METHODS IN MOLECULAR BIOLOGY™

Series Editor
John M. Walker
School of Life Sciences
University of Hertfordshire
Hatfield, Hertfordshire, AL10 9AB, UK

For further volumes:
http://www.springer.com/series/7651

Fluorescent Protein-Based Biosensors

Methods and Protocols

Edited by

Jin Zhang and Qiang Ni

Department of Pharmacology and Molecular Sciences,
The Johns Hopkins University School of Medicine, Baltimore, MD, USA

Robert H. Newman

Department of Pharmacology and Molecular Sciences,
The Johns Hopkins University School of Medicine, Baltimore, MD, USA

Department of Biology, North Carolina A&T State University, Greensboro, NC, USA

Editors
Jin Zhang
Department of Pharmacology
and Molecular Sciences
The Johns Hopkins University
School of Medicine
Baltimore, MD, USA

Qiang Ni
Department of Pharmacology
and Molecular Sciences
The Johns Hopkins University
School of Medicine
Baltimore, MD, USA

Robert H. Newman
Department of Pharmacology
and Molecular Sciences,
The Johns Hopkins University School
of Medicine, Baltimore, MD, USA
Department of Biology
North Carolina A&T State University
Greensboro, NC, USA

ISSN 1064-3745 ISSN 1940-6029 (electronic)
ISBN 978-1-62703-621-4 ISBN 978-1-62703-622-1 (eBook)
DOI 10.1007/978-1-62703-622-1
Springer New York Heidelberg Dordrecht London

Library of Congress Control Number: 2013946935

Printed on acid-free paper

Humana Press is a brand of Springer
Springer is part of Springer Science+Business Media (www.springer.com)

Preface

The past decade has witnessed a revolution in the field of live cell imaging. This revolution was sparked, in large part, by the discovery of a family of autofluorescent proteins (FPs) from several bioluminescent marine organisms, the most notable of which is the green fluorescent protein (GFP) from *Aequorea victoria*. The subsequent cloning of GFP and its color variants made possible the development of genetically encodable fluorescent biosensors designed to probe biological processes in living cells and animals. These biosensors, which can be targeted to specific subcellular regions using standard cell and molecular biology techniques, have allowed researchers to observe biochemical processes within the endogenous cellular environment with unprecedented spatiotemporal resolution. As the number and diversity of available biosensors grow, we believe that it is increasingly important to equip researchers (both present and future) with an understanding of the key concepts underlying the design and application of genetically encodable fluorescent biosensors to live cell imaging. To this end, we have assembled a series of detailed protocols, written by experts in the field, describing several ways in which FP-based reporters can be used to gain unique insights into the regulation of cellular signal transduction. It is our hope that this work will serve as a user's guide for researchers who wish to (1) apply current biosensors to the study of interesting biological questions and (2) develop novel fluorescent biosensors to monitor additional biochemical processes in living cells.

In this textbook, we will first briefly explore the design of genetically encodable fluorescent biosensors as well as some of the fluorescence imaging techniques that have been developed to visualize them. We will then turn our attention to the development of genetically encodable biosensors used to probe the cellular environment (e.g., redox state), track diverse signaling molecules (e.g., calcium and cyclic nucleotides), and visualize discrete cellular signaling events (e.g., G-protein activation and protein kinase activity). Finally, we will explore the diverse ways in which fluorescent biosensors have been adapted to different biological applications (e.g., high throughput drug screening and computational modeling). In so doing, we hope to provide the reader with a sense of the vast potential, as well as some of the current limitations, of fluorescent biosensor technology for studying "biochemistry in the wild."

Baltimore, MD, USA	*Jin Zhang*
Baltimore, MD, USA	*Qiang Ni*
Greensboro, NC, USA	*Robert H. Newman*

Contents

Contributors

NWE-NWE AYE-HAN • *Department of Pharmacology and Molecular Sciences, The Johns Hopkins University School of Medicine, Baltimore, MD, USA*
ADAM E. COHEN • *Department of Chemistry and Department of Chemical Biology and Physics, Harvard University, Cambridge, MA, USA*
SHUBHAM DIPT • *Johann-Friedrich-Blumenbach-Institute for Zoology and Anthropology and Molecular Neurobiology of Behavior, Georg-August-Universität Göttingen, Göttingen, Germany*
ANDRÉ FIALA • *Johann-Friedrich-Blumenbach-Institute for Zoology and Anthropology and Molecular Neurobiology of Behavior, Georg-August-Universität Göttingen, Göttingen, Germany*
LAURA ASHLEY FIELDS • *College of Medical, Veterinary & Life Sciences, Institute of Neuroscience and Psychology, University of Glasgow, Glasgow, UK*
J. GOEDHART • *Section of Molecular Cytology, Swammerdam Institute for Life Sciences, University of Amsterdam, Amsterdam, The Netherlands*
ERIC C. GREENWALD • *Department of Biomedical Engineering, University of Virginia, Charlottesville, VA, USA*
ELVIRE GUIOT • *Centre National de la Recherche Scientifique, Unité Mixe de Recherche and Université Pierre et Marie Curie, Paris, France*
KATIE HERBST ROBINSON • *Department of Pharmacology and Molecular Sciences, The Johns Hopkins University School of Medicine, Baltimore, MD, USA*
MARK A. HINK • *Section Molecular Cytology, Swammerdam Institute for Life Sciences, University of Amsterdam, Amsterdam, The Netherlands*
DANIEL R. HOCHBAUM • *Department of Chemistry and Department of Chemical Biology and Physics, Harvard University, Cambridge, MA, USA*
YIN PUN HUNG • *Department of Neurobiology, Harvard Medical School, Cambridge, MA, USA*
KEES JALINK • *Section of Molecular Cytology, Swammerdam Institute for Life Sciences, University of Amsterdam and The van Leeuwenhoek Centre of Advanced Microscopy, Amsterdam, The Netherlands*
HE JIANG • *College of Medical, Veterinary & Life Sciences, Institute of Neuroscience and Psychology, University of Glasgow, Glasgow, UK*
J. KLARENBEEK • *Division of Cell Biology, The Netherlands Cancer Institute, Amsterdam, The Netherlands*
ANDREAS KOSCHINSKI • *College of Medical, Veterinary & Life Sciences, Institute of Neuroscience and Psychology, University of Glasgow, Glasgow, UK*
MAYA T. KUNKEL • *Department of Pharmacology, University of California at San Diego, San Diego, CA, USA*
SHAOYING LU • *Department of Bioengineering, University of Illinois at Urbana-Champaign, Champaign, IL, USA*

SOHUM MEHTA • *Department of Pharmacology and Molecular Sciences, The Johns Hopkins University School of Medicine, Baltimore, MD, USA*
ROBERT H. NEWMAN • *Department of Pharmacology and Molecular Sciences, The Johns Hopkins University School of Medicine, Baltimore, MD, USA; Department of Biology, North Carolina A&T State University, Greensboro, NC, USA*
ALEXANDRA C. NEWTON • *Department of Pharmacology, University of California at San Diego, San Diego, CA, USA*
ANA F. OLIVEIRA • *Department of Neurobiology, Duke University Medical Center, Durham, NC, USA*
MINGXING OUYANG • *Department of Bioengineering, University of Illinois at Urbana-Champaign, Champaign, IL, USA*
AMY E. PALMER • *Department of Biochemistry and Chemistry, BioFrontiers Institute, University of Colorado Boulder, Boulder, CO, USA*
J. GENEVIEVE PARK • *Department of Biochemistry and Chemistry, BioFrontiers Institute, University of Colorado Boulder, Boulder, CO, USA*
RENATA K. POLANOWSKA-GRABOWSKA • *Department of Biomedical Engineering, University of Virginia, Charlottesville, VA, USA*
MARINA POLITO • *Centre National de la Recherche Scientifique, Unité Mixe de Recherche and Université Pierre et Marie Curie, Paris, France*
THOMAS RIEMENSPERGER • *Johann-Friedrich-Blumenbach-Institute for Zoology and Anthropology and Molecular Neurobiology of Behavior, Georg-August-Universität Göttingen, Göttingen, Germany*
KAZUKI SASAKI • *Chemical Genetics Laboratory/Chemical Genomics Research Group, RIKEN Advanced Science Institute, Wako, Japan*
MORITOSHI SATO • *Graduate School of Arts and Sciences, The University of Tokyo, Tokyo, Japan*
JEFFREY J. SAUCERMAN • *Department of Biomedical Engineering, University of Virginia, Charlottesville, VA, USA*
ALESSANDRA STANGHERLIN • *College of Medical, Veterinary & Life Sciences, Institute of Neuroscience and Psychology, University of Glasgow, Glasgow, UK*
ANNA TERRIN • *College of Medical, Veterinary & Life Sciences, Institute of Neuroscience and Psychology, University of Glasgow, Glasgow, UK*
PIERRE VINCENT • *Centre National de la Recherche Scientifique, Unité Mixe de Recherche and Université Pierre et Marie Curie, Paris, France*
YINGXIAO WANG • *Department of Bioengineering, Neuroscience Program, Center of Biophysics and Computational Biology, Beckman Institute for Advanced Science and Technology, University of Illinois at Urbana-Champaign, Champaign, IL, USA; Department of Molecular and Integrative Physiology, University of Illinois at Urbana-Champaign, Champaign, IL, USA*
JESSICA R. YANG • *Department of Pharmacology and Molecular Sciences, The Johns Hopkins University School of Medicine, Baltimore, MD, USA*
RYOHEI YASUDA • *Departments of Neurobiology, Cell Biology, and Physics, Duke University Medical Center, Durham, NC, USA; Max-Planck Florida Institute for Neuroscience, Jupiter, FL, USA*
GARY YELLEN • *Department of Neurobiology, Harvard Medical School, Cambridge, MA, USA*

MINORU YOSHIDA • *Chemical Genetics Laboratory/Chemical Genomics Research Group, RIKEN Advanced Science Institute, Graduate School of Science and Engineering, Saitama University, Saitama, Japan; CREST Research Project, Japan Science and Technology Corporation, Kawaguchi, Saitama, Japan*
MANUELA ZACCOLO • *College of Medical, Veterinary & Life Sciences, Institute of Neuroscience and Psychology, University of Glasgow, Glasgow, UK; Department of Physiology, Anatomy & Genetics, Oxford University, Oxford, UK*
JIN ZHANG • *Department of Pharmacology and Molecular Sciences, The Johns Hopkins University School of Medicine, Baltimore, MD, USA*
ANNA ZOCCARATO • *College of Medical, Veterinary & Life Sciences, Institute of Neuroscience and Psychology, University of Glasgow, Glasgow, UK*

Chapter 1

The Design and Application of Genetically Encodable Biosensors Based on Fluorescent Proteins

Robert H. Newman and Jin Zhang

Abstract

To track the activity of cellular signaling molecules within the endogenous cellular environment, researchers have developed a diverse set of genetically encodable fluorescent biosensors. These sensors, which can be targeted to specific subcellular regions to monitor specific pools of a given signaling molecule in real time, rely upon conformational changes in a sensor domain to alter the photophysical properties of green fluorescent protein (GFP) family members. In this introductory chapter, we first discuss the properties of GFP family members before turning our attention to the design and application of genetically encodable fluorescent biosensors to live cell imaging.

Key words Fluorescent proteins, Fluorescent biosensor, Biosensor design, FRET, Cell signaling, Live cell imaging

1 Introduction

Cells do not exist in a static environment. Instead, they must simultaneously sense multiple extracellular cues and respond accordingly. This task is accomplished by a host of cellular signaling molecules whose activities are precisely coordinated by endogenous regulatory factors. Together, signaling molecules and their cofactors form tightly regulated signaling networks that impact nearly every aspect of cellular physiology. Therefore, to better understand how individual signaling molecules are controlled inside cells, it is necessary to develop tools that are able to monitor their activities within the context of intact, fully functional signaling networks. To this end, researchers have recently developed a diverse set of genetically encodable fluorescent biosensors designed to probe dynamic cellular events in living cells with high spatial and temporal resolution. These sensors, which typically involve the fusion of green fluorescent protein (GFP) or a related fluorescent protein (FP) color variant(s) into the primary amino acid sequence of a protein or a selected protein domain, have enabled researchers

Jin Zhang et al. (eds.), *Fluorescent Protein-Based Biosensors: Methods and Protocols*, Methods in Molecular Biology, vol. 1071, DOI 10.1007/978-1-62703-622-1_1,

to track various components of intracellular signaling networks in real time within the native cellular environment.

In this chapter, we provide a general overview of genetically encodable fluorescent biosensors. To this end, we will briefly examine the properties of GFP and several FP family members commonly used to construct fluorescent biosensors before turning our attention to the development and application of fluorescent biosensors for studying dynamic signaling processes in living cells.

2 Fluorescent Proteins for Biosensor Development

Aequorea victoria GFP (avGFP) is the founding member of a large family of autofluorescent proteins isolated from bioluminescent marine organisms such as hydrozoa and reef-building corals [1, 2]. FP family members, which require only molecular oxygen to generate their intrinsically derived fluorophores, form a highly stable 11-stranded β-barrel structure whose architecture aids in both the formation and stabilization of the conjugated ring systems that account for their spectral properties (Fig. 1a) [1, 2]. Importantly, in addition to the extent of conjugation within the ring system, the fluorescent properties of FP fluorophores can also be substantially influenced by their local protein microenvironment [1, 3]. For instance, the *p*-hydroxylbenzylideneimidazolinone (*p*-HBI) fluorophore generated by wild-type avGFP is formed through a three-step autocatalytic cyclization reaction involving a triplet of amino acids, Ser65-Tyr66-Gly67, located along a central α-helix that runs through the interior of the β-barrel (Fig. 1a) [1–3]. In the context of the popular avGFP derivative, enhanced GFP (EGFP), the *p*-HBI chromophore emits green light (507 nm) following excitation with 484 nm light. Meanwhile, the excitation and emission spectra of this same *p*-HBI species is red-shifted by ~20 nm in the closely related color variant, enhanced yellow FP (EYFP) [4, 5]. The dramatic changes observed in the spectral properties of the EYFP fluorophore are attributed to π-stacking interactions caused by the introduction of a Tyr residue (T203Y) in the vicinity of the chromophore (Fig. 1a). In fact, together with the surprising tolerance of the avGFP fluorophore to mutation (only Gly67 appears to be essential for its formation), modification and refinement of the protein microenvironment surrounding the fluorophore is a primary focus in the development of many of the avGFP color variants used to construct fluorescent biosensors [6, 7].

The FP toolbox contains a broad range of color variants whose emission profiles span most of the visible spectrum. Though in reality there is no clear line of demarcation between them, FP family members are often divided into seven spectral classes according to their emission maxima. These include blue FPs (BFPs), which

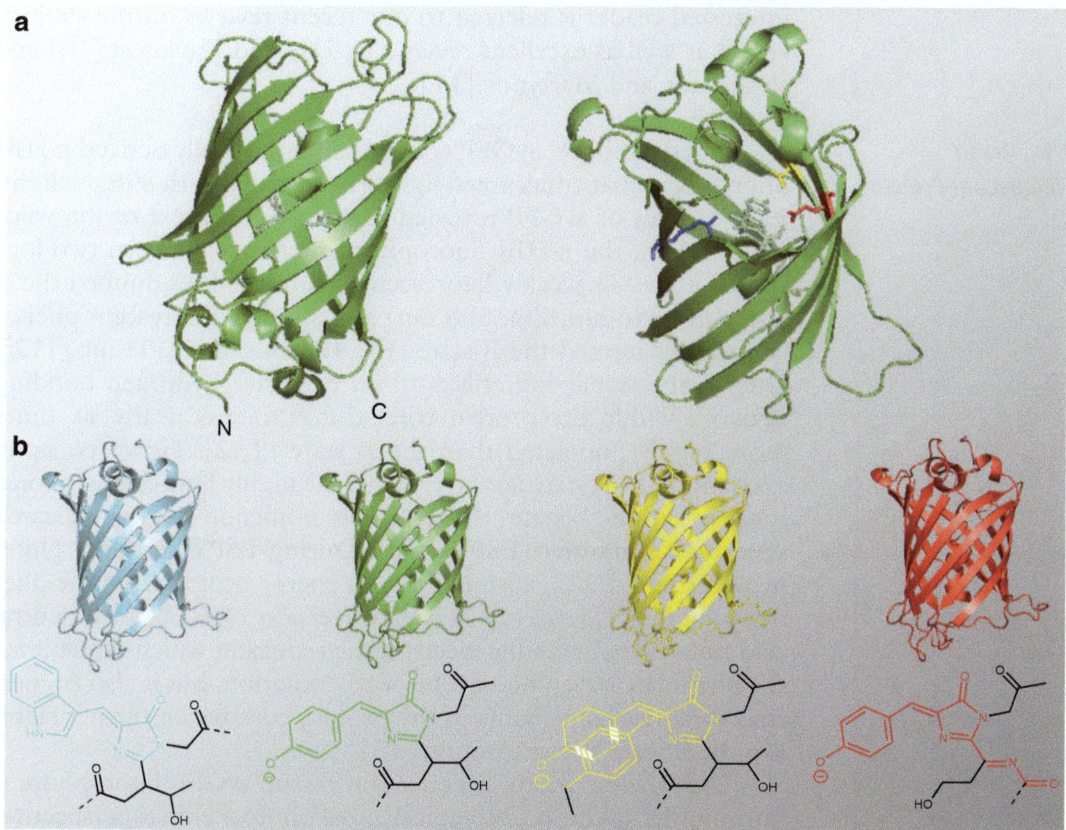

Fig. 1 Structure and function of GFP. (**a**) Structure of *A. victoria* GFP, highlighting the protein's β-barrel architecture. In the structure on the *right*, which is rotated approximately 180° along both the *X* and *Y* axes relative to the structure on the *left*, four residues (V68 through F71) have been removed in order to highlight the location of the fluorophore (*pale green*) on the alpha helix that runs through the core of the protein. Two residues, E222 (*red*) and R96 (*blue*), involved both in fluorophore maturation and the formation of the hydrogen bonding network are shown. In addition, the position of T203 (*yellow*), which is mutated to Tyr in YFP variants, is shown. (**b**) FP color variants frequently used in biosensor development, along with the structure of their fluorophores, are shown. The conjugated ring structure responsible for the spectral properties of each fluorophore is colored accordingly. The fluorophore structures correspond to those found in ECFP (Ex: 439 nm; Em: 476 nm), EGFP (Ex: 484 nm; Em: 507 nm), EYFP (Ex: 514 nm; Em: 527 nm), and mCherry (Ex: 587 nm; Em: 610 nm). All protein structures were generated using Pymol using coordinates for avGFP (PDB ID: 1gfl) originally published by Yang et al. [57]

emit between 440–470 nm, cyan FPs (CFPs; 471–500 nm), GFPs (501–520 nm), YFPs (521–550 nm), orange FPs (OFPs; 551–575 nm), red FPs (RFPs: 576–610 nm) and far-red FPs (FRFPs; 611–660 nm). In this chapter, we will examine the molecular characteristics of representative members from each of the spectral classes most commonly used in biosensor development, namely CFP, GFP, YFP, and RFP family members (Fig. 1a). Though an in-depth discussion of the many FP color variants currently available for live cell imaging is beyond the scope of this chapter, the interested reader is referred to our recent reviews on the subject, [8, 9] as well as excellent reviews by Day and Davidson [10] and Pakhomov and Martynov [11].

2.1 Green Fluorescent Protein

As alluded to above, avGFP contains an intrinsically derived *p*-HBI fluorophore that emits green light when excited with 396 nm light. Early studies of avGFP revealed that, in the context of the wild-type protein, the *p*-HBI fluorophore alternates between two ionization states—a weakly fluorescent, neutral species dubbed the A state (Ex: 396 nm, Em: 508 nm) and a highly fluorescent phenolate species termed the B state (Ex: 475 nm; Em: 503 nm) [12]. Due to the stabilizing effects of an extensive hydrogen bonding network within the protein core, the A state is nearly six times more highly populated than the B state [1, 2]. However, upon excitation, the A state is converted to the highly fluorescent anionic species (i.e., the B state) through a phenomenon known as excited state proton transfer (ESPT) [12]. During ESPT, a nearby glutamate residue, E222, abstracts a high energy proton from the phenol moiety of Tyr66 [13]. The conversion of E222 to a neutral acid not only relieves the electrostatic repulsion which destabilizes the phenolate oxyanion of B prior to excitation, but it also triggers structural rearrangements in the protein core which further stabilize the B state after excitation [13].

Though the photoisomerization of the avGFP fluorophore is intriguing from both a biological and a photochemical perspective, this behavior is not ideal for biosensor development. For instance, the existence of two isomerization states with distinct excitation/emission profiles can complicate the interpretation of experimental data. Moreover, the ultraviolet (UV) light required to excite the A state not only induces cellular autofluorescence, but it can also adversely affect the cellular system under study. Therefore, several mutations have been introduced into the wild-type protein that simplify its excitation/emission spectra and improve its spectral properties for live cell imaging. One of these mutations, the substitution of a Thr residue for Ser at position 65 (S65T), introduces a single methyl group into the fluorophore [14]. Though this mutation does not dramatically affect the structure of the fluorophore itself, it has a profound impact on the spectral properties of the resultant GFP variant. This is because the methyl group causes local conformational changes in the protein core that disrupt the ground state hydrogen bonding network originally involved in stabilization of the A state [13, 14]. As a consequence, the resulting fluorophore is converted almost exclusively to the phenolate species in the unexcited state, simplifying its excitation/emission spectra and making it better suited for live cell imaging. For this reason, the S65T mutation, along with a series of mutations that improve the expression and maturation of avGFP in mammalian cells, has been incorporated into EGFP (avGFP/F64L/S65T).

For both practical and historical reasons, EGFP is the most widely used FP in the biological sciences and the starting point for the development of many FP-based biosensors.

2.2 Yellow Fluorescent Proteins

Due to their relatively high extinction coefficients (ε) and quantum yields (ϕ), YFPs represent some of the brightest monomeric FPs reported to date [10, 14]. As alluded to above, FPs from this class are generated by the introduction of an aromatic Tyr residue near the phenolate ion of the *p*-HBI fluorophore (T203Y) [1, 14]. This mutation promotes π–π stacking interactions that stabilize the excited state dipole moment of the fluorophore, shifting the excitation and emission spectra of YFP variants by ~20 nm toward the red end of the spectrum.

The introduction of a bulky Tyr residue within the protein core is believed to disrupt internal hydrogen bonding networks and reduce steric packing interactions around the fluorophore [5]. As a consequence, YFP family members often exhibit acute sensitivities to several cellular parameters. Though not ideal for general biosensor development, the environmental sensitivities of YFP family members have been exploited to probe changes in the cellular environment. For example, EYFP, a first generation YFP derived from EGFP, exhibits a relatively high pK_a ($pK_a = 6.2$) and is highly sensitive to halide ions [5, 15]. The acute sensitivity of EYFP to these cellular parameters appears to be related to the stabilization of the weakly fluorescent neutral form of *p*-HBI at low pH or in the presence of a bound halide ion in the vicinity of the fluorophore. By exploiting these sensitivities, researchers have designed a series of EYFP-based biosensors capable of tracking dynamic changes in the concentration of halide ions and/or fluctuations in pH within different subcellular compartments [16–18].

To produce a YFP variant that is useful for general biosensor development, Griesbeck et al. employed a molecular evolution strategy to produce mCitrine (EYFP-V68L, Q69M) [19]. Compared to the parent EYFP species, mCitrine exhibits a markedly reduced sensitivity to halide ions, better photostability, and a lower pK_a ($pK_a = 5.7$). Structural analysis revealed that the reduced halide sensitivity exhibited by mCitrine is likely conferred by a M69Q point mutation that plugs a large halide-binding pocket next to the fluorophore [19]. By preventing the binding of halide ions in the vicinity of the fluorophore, this substitution relieves repulsive forces caused by the binding of a negatively charged halide ion in close proximity to the fluorophore which destabilizes the highly fluorescent phenolate anion in the parent species. As a consequence, the phenolate form of *p*-HBI is more highly populated than it is in EYFP. Moreover, the tight packing interactions caused by the introduction of the neutral methionine side chain within the core of the protein are believed to further stabilize the phenolate species. Together, these modifications help to shift the equilibrium away from the neutral form of the *p*-HBI chromophore and toward the phenolate anion, leading to the enhanced fluorescence exhibited by this variant. In addition to mCitrine, another popular EYFP variant, Venus, also exhibits enhanced photostability and reduced sensitivity to both pH and fluctuations in halide ion concentrations [20]. As a consequence, mCitrine and Venus have been incorporated into a large number of biosensors.

2.3 Cyan Fluorescent Proteins

Perhaps surprisingly, the SYG tripeptide used to generate the *p-HBI* fluorophore found in wild-type avGFP is quite tolerant of substitutions. This conformational flexibility permits the spontaneous formation of several different fluorophore structures, each of which exhibits unique photophysical properties. For instance, in the case of CFP family members derived from avGFP, substitution of Trp for Y66 (Y66W) results in the formation of a fluorophore with an indole moiety in place of the phenol ring [21]. As a consequence, the excitation and emission wavelengths of CFPs are blue-shifted relative to the parent protein, giving members of this family a bluish-green appearance following excitation with ~450 nm light. However, because the protein core of the parent avGFP species is designed to accommodate the phenol-containing *p*-HBI species, modification of the fluorophore structure often comes at a price. For instance, the indole-containing fluorophore present in the enhanced CFP (ECFP) variant behaves as two distinct species, each characterized by a unique fluorescence lifetime decay curve and exhibiting distinct excitation and emission spectra. Using the crystal structure of ECFP as a guide, Rizzo and colleagues hypothesized that the distinct fluorogenic species observed in ECFP arise from isostable conformations of the protein caused by the dynamic exchange of two bulky residues, Tyr145 and His148, in the region surrounding the rigid fluorophore [22]. Therefore, to eliminate the biphasic character of ECFP, a combination of site-directed and random mutagenesis was employed to produce mCerulean (ECFP-S72A/H148D/Y154A), a 2.5-fold brighter ECFP variant that exhibits a monophasic decay curve [22]. The key mutation in mCerulean is a H148D substitution that eliminates the exchange between Tyr145 and His148, thus stabilizing the first conformational state and improving the fluorescence lifetime characteristics of the protein. Recently, the photophysical properties of mCerulean were further improved by (1) sealing a gap between β-strands 7 and 8 in the β-barrel and (2) converting T65 in ECFP back to the Ser residue found in wild-type avGFP [22]. By improving the stability and ϕ of the fluorophore, these mutations increase the relative brightness of the resulting CFP variant, mCerulean3, by approximately 65 % compared to mCerulean. Likewise, the recently described CFP variant, mTurquoise, which contains the same T65S mutation as mCerulean3, is 1.5 times brighter than mCerulean and exhibits a monoexponential decay curve [23].

Together, their photophysical properties suggest that these new CFP variants will be broadly useful in biosensor development (e.g., *see* Chapter 4 in this textbook and ref. 24).

2.4 Red Fluorescent Proteins

In the same way that the spectral properties of CFP color variants can be altered by direct chemical modification of their fluorophores, the fluorophores generated by RFP family members also exhibit altered chemical structures whose excitation and emission spectra are shifted relative to the *p*-HBI species utilized by avGFP. However, unlike CFP variants and other avGFP derivatives, red-emitting FPs undergo a second oxidation step along the Cα-N bond of residue 65 (according to the GFP numbering system) to form an acylimine linkage with the polypeptide backbone [25, 26]. As a consequence, the conjugated π-system of RFP-derived fluorophores are extended over a greater distance, thereby lowering the energy barrier separating the ground (S_0) and excited states (S_1) of the fluorophore and leading to red-shifted excitation and emission spectra [6, 27, 28].

Importantly, the longer wavelength light used to excite RFPs (>550 nm) substantially reduces the effects of cellular autofluorescence caused by the excitation of abundant fluorescent biomolecules inside the cell. Moreover, the light emitted by RFPs (>580 nm) is less susceptible to interference, such as light scattering, that can reduce the signal-to-noise ratio in an imaging experiment. Together, these spectral properties make RFPs particularly attractive for deep tissue and whole-body imaging, where the excitation and emission light must traverse several layers of cells. As a consequence, much effort has been devoted to the development of monomeric RFP variants suitable for biosensor development. These efforts have culminated in the evolution of two major families of mRFPs: the mFruits and a series of eqFP578 derivatives consisting of TagRFP [29], TagRFP-T [30], mKate [31], mKate2 [32], and Neptune [33]. Together, these proteins provide researchers with several monomeric FPs that are excited at wavelengths >550 nm. Though no single variant is optimal for all applications, the photophysical properties of mCherry, mRuby2 [34], and TagRFP-T make them the best general-purpose mRFPs for biosensor development.

3 Design and Application of Fluorescent Protein-Based Biosensors

Aside from allowing multiple cellular parameters, such as protein dynamics or gene expression profiles, to be probed simultaneously, the availability of spectrally distinct FP variants has also prompted the development of novel imaging techniques, such as fluorescence resonance energy transfer (FRET)-based approaches. Such techniques extend the application of FP technologies, enabling researchers to visualize other dynamic cellular events, such as the turnover of small molecule second messengers like Ca^{2+} and cyclic AMP (cAMP) or the activity profiles of cellular signaling enzymes. In some cases, two or more cellular parameters can even be measured simultaneously in the same cell using biosensors that employ three spectrally distinct FPs, such as CFP, YFP, and RFP (e.g., *see* Chapter 16 in this textbook). Below, we explore several of the design principles used to construct FP-based biosensors and briefly discuss some of the ways in which these sensors have been used to gain unique insights into the activation and regulation of cellular signaling molecules within their native cellular environment. During the course of this discussion, we will also highlight several of the key parameters that must be considered when utilizing fluorescent biosensors for live cell imaging. These include parameters that are intrinsic to the sensors themselves, such as sensitivity, reversibility, response kinetics and dynamic range, as well as cellular factors, such as interference from endogenous proteins or small molecules, which can impact sensor performance.

Under certain circumstances, biosensors can be constructed based on the intrinsic environmental sensitivities of FP color variants, as illustrated by the EYFP-based probes discussed above. However, in addition to environmental parameters, such as pH and halide ion concentration, FP-based biosensors can also be engineered to directly sense other important cellular parameters. In general, engineered fluorescent biosensors contain two basic components: (1) a "sensor unit", which undergoes a conformational change in response to a given cellular parameter and (2) a "reporter unit", which converts the induced conformational change into a fluorescent readout (Fig. 2) [8]. In the case of FP-based sensors, the reporter unit usually consists of a pair of FPs that undergo FRET or a single FP whose fluorescent properties are altered in response to a conformational change in the sensor unit. Meanwhile, the molecular switch utilized by the sensor unit can be generated in many different ways, provided that it promotes a conformational change in response to the cellular parameter under study. For instance, a relatively simple molecular switch is employed by two related classes of redox-sensitive FP indicators known as roGFPs and rxYFPs [35]. These biosensors, which change either their spectral properties (roGFPs) or fluorescence intensity (rxYFPs) in response to changes in the cellular redox potential, use disulfide bond formation between pairs of carefully positioned cysteine residues on the surface of their respective FPs to induce conformational changes in the FP [35]. In the case of rxYFP, structural analysis suggests that disulfide bond formation leads to the reorganization of residues in close proximity to the fluorophore. Reorganization of the fluorophore microenvironment, particularly the hydrogen bonding network, shifts the equilibrium between the neutral (A) state and the phenolic (B) state, leading to weak

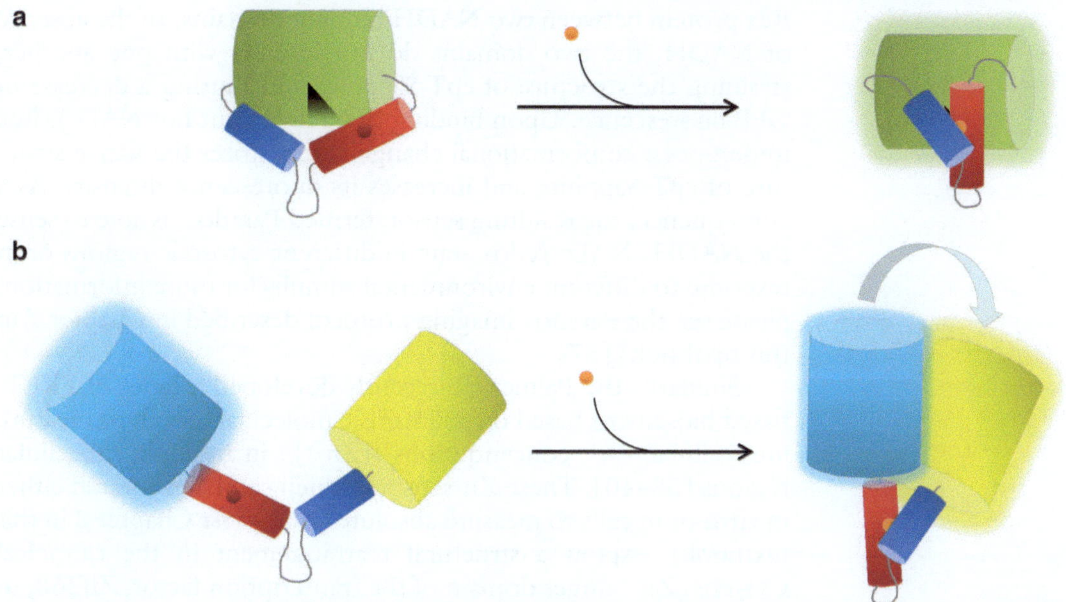

Fig. 2 FP-based biosensor designs. (**a**) Cartoon depicting a single FP biosensor based on an engineered molecular switch. In this design, a sensor domain composed of a receiver unit (*red*) and a switching unit (*blue*) is grafted into a circularly permuted version of GFP (*green cylinder*). In the unbound "open" state, a hole in the β-barrel created by circular permutation allows bulk solvent access to the fluorophore, quenching its fluorescence. In the presence of ligand (*orange sphere*), the sensor domain undergoes a conformational change that plugs the hole and increases fluorescence (*green glow*). (**b**) Cartoon depicting a FRET-based FP biosensor using the same engineered molecular switch described in (**a**). In this case, conformational changes induced by ligand binding alter the relative distance and orientation of the donor CFP (*cyan cylinder*) and the acceptor YFP (*yellow cylinder*), increasing FRET between them

fluorescence in oxidizing environments and increased fluorescence under reducing conditions [36].

Though roGFPs and rxYFPs employ a relatively simple molecular switch to convert changes in the cellular environment into a fluorescence output, in general, more complex switches are necessary to detect changes in other cellular parameters. Under these circumstances, the molecular switch can be derived from a conformational change intrinsic to an endogenous protein or it can be generated via an engineered switch [8]. By combining the molecular switch with an appropriate reporter unit—either by flanking the switch region with complementary FRET pairs or by grafting it into the FP itself—conformational changes in the sensor unit can be translated into a fluorescence readout from the reporter unit (Fig. 2). For instance, to construct a biosensor that monitors the NADH/NAD$^+$ redox state, Hung et al. employed an intrinsic molecular switch based on the bacterial NADH binding protein, Rex [37]. In this sensor design, a circularly permuted version of the GFP variant, T-Sapphire (cpT-Sapphire), is grafted into the

Rex protein between two NADH binding domains. In the absence of NADH, the two domains do not interact with one another, straining the structure of cpT-Sapphire and causing a decrease in GFP fluorescence. Upon binding of NADH (but not NAD^+), Rex undergoes a conformational change that restores the native structure of cpT-Sapphire and increases its fluorescence intensity. As a consequence, the resulting sensor, termed Paradox, is able to sense the NADH/NAD^+ redox state in different cytosolic regions or in response to different environmental stimuli (for more information, please see the Paradox imaging protocol described in Chapter 7 in this textbook) [37].

Similarly, the Palmer lab recently developed a series of FRET-based biosensors based on an intrinsic molecular switch to measure intracellular Zn^{2+} concentrations ($[Zn^{2+}]_i$) in multiple subcellular regions [38–40]. These Zn-sensors, which can be calibrated either in vitro or in cells to measure absolute $[Zn^{2+}]_i$ (*see* Chapter 3 in this textbook), exploit a structural rearrangement in the canonical Cys_2His_2 Zn^{2+} finger domain of the transcription factor, Zif268, to change the relative distance/orientation of flanking FP color variants [38]. In the absence of Zn^{2+}, the Zn^{2+}-finger domain is largely unstructured, leading to large degree of separation between the FPs in space. Zn^{2+} binding causes the domain to fold into a compact structure that brings the FPs into close proximity with one another, facilitating FRET between them. Importantly, the Zn^{2+} binding affinity of Zn-sensor can be reduced ~100-fold by substituting His for Cys in the Cys_2His_2 motif [38]. Due to its reduced sensitivity for Zn^{2+}, the resulting His_4 Zn-sensor is able to detect changes in intracellular Zn^{2+} at concentrations that would otherwise saturate biosensors based on the canonical Cys_2His_2 motif. Thus, by tuning the sensitivity of the sensor, researchers have effectively expanded the range of $[Zn^{2+}]_i$ that can be reliably measured in cells [38]. Together, this family of Zn-sensors has been used to gain a better understanding of $[Zn^{2+}]_i$ regulation in several organelles, including the cytosol [38], ER [40], Golgi [40], and mitochondria [39].

In contrast to the intrinsic molecular switches utilized by Paradox and Zn-sensor, which each rely upon a naturally occuring conformational change to alter the fluorescent properties of the reporter unit, engineered switches can be constructed based on a modular design. Engineered molecular switches typically consist of a "receiver" module and a "switching" module. The receiver module "senses" the cellular parameter under study (be it through the binding of a small molecule or via posttranslational modification by a signaling enzyme) while the switching module converts changes in the receiver module into a conformational change that alters the fluorescent properties of the FPs in the reporter unit (Fig. 2). For instance, two of the most popular families of genetically targetable Ca^{2+} sensors, the GCaMPs and yellow cameleons (YC's), both rely upon an engineered molecular switch composed of the Ca^{2+} binding protein, calmodulin (CaM), and the CaM-binding peptide, M13 [5, 41–46]. In the case of the GCaMPs, the CaM-M13 switch is fused to a circularly permuted version of EGFP. In the presence of Ca^{2+}, CaM binds M13 and induces a conformational change in the reporter that leads to an increase in its fluorescence intensity. In the context of the FRET-based YC's, in which the same CaM-M13 switch is sandwiched between a CFP-YFP FRET pair, the Ca^{2+}-dependent conformational change in CaM-M13 alters the distance and orientation of the CFP donor relative to the YFP acceptor, altering their FRET emission ratio. Importantly, because the CaM-M13 interaction is readily reversible, both YC's and GCaMPs can report transient changes in intracellular Ca^{2+} levels ($[Ca^{2+}]_i$) in real-time. Moreover, since the CaM-based receiver module is tethered directly to the M13-based switching module, the response kinetics of the sensors is largely determined by the intrinsic rates of association and dissociation of Ca^{2+}-CaM for M13.

Aside from simplifying the response kinetics, the modular design of the sensor domain also ensures that the effective concentrations of the two halves of the molecular switch are relatively high in the context of the sensor. As a consequence, in most cases, the performance of GCaMP and YC family members is not dramatically influenced by nonproductive interactions with endogenous binding partners. Nonetheless, there are circumstances in which interference becomes a concern. For instance, in cellular environments where the concentration of endogenous CaM is extremely high, such as at the plasma membranes of hippocampal neurons, the ability of YC biosensors to report changes in $[Ca^{2+}]_i$ is hindered by interactions with endogenous CaM. Thus, to detect changes in $[Ca^{2+}]_i$ in regions of high endogenous CaM, the Tsien lab used a "bump and hole" strategy to create CaM and M13 variants that exhibit a very low affinity for wild-type CaM but were still able to interact with one another efficiently [45].

A modular design similar to that described above has been used to construct engineered molecular switches for a wide range of cellular parameters. For instance, in order to monitor changes in the activity profiles of different protein kinases inside the cell, a series of FRET-based kinase activity reporters have been developed based on an engineered molecular switch [47]. In this type of sensor, a short peptide sequence that is specifically phosphorylated by the kinase-of-interest serves as the receiver module while a phospho-amino acid binding domain (PAABD) that reversibly associates with the phosphorylated form of the substrate region serves as the switching module. In order to detect both activation *and* attenuation of kinase activity, the PAABD must be carefully selected when designing kinase activity reporters. Ideally, the PAABD would associate with the phosphorylated form of the substrate region rapidly but not bind so tightly as to prevent dephosphorylation by cellular phosphatases. For instance, though the first protein kinase A (PKA) activity sensor, AKAR1, accurately reported increases in PKA activity, the high affinity of the 14-3-3τ PAABD employed by this sensor blocked access of cellular phosphophosphatases to the substrate domain, preventing its dephosphorylation once the activity of PKA decreased [48]. Therefore, to improve the reversibility of subsequent AKAR variants, 14-3-3τ was replaced with the forkhead associated 1 (FHA1) domain from Rad53p [49–51]. The FHA1 domain exhibits a lower K_d for the phosphorylated substrate domain than 14-3-3τ, allowing both the activation and attentuation of PKA to be visualzed in cells. As described in Chapter 10, kinase activity reporters can be targeted to specific subcellular compartments through the incorporation of short targeting sequences, allowing researchers to gain important insights into the regulation of specific pools of a given kinase under various cellular conditions.

It is important to note that, due to the modular design of their sensor unit, unimolecular FRET-based probes that utilize an engineered molecular switch can be converted into bimolecular probes simply by removal of the flexible linker between the receiver domain and the switching domain (Fig. 3). One of the primary advantages of a bimolecular design is that the reporter typically exhibits a larger dynamic range than its unimolecular counterpart [52]. Presumably, this is due to lower basal FRET caused by a large degree of separation between the donor and acceptor fluorophores in the uninduced state [41, 53, 54]. However, despite this potential advantage, bimolecular FRET-based reporter systems also present researchers with unique challenges not typically encountered when using unimolecular sensors. For instance, because both halves of the sensor are required to generate a measurable FRET response, the stoichiometric ratio between them must be strictly regulated when using bimolecular reporter systems. This task is nontrivial considering the variability that often exists in parameters such as DNA transfection efficiency, transcriptional regulation, and protein translation, to name a few. As a consequence, changes in FRET are often measured differently when utilizing unimolecular versus bimolecular biosensor designs. For instance, when the stoichiometry between the donor and acceptor fluorophores is fixed, as it is in the case of unimolecular probes, the donor-to-acceptor emission ratio is generally the easiest and most convenient means of measuring changes in FRET [55]. On the other hand, if the stoichiometries between the donor and acceptor are variable, as is often the case when using bimolecular reporter systems, more sophisticated measures of FRET efficiency, such as donor fluorescence recovery after acceptor photobleaching and fluorescence lifetime imaging (FLIM), need to be used (for a more detailed discussion about commonly used FRET imaging techniques, please *see* Chapter 2 in this textbook) [55]. Moreover, because the two sensor halves exist independently of one another inside the cell,

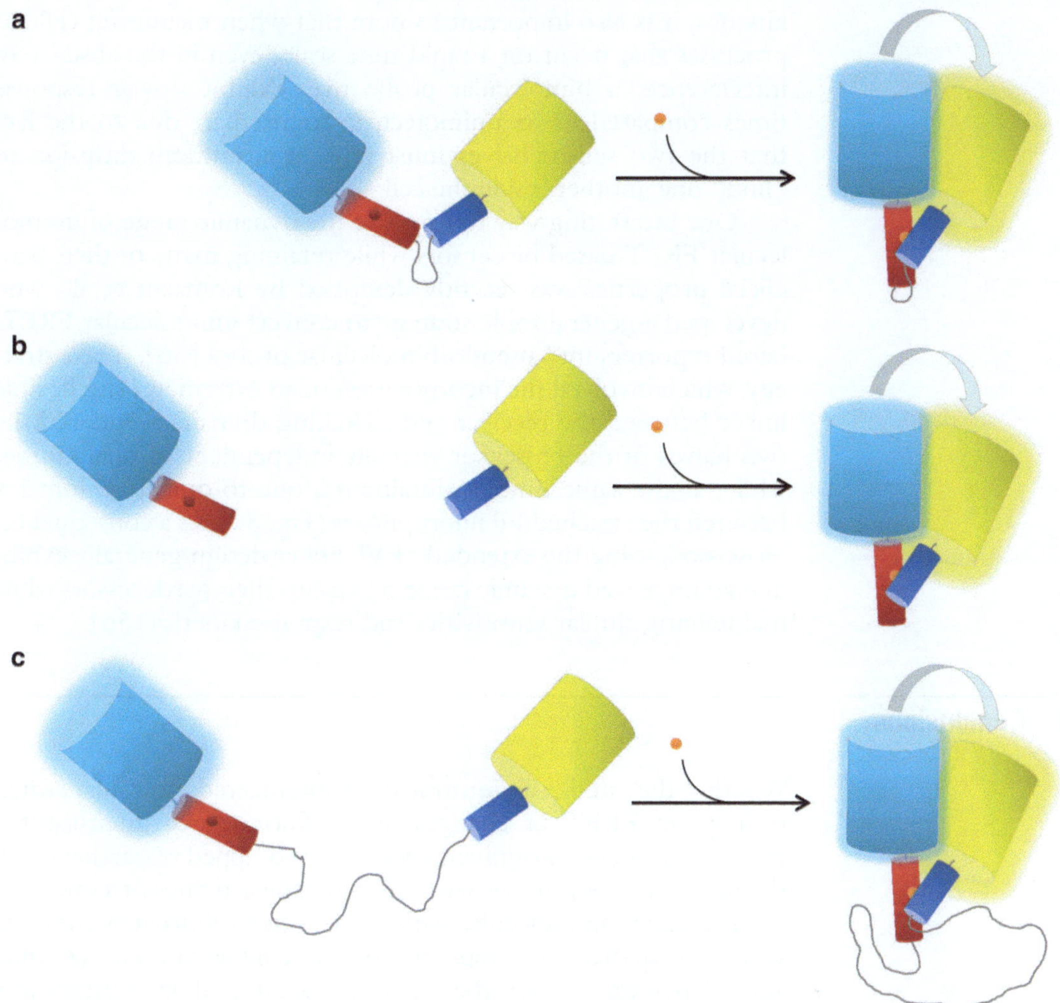

Fig. 3 Variations on an engineered molecular switch. (**a**) The "classic" unimolecular biosensor design. According to this design, the receiver unit (*red*) is tethered to the switching unit (*blue*) by a short flexible linker that is typically between 5 and 15 residues in length. In the presence of ligand (*orange sphere*), the sensor domain undergoes a conformational change that alters the relative distance and orientation of the donor CFP (*cyan cylinder*) and the acceptor YFP (*yellow cylinder*), increasing FRET between them. Most current FRET-based FP biosensors that employ an engineered molecular switch are constructed according to this design. (**b**) A bimolecular reporter design. By removing the flexible linker between the receiver and switching units, the unimolecular biosensor shown in (**a**) is converted to a bimolecular reporter. In the absence of ligand, each half of the sensor behaves independently of the other, leading to lower basal FRET and a higher dynamic range for the sensor (see text for details). (**c**) A pseudo-bimolecular biosensor design. In this design, the receiver unit and the sensor unit are separated by a very long flexible linker that is between 52 and 244 amino acids in length. As a consequence, pseudo-bimolecular reporters exhibit features of both unimolecular and bimolecular probes (see text for details)

they are more susceptible to artifacts such as interactions with endogenous, untagged binding partners. As alluded to above, such nonproductive interactions can both reduce the sensitivity and slow the response kinetics of the sensor. With regard to response kinetics, it is also important to note that when measuring cellular processes that occur on a rapid time scale, even in the absence of interference, a bimolecular probe may exhibit slower response times compared to its unimolecular counterpart due to the fact that the two sensor halves must rely upon random diffusion to "find" one another inside the cell.

One interesting way to improve the dynamic range of unimolecular FRET-based biosensors while retaining many of their beneficial properties was recently described by Komatsu et al., who developed a generalizable strategy to convert unimolecular FRET-based reporters into pseudo-bimolecular probes [56]. Their strategy, which involved the incorporation of an extremely long flexible linker between the receiver and switching domains, rendered the two halves of the biosensor virtually independent of one another while, at the same time, maintaining a one-to-one stoichiometry between the attached FP fluorophores (Fig. 3c). As a consequence, biosensors using the extended "EV" linker design generally exhibited an improved dynamic range relative to their predecessors while maintaining similar sensitivities and response kinetics [56].

4 Conclusions

Whether they utilize an intrinsic or an engineered molecular switch or rely on FRET or changes in FP fluorescence intensity, the FP-based biosensors outlined above have equipped researchers with the molecular tools necessary to monitor the activities of a variety of cellular signaling molecules within the native cellular environment with high spatial and temporal resolution. Due to their versatility and relative ease of use, these sensors have found applications in a variety of areas, including in situ and in vivo imaging (e.g., *see* Chapters 14 and 15), compound screening (e.g., *see* Chapter 17), and computational modeling (e.g., *see* Chapter 18). In the following chapters, we provide in-depth protocols describing not only how to conduct imaging experiments using these sensors, but also how the quantitative data generated from these experiments can be used to better understand the inner-workings of the cell.

References

1. Tsien RY (1998) The green fluorescent protein. Annu Rev Biochem 67:509–544
2. Zimmer M (2002) Green fluorescent protein (GFP): applications, structure, and related photophysical behavior. Chem Rev 102:759–781
3. Remington SJ (2006) Fluorescent proteins: maturation, photochemistry and photophysics. Curr Opin Struct Biol 16:714–721
4. Heim R, Tsien RY (1996) Engineering green fluorescent protein for improved brightness, longer wavelengths and fluorescence resonance energy transfer. Curr Biol 6:178–182
5. Miyawaki A, Griesbeck O, Heim R, Tsien RY (1999) Dynamic and quantitative Ca2+ measurements using improved cameleons. Proc Natl Acad Sci U S A 96:2135–2140
6. Davidson MW, Campbell RE (2009) Engineered fluorescent proteins: innovations and applications. Nat Methods 6: 713–717

7. Shaner NC, Patterson GH, Davidson MW (2007) Advances in fluorescent protein technology. J Cell Sci 120:4247–4260
8. Newman RH, Fosbrink MD, Zhang J (2011) Genetically encodable fluorescent biosensors for tracking signaling dynamics in living cells. Chem Rev 111:3614–3666
9. Sample V, Newman RH, Zhang J (2009) The structure and function of fluorescent proteins. Chem Soc Rev 38:2852–2864
10. Day RN, Davidson MW (2009) The fluorescent protein palette: tools for cellular imaging. Chem Soc Rev 38:2887–2921
11. Pakhomov AA, Martynov VI (2008) GFP family: structural insights into spectral tuning. Chem Biol 15:755–764
12. Chattoraj M, King BA, Bublitz GU, Boxer SG (1996) Ultra-fast excited state dynamics in green fluorescent protein: multiple states and proton transfer. Proc Natl Acad Sci U S A 93: 8362–8367
13. Brejc K, Sixma TK, Kitts PA, Kain SR et al (1997) Structural basis for dual excitation and photoisomerization of the Aequorea victoria green fluorescent protein. Proc Natl Acad Sci U S A 94:2306–2311
14. Ormo M, Cubitt AB, Kallio K, Gross LA et al (1996) Crystal structure of the Aequorea victoria green fluorescent protein. Science 273: 1392–1395
15. Shaner NC, Steinbach PA, Tsien RY (2005) A guide to choosing fluorescent proteins. Nat Methods 2:905–909
16. Abad MF, Di Benedetto G, Magalhaes PJ, Filippin L et al (2004) Mitochondrial pH monitored by a new engineered green fluorescent protein mutant. J Biol Chem 279: 11521–11529
17. Kneen M, Farinas J, Li Y, Verkman AS (1998) Green fluorescent protein as a noninvasive intracellular pH indicator. Biophys J 74: 1591–1599
18. Llopis J, McCaffery JM, Miyawaki A, Farquhar MG et al (1998) Measurement of cytosolic, mitochondrial, and Golgi pH in single living cells with green fluorescent proteins. Proc Natl Acad Sci U S A 95:6803–6808
19. Griesbeck O, Baird GS, Campbell RE, Zacharias DA et al (2001) Reducing the environmental sensitivity of yellow fluorescent protein. Mechanism and applications. J Biol Chem 276:29188–29194
20. Nagai T, Ibata K, Park ES, Kubota M et al (2002) A variant of yellow fluorescent protein with fast and efficient maturation for cell-biological applications. Nat Biotechnol 20: 87–90
21. Heim R, Prasher DC, Tsien RY (1994) Wavelength mutations and posttranslational autoxidation of green fluorescent protein. Proc Natl Acad Sci U S A 91:12501–12504
22. Rizzo MA, Springer GH, Granada B, Piston DW (2004) An improved cyan fluorescent protein variant useful for FRET. Nat Biotechnol 22:445–449
23. Goedhart J, van Weeren L, Hink MA, Vischer NO et al (2010) Bright cyan fluorescent protein variants identified by fluorescence lifetime screening. Nat Methods 7:137–139
24. Klarenbeek JB, Goedhart J, Hink MA, Gadella TW et al (2011) A mTurquoise-based cAMP sensor for both FLIM and ratiometric read-out has improved dynamic range. PLoS One 6: e19170
25. Verkhusha VV, Lukyanov KA (2004) The molecular properties and applications of Anthozoa fluorescent proteins and chromoproteins. Nat Biotechnol 22:289–296
26. Shu X, Shaner NC, Yarbrough CA, Tsien RY et al (2006) Novel chromophores and buried charges control color in mFruits. Biochemistry 45:9639–9647
27. Chudakov DM, Lukyanov S, Lukyanov KA (2005) Fluorescent proteins as a toolkit for in vivo imaging. Trends Biotechnol 23:605–613
28. Wachter RM, Watkins JL, Kim H (2010) Mechanistic diversity of red fluorescence acquisition by GFP-like proteins. Biochemistry 49:7417–7427
29. Merzlyak EM, Goedhart J, Shcherbo D, Bulina ME et al (2007) Bright monomeric red fluorescent protein with an extended fluorescence lifetime. Nat Methods 4:555–557
30. Shaner NC, Lin MZ, McKeown MR, Steinbach PA et al (2008) Improving the photostability of bright monomeric orange and red fluorescent proteins. Nat Methods 5:545–551
31. Shcherbo D, Merzlyak EM, Chepurnykh TV, Fradkov AF et al (2007) Bright far-red fluorescent protein for whole-body imaging. Nat Methods 4:741–746
32. Shcherbo D, Murphy CS, Ermakova GV, Solovieva EA et al (2009) Far-red fluorescent tags for protein imaging in living tissues. Biochem J 418:567–574
33. Lin MZ, McKeown MR, Ng HL, Aguilera TA et al (2009) Autofluorescent proteins with excitation in the optical window for intravital imaging in mammals. Chem Biol 16: 1169–1179
34. Lam AJ, St Pierre F, Gong Y, Marshall JD et al (2012) Improving FRET dynamic range with bright green and red fluorescent proteins. Nat Methods 9:1005–1012
35. Meyer AJ, Dick TP (2010) Fluorescent protein-based redox probes. Antioxid Redox Signal 13:621–650
36. Hanson GT, Aggeler R, Oglesbee D, Cannon M et al (2004) Investigating mitochondrial redox potential with redox-sensitive green fluorescent protein indicators. J Biol Chem 279: 13044–13053

37. Hung YP, Albeck JG, Tantama M, Yellen G (2011) Imaging cytosolic NADH-NAD(+) redox state with a genetically encoded fluorescent biosensor. Cell Metab 14:545–554
38. Dittmer PJ, Miranda JG, Gorski JA, Palmer AE (2009) Genetically encoded sensors to elucidate spatial distribution of cellular zinc. J Biol Chem 284:16289–16297
39. Park JG, Qin Y, Galati DF, Palmer AE (2012) New sensors for quantitative measurement of mitochondrial Zn(2+). ACS Chem Biol 7: 1636–1640
40. Qin Y, Dittmer PJ, Park JG, Jansen KB et al (2011) Measuring steady-state and dynamic endoplasmic reticulum and Golgi Zn2+ with genetically encoded sensors. Proc Natl Acad Sci U S A 108:7351–7356
41. Miyawaki A, Llopis J, Heim R, McCaffery JM et al (1997) Fluorescent indicators for Ca2+ based on green fluorescent proteins and calmodulin. Nature 388:882–887
42. Nagai T, Sawano A, Park ES, Miyawaki A (2001) Circularly permuted green fluorescent proteins engineered to sense Ca2+. Proc Natl Acad Sci U S A 98:3197–3202
43. Nagai T, Yamada S, Tominaga T, Ichikawa M et al (2004) Expanded dynamic range of fluorescent indicators for Ca(2+) by circularly permuted yellow fluorescent proteins. Proc Natl Acad Sci U S A 101:10554–10559
44. Ohkura M, Matsuzaki M, Kasai H, Imoto K et al (2005) Genetically encoded bright Ca2+ probe applicable for dynamic Ca2+ imaging of dendritic spines. Anal Chem 77: 5861–5869
45. Palmer AE, Giacomello M, Kortemme T, Hires SA et al (2006) Ca2+ indicators based on computationally redesigned calmodulin-peptide pairs. Chem Biol 13:521–530
46. Tallini YN, Ohkura M, Choi BR, Ji G et al (2006) Imaging cellular signals in the heart in vivo: cardiac expression of the high-signal Ca2+ indicator GCaMP2. Proc Natl Acad Sci U S A 103:4753–4758
47. Zhang J, Allen MD (2007) FRET-based biosensors for protein kinases: illuminating the kinome. Mol Biosyst 3:759–765
48. Zhang J, Ma Y, Taylor SS, Tsien RY (2001) Genetically encoded reporters of protein kinase A activity reveal impact of substrate tethering. Proc Natl Acad Sci U S A 98: 14997–15002
49. Allen MD, Zhang J (2006) Subcellular dynamics of protein kinase A activity visualized by FRET-based reporters. Biochem Biophys Res Commun 348:716–721
50. Depry C, Allen MD, Zhang J (2011) Visualization of PKA activity in plasma membrane microdomains. Mol Biosyst 7:52–58
51. Zhang J, Hupfeld CJ, Taylor SS, Olefsky JM et al (2005) Insulin disrupts beta-adrenergic signalling to protein kinase A in adipocytes. Nature 437:569–573
52. Zhou X, Herbst-Robinson KJ, Zhang J (2012) Visualizing dynamic activities of signaling enzymes using genetically encodable FRET-based biosensors from designs to applications. Methods Enzymol 504:317–340
53. Knopfel T, Tomita K, Shimazaki R, Sakai R (2003) Optical recordings of membrane potential using genetically targeted voltage-sensitive fluorescent proteins. Methods 30: 42–48
54. Lundby A, Mutoh H, Dimitrov D, Akemann W et al (2008) Engineering of a genetically encodable fluorescent voltage sensor exploiting fast Ci-VSP voltage-sensing movements. PLoS One 3:e2514
55. Zhang J, Campbell RE, Ting AY, Tsien RY (2002) Creating new fluorescent probes for cell biology. Nat Rev Mol Cell Biol 3:906–918
56. Komatsu N, Aoki K, Yamada M, Yukinaga H et al (2011) Development of an optimized backbone of FRET biosensors for kinases and GTPases. Mol Biol Cell 22:4647–4656
57. Yang F, Moss LG, Phillips GN Jr (1996) The molecular structure of green fluorescent protein. Nat Biotechnol 14:1246–1251

Chapter 2

An Introduction to Fluorescence Imaging Techniques Geared Towards Biosensor Applications

J. Goedhart, Mark A. Hink, and Kees Jalink

Abstract

After providing a brief overview of the basics of fluorescence and FRET, this chapter discusses the most commonly used methods to record FRET. Emphasis is on microscopy methods that are widely used for biosensor imaging. We cover choice of instruments, describe various ways to detect FRET based on intensity as well as on donor lifetime, and provide some guidelines to match particular recording methods with specific scientific experiments. We end with an extensive discussion on further practical considerations that may greatly affect the success of the experiments.

Key words Wide-field fluorescence microscopy, Laser scanning microscopy, Ratio-imaging, Fluorescence lifetime imaging

1 Introduction

After providing a brief overview of the basics of fluorescence and FRET, this chapter discusses the most commonly used methods to record FRET. Emphasis is on microscopy methods that are widely used for biosensor imaging. We provide some guidelines to match particular recording methods with specific scientific experiments, and we end by discussing further practical considerations that may greatly affect the success of the experiments.

2 Fluorescence Basics

The term fluorescence was coined by Sir George Gabriel Stokes in the nineteenth century. It describes the process in which a material (the so-called fluorophore) is excited by absorbing light which is subsequently emitted at a longer wavelength. After absorbing light, the molecule is in an excited state. The excited state is short-lived, typically in the nanosecond (10^{-9} s) time range.

Jin Zhang et al. (eds.), *Fluorescent Protein-Based Biosensors: Methods and Protocols*, Methods in Molecular Biology, vol. 1071, DOI 10.1007/978-1-62703-622-1_2,

The light emitted when the molecule returns to the ground state is known as fluorescence emission light or fluorescence for short.

Typically, the excitation light is several orders of magnitude stronger than the emission. To obtain pure fluorescence emission signals, the fluorescence signal needs to be isolated somehow. While this is easy to achieve in a fluorimeter by detecting fluorescence emission from a cuvette at an angle perpendicular to the excitation source, it is more complicated in a microscope. The most popular configuration to observe fluorescence with a microscope is the epifluorescence mode, in which both the excitation light and emission light pass through the objective (Fig. 1). Excitation light is typically obtained by filtering suitable wavelength from a white-light source using a interference filter. A dichroic mirror reflects the excitation light onto the sample while allowing the fluorescence emission light to freely pass. An emission filter is required to completely block any excitation light that is reflected (scattered) by the sample and might leak through the dichroic mirror. Emission filters typically have a transmittance of 10^{-4}–10^{-5} % at the excitation wavelengths whereas they allow light to pass with an efficiency close to 100 % at the emission wavelengths.

Although the epifluorescence mode provides high signal to noise, it is important to note that only a fraction of the emission light is detected (about 10–25 %). The detection efficiency scales with the square of the numerical aperture (NA) of the objective. Therefore, the use of high NA objectives (NA 1.2–1.4) is recommended when high sensitivity is desired.

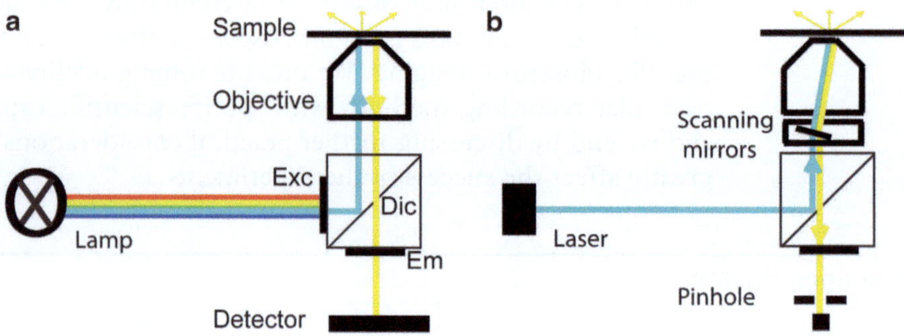

Fig. 1 (**a**) Basic setup of filters in fluorescence microscopy. The excitation source, here a mercury lamp, produces a broad range of light of various colors. A restricted range of wavelengths (here *blue*) is selected by the excitation filter (Exc) and reflected towards the sample by the dichroic mirror (Dic), placed under an angle of 45°. The light is focused by an objective lens into the sample resulting in a full illumination of the sample, resulting in fluorescence (*yellow*). Since the fluorescence has a different wavelength, light passes the dichroic mirror. An emission filter (Em) allows detection of the fluorescent signal while providing an additional block for the intense excitation light. In the case of a laser scanning confocal microscope (**b**), a monochromatic laser beam is focused into a single point in the sample via the scanning mirrors. By moving the scan-mirrors and therefore moving the excitation spot, the whole sample is imaged point-by-point. Detection of out-of-focus light is reduced by placing an aperture, the pinhole, in the detection path resulting in optical sectioning

3 FRET

Förster resonance energy transfer (FRET) is the radiationless transfer of energy from the excited state of a donor fluorophore to an acceptor. FRET occurs when a donor and acceptor are in close proximity, usually within 1–10 nm of each other. Importantly, this range matches the scale at which protein–protein interactions take place. Since the FRET efficiency depends on the inverse sixth power of distance ($E \equiv 1/r^6$), it is very sensitive to distance. Other than distance, the efficiency of FRET also depends strongly on the orientation between the donor and acceptor dipole moments, which allows FRET-based detection of conformational changes.

The applications of FRET include detection of protein–protein interactions, detecting protein conformational changes, and, most relevant to this volume, as readout of so-called "sensors", i.e., genetically encoded constructs that are designed to spy on cellular biochemistry, such as the activity of proteases or the concentration of second messengers. Literally hundreds of such sensors have been described and the development and optimization of FRET sensors is a very active field. The application of FRET sensors allows us to detect any process with high spatial and temporal resolution in single living cells, giving us an unprecedented view into cellular functioning.

The basic spectroscopic changes that occur upon FRET are quenching of donor fluorescence, a decrease in excited state lifetime of the donor and, if the acceptor is a fluorophore, emission of fluorescence by the acceptor. The latter is known as sensitized emission. All these parameters can be used to detect and quantify FRET by fluorescence microscopy. Other spectroscopic changes, including fluorescence anisotropy and a change in photobleaching kinetics, can also be used to measure FRET but will not be discussed here. Further details on FRET recording and the required instrumentation will be described in next section.

FRET requires matched fluorophores, which means that the spectra of donor emission and acceptor absorbance should have substantial overlap. In addition the quantum yield of the donor and extinction coefficient of the acceptor should be sufficiently high. The cyan and yellow fluorescent protein pair is currently the most widely employed pair for FRET based genetically encoded biosensors, both for historical reasons as well as favorable spectroscopic characteristics. For more information about suitable fluorophores for FRET we refer to Chapter 1. It has been pointed out that the method of FRET detection also affects the choice of donor and acceptor fluorophores, see for instance [1].

4 Fluorescence Microscopes

Two kinds of fluorescence microscopy setups are commonly available in most cell biology laboratories: the wide-field fluorescence microscope and the confocal (laser scanning) microscope. Both instruments can be readily used to perform FRET imaging. We will discuss some of the basic principles and differences.

4.1 The Wide-Field Microscope

In wide-field microscopy, the sample is illuminated uniformly over the field of view. The emitted fluorescence is captured by a camera. In contrast to confocal microscopy, there is no optical sectioning, resulting in the image being somewhat deteriorated by blur from out-of-focus parts of the cells. The typical resolution for a wide-field microscope is about half the wavelength of the used light, or ~250 nm, in the *XY*-plane and whereas the resolution along the *Z*-axis is about 700 nm.

High power Mercury and Xenon lamps are often used as excitation sources. These lamps emit broad-spectrum (white) light, which needs to be filtered to obtain the correct wavelength for excitation. High quality excitation filters are used that only transmit a small bandwidth of light (Fig. 1a). More recently, colored Light Emitting Diodes (LEDs) are replacing the arc lamps. For quantitative microscopy, sensitive cameras that linearly detect fluorescence are crucial. Cooled CCD (Charge-Coupled Device) cameras provide good sensitivity and high resolution. Current state-of-the-art imaging is performed with Electron-Multiplied CCD cameras which combine optimal sensitivity (95 % quantum efficiency for back-illuminated CCD cameras), fast acquisition (>1,000 frames/s) and high resolution (>1 megapixels).

4.2 The Confocal Laser Scanning Microscope

In laser scanning microscopy, for excitation a focused light spot is used to scan the sample point-by-point (Fig. 1b). The emission light is passed through a pinhole and fluorescence intensity of each point, the pixel, is recorded with a detector. The pinhole rejects out-of-focus fluorescence, which results in "optical sectioning" Thus, the resolution is increased, especially along the *z*-axis. The confocal resolution is only marginally better than that of the wide-field microscope, but since it lacks out-of-focus blur it yields crisp optical sections.

A laser is usually employed as the excitation source, since it produces intense light that can be focused to a small so-called diffraction-limited spot. Since lasers are monochromatic, one is limited to the available lasers for excitation but excitation filters are not necessary. Detection of fluorescence is usually by a photomultiplier tube (PMT). The avalanche photodiode (APD) is a more sensitive alternative, but it requires more careful handling. Important parameters when purchasing a PMT are the sensitivity and speed needed in the experiment. The detection efficiency of the classical PMTs that are included in most confocal microscopes is 10–15 % and it decreases sharply in the red part of the spectrum. Recent, more sensitive variants may have efficiencies up to 50 %. PMTs and APDs may be operated either in linear mode or in photon-counting (Geiger) mode. It is important to note that, at very low and at high photon fluxes, detectors may not be linear which would of course hamper quantitative measurements. In photon counting mode, detector dead-time (the time period required to process the signal of the photon, during which it is insensitive for further photons) is an important consideration because it limits detection speed as well as efficiency.

4.3 Alternative Excitation Modes

Several alternative approaches to image fluorescence are available. For wide-field microscopes, TIRF excitation [2] or Nipkow spinning disk scanning [3] can be implemented to increase *z-resolution*. Laser scanning confocal microscopes can be equipped with a source for two-photon excitation, providing better tissue penetration for intravital imaging. Although useful for FRET, a detailed explanation of these techniques is outside of the scope of this introductory chapter.

5 FRET Detection Methods

For a detailed description of FRET imaging we refer to the authoritative volume edited by Gadella Jr. [4] and to Jares-Erijman [5]. Useful suggestions on selecting the right imaging approach for an application are in ref. 6. Here, we will describe the most common techniques to measure FRET in single living cells, emphasizing those most useful for fluorescent biosensors. Besides intensity-based methods (ratio-imaging and acceptor photobleaching) which are readily performed on ordinary fluorescence microscopes we will also discuss fluorescence lifetime imaging as a robust method to detect FRET. FLIM detection equipment is now commercially available, either as integrated "push-button" system or as add-on for wide-field or laser scanning microscopes.

5.1 Intensity-Imaging

In *ratio-imaging* (Fig. 2a), the intensity of both donor emission and sensitized emission (i.e., the emission detected from the acceptor upon excitation of the donor) are detected. This is by far the most popular method for (dynamic) biosensor imaging due to its simplicity, sensitivity and speed. Ratio-imaging is often not carried out in a quantitative manner, although addition of extra corrections or use of an end-point calibration can easily remedy that. This method is very photon-efficient and most useful to record FRET changes. As said, ratiometry is not, in general, quantitative because the recorded sensitized emission image is contaminated by bleedthrough of the donor molecules, and because it also contains

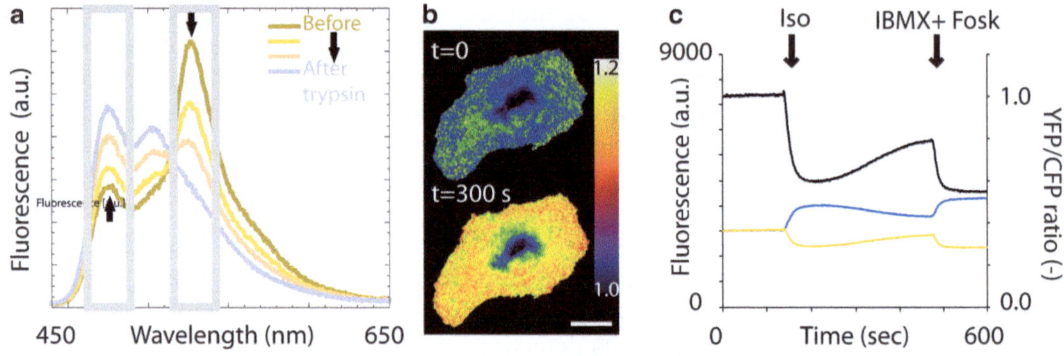

Fig. 2 (**a**) Fluorescence spectra of a purified CFP-YFP fusion protein After addition of trypsin, the fusion protein is progressively cleaved, resulting in a loss of FRET that can be monitored over time as a loss in sensitized emission (~530 nm) and an increase in CFP donor fluorescence (~480 nm). A fast readout of this sensor may be achieved by monitoring the 530/480 intensity ratio using emission filters specific for these spectral regions, as indicated by the *gray* blocks. (**b**) YFP/CFP ratio image of HUVEC cells transfected with a FRET sensor. A significant ratio change was observed 300 s after stimulation of the cells. (**c**) Dual-PMT ratio experiment of ^{T}EPAC^{VV} expressed in HeLa cells. Upon stimulation with isoproterenol or IBMX and forskolin, the FRET efficiency is lowered thereby increasing the CFP (*blue*) and lowering the YFP (*yellow*). *Black* denotes the ratio (YFP/CFP)

emission from acceptors that are directly excited by the excitation at donor wavelength, so-called cross-excitation.

Experiments usually start with recording of a baseline from a cell or population of cells. Cells are then stimulated with an agonist that alters FRET, and the changes are expressed as deviation (in percentage) of that change from the baseline, or as a percentage of the change induced by a control stimulus at the end of the experiment (Fig. 2b).

An extension of this technique, *filterFRET*, allows quantitative FRET recording from emission intensities. To this goal, ratiometry is upgraded by simply recording an additional image that reflects the emission of acceptors when excited at their own characteristic wavelength. It can be shown that from such image triplets, FRET can be quantified in a straight-forward manner [4, 7]. Detailed description is beyond the scope of this chapter but provided in the literature.

For ratio-imaging, the speed of acquisition requires some consideration. Fast-moving structures such as cellular vesicles or the leading edge of the plasma membrane of a moving cell may lead to an artifact when the ratio of the two images is taken. Thus, the acquisition of donor and acceptor channels should be faster than those changes to avoid movement artifacts. In confocal microscopes, simultaneous detection of two or more channels is a basic feature, allowing donor and acceptor channels to be imaged without delay. In wide-field microscopes, the emission filters are often located in a filter cube or emission filter wheel, and the images have to be taken consecutively. The integration time and time added by changing the filter will limit the application for the fastest biological processes. Parallel imaging can, however, be achieved in several ways. It is relatively straightforward to incorporate a secondary dichroic mirror and two CCD cameras, but this is rather expensive. An excitation-economic alternative is implementation of an image splitter (marketed under the name of optosplit or dualview), which projects two emission channels side-by-side on a single CCD camera. Another option is to combine an emission beamsplitter with two point detectors (PMTs), a method known as "dual-photometer detection". At the expense of spatial information, this allows the fastest and highest sensitivity detection by far. Dual photometer detection is often used when very dim excitation intensities are necessary to avoid photodamage to sensitive cells (Fig. 2c).

5.2 Acceptor Photobleaching

The basic idea behind acceptor photobleaching is that the donor is quenched due to FRET. By destroying the acceptor, FRET is no longer taking place and the donor becomes dequenched (Fig. 3). The FRET efficiency can be directly quantified from the intensity increase of the donor.

Laser scanning systems are popular for this technique as these have high-intensity monochromatic excitation lasers which can be

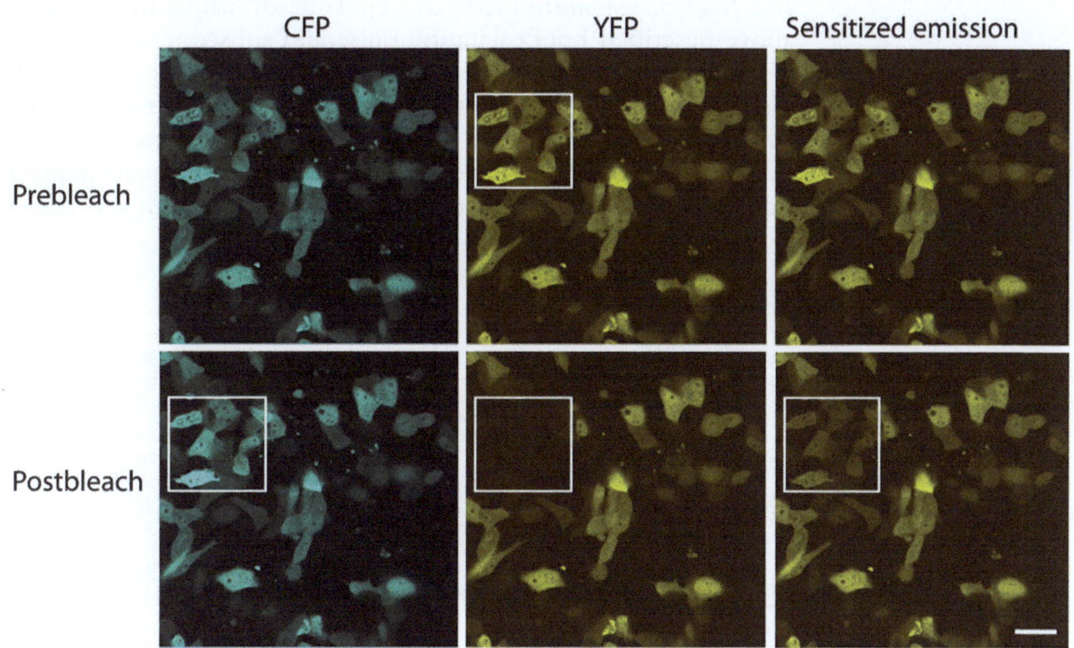

Fig. 3 Acceptor photobleaching experiment of a FRETing CFP-YFP fusion protein expressed in HeLa cells. By selectively exciting the YFP acceptor protein in a region of interest (*box*) using the full intensity of a 514 nm laser line, the YFP protein was destroyed locally. In this ROI the FRET was abolished resulting in an increase of the CFP donor fluorescence and a reduction of the sensitized emission. The remaining fluorescence in the sensitized emission panel is CFP cross-talk (leak-through) in the YFP detector

used to excite/bleach the acceptor with high specificity without exciting and possibly destroying the donor. For instance, the 514 nm laser line selectively bleaches YFP without exciting CFP, and 561/568 nm laser lines can be used to specifically bleach red fluorescent proteins that act as acceptors to green-shifted variants. Note that this technique is destructive so it cannot provide temporal information; it is, however, very useful as an extra control (end-point calibration) for other methods.

5.3 Lifetime Imaging

Imagine a population of fluorophores that is excited by a very brief (ps) pulse of light. These molecules will not emit fluorescence all at the same time; rather, they emit randomly and the population will thus send out light in an exponentially decaying manner (Fig. 4a). The decay time or characteristic fluorescence lifetime, τ, for common fluorophores in biology is between ~0.1 and 5 ns. FRET can be quantified with high precision by detecting lifetimes because it causes a reduction of the lifetime. Since the lifetime is a kinetic parameter, it is independent of excitation or emission intensity, and not affected by expression level, cell thickness and local concentration. Fluorescence lifetime imaging microscopy (FLIM) requires dedicated and expensive equipment but delivers trouble-free "push-button" FRET pictures.

Two acquisition modes are popular, the so-called *time-domain* and *frequency-domain* methods [4]. Time-domain FLIM uses the above-described brief excitation pulse and subsequently measures (many) photon arrival times to construct a decay curve in a technique known as time-correlated singles photon counting (TCSPC). The kinetics of the decay curve can be analyzed to obtain the excited state lifetime. This technique presents the ultimate in precision, but since it depends on individual photons being collected

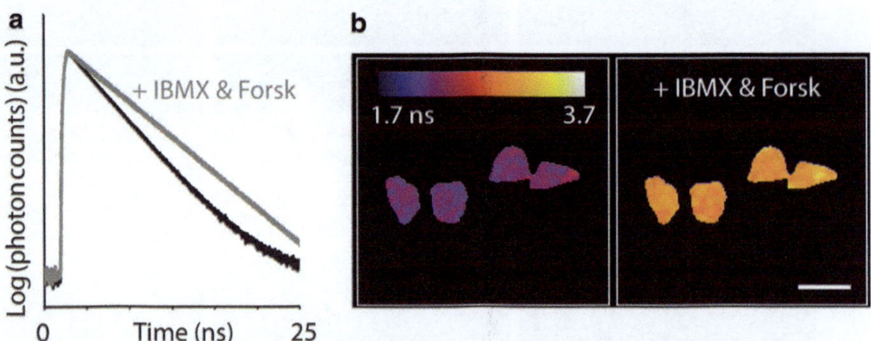

Fig. 4 (**a**) Fluorescence lifetime decay histogram of purified cAMP sensor ^{T}EPAC^{VV} (*black*). Upon stimulation with IBMX and forskolin, FRET is lowered, resulting in increased fluorescence lifetime and thus a slower decay of the curve (*gray*). The amplitudes of the curves were normalized for clarity. (**b**) Frequency-domain FLIM analysis of ^{T}EPAC^{VV} expressed in HeLa cells before and after stimulation. The images were intensity thresholded to reduce background contribution and pseudo-colored to display the phase lifetimes

sequentially for each pixel, image acquisition times of several minutes are common.

In frequency domain FLIM, the sample is irradiated with light that is intensity-modulated, typically at 40–80 MHz [4]. The fluorescence emitted by the cells is therefore also modulated with the same frequency, but it will be shifted in phase and its amplitude will be demodulated, depending on the lifetime. Calculation of the lifetime based on phase-shift and demodulation yields not one, but two parameters with usable information. For mono-exponentially decaying fluorophores, both lifetimes should be identical, but when the decay is multi-exponential, the modulation lifetime is higher than the phase lifetime.

Acquisition of lifetime images (Fig. 4b) with frequency-FLIM is much faster than with TCSPC but still slower than with intensity-based methods (ratio-imaging) and therefore this approach is less suitable for fast FRET imaging (<1 s). Care should be taken that the acquisition time is short relative to possible movement of structures in the sample in order to obtain reliable data. Quantification of FRET by FLIM may be complicated if the donor has a multi-exponential decay, like most CFP variants. EGFP [8] and the new CFP variants mTurquoise and mTurquoise2 [9] are mono-exponential and therefore popular for FLIM.

Both frequency-domain and time-domain FLIM can be implemented on fluorescence microscopes. For technical reasons, frequency-domain FLIM is usually implemented on wide-field systems, while the time-domain method is more easily compatible with confocal laser scanning microscopes.

6 Spying on Intracellular Messengers by FRET: Practical Considerations

Many practical considerations guide readout of FRET sensors by microscopy. These can be loosely grouped in issues relating to biology, to the fluorophores and, most important in the context of this chapter, to the choice of acquisition method.

6.1 Biology

Biosensors should interfere minimally with normal cell physiology. It is important to carefully ensure in the cells to be studied that the biosensor has minimal effects of the readout, for example by comparing responses in cells with high expression levels to those with (very) low levels of the sensor. Whereas the noise levels may be high in the latter case, both experiments should reveal at least similar kinetics of the process under study. This ensures that the sensor neither buffers the analyte nor affects the reaction steps in another manner. Any unwanted activity of the sensor, such as catalytic activity [10] or binding to endogenous components [11], must be removed. Note that the choice of fluorophores may be important too. For example, we noticed that FRET sensors employing monomeric

GFP variants may behave remarkably different from similar sensors that incorporate "dimerizing" fluorophores in, for example, their tendency to dimerize and localize to organelles (unpublished observations and ref. 1).

The various techniques described in the previous paragraphs require widely differing amounts of excitation light to achieve a good signal-to-noise ratio of the readout. In order not to disturb sensitive cellular processes such as mitosis, it may be necessary to be less demanding on the required spatiotemporal resolution and precision. Simple ratiometric imaging is very photon-efficient, and both sensitized emission and FLIM may be sped up by summing, so-called binning, camera pixels.

6.2 Fluorophores

Stable fluorescence signals are essential for quantitative imaging, regardless of the approach. The sensors should exhibit stable (i.e., no photobleaching and no photochromism) and bright signals. Large differences exist in the performance of individual FPs, in this respect. Therefore, photostable FPs should be used at the lowest practical intensity of illumination.

Theoretically, it is preferable to employ fluorescent proteins that emit in the red part of the visible spectrum, since cells show less autofluorescence in this region. In addition, the Förster radius, a parameter that quantifies the FRET efficiency of a certain donor–acceptor pair, is higher while the excitation light is of lower energy, causing it to be less harmful for the cells. However, several issues impede the wide application of red acceptors. First of all, the red acceptors reported thus far have rather low quantum yield.

Secondly, some red fluorescent proteins mature via green intermediates. Therefore, maturation should be fast, a requirement which is met by only a few variants [1, 12]. For example, we find that the majority of inter-cell variability when reading out resonance levels with, e.g., the GFP-tdTomato FRET pair, is caused by poor maturation of tdTomato. Unmatured acceptors will cause erroneous readout of FRET in ratio measurements, as well as in FLIM. We anticipate that red-shifted FRET biosensors will become more popular once maturation- and brightness issues have been solved. Fluorophore choice is discussed in much more detail in Chapter 4 of this textbook.

6.3 Acquisition

Not all biosensors are equally suited for the different acquisition methods described in this chapter. Ratio-imaging and *filterFRET* rely on detection of sensitized emission and therefore profit from bright (high quantum yield) acceptors. An elegant way to improve acceptor brightness is by increasing the number of acceptors [1]. FLIM, on the other hand, is based on donor fluorescence only and thus the light yield of the acceptor is not important. Rather, FLIM may benefit from so-called "dark" acceptors in biosensors [13]. These acceptors are chromophores, rather than fluorophores, and absorb but do not emit light. Therefore, the acceptor channel is not occupied by the biosensor and can be used for another probe or sensor for multiplexing applications. A disadvantage may be that no information on the actual distribution of the acceptor can be obtained. Furthermore, since the donor is quenched when FRET occurs, the signal obtained by FLIM decreases when FRET increases. This means that the FRET efficiency should not be too high for reliable analysis by FLIM; in practice, this is hardly ever a problem since, in fluorescent protein-based FRET, efficiencies are usually well below 50 %.

Other considerations concern the quantitation of the response. Simple ratiometry effectively cancels out excitation intensity fluctuations, image shading (lateral intensity variations present in the image) and local abundance of the sensor. However, in triple-image *filterFRET* images, the acquired acceptor image may contain a differing amount of shading, and this should be corrected in order to get the best possible quantitation, particularly on older confocal microscopes. For an excellent review on details of quantitative microscopy and necessary corrections, see Waters [14].

Pixel-shift between donor and acceptor channel may be introduced by image-splitting devices on wide-field microscopes and needs to be meticulously corrected for, e.g., by imaging submicron beads. Post-acquisition, the pixel-shift can be corrected using image processing (for a detailed procedure see Kardash et al. [15]). A detailed procedure for sensitized emission–based FRET imaging using confocal microscopy, including all corrections has also been reported [4, 7]. Note that FLIM measurements are essentially free of shading and pixel-shift issues.

Finally, it must be emphasized again that the various acquisition methods differ widely in sensitivity, speed and photon efficiency. TCSPC represents the ultimate in precision, requiring no calibration at all, but the acquisition is so slow that for imaging live-cell dynamics it is usually not an option. However, when coupled to a confocal microscope, high-resolution FRET images can be obtained. On the other hand, depending on sensor expression levels, frequency-domain FLIM yields quantitative results in ~1 to a few seconds. With current instrumentation, this technique is not very photon-efficient, requiring an estimated 5–20 times more excitation to achieve a good quality image. In addition, in our experience, very small FRET changes usually escape detection by these FLIM methods. Intensity-based FRET readings, on the other hand, are not quantitative unless additional calibrations and corrections are carried out, but they are very sensitive, even to small changes in FRET. The photon efficiency is high and readout rate is, in principle, only limited by the emission intensity; for example, using a resonant scanning confocal, full-frame emission ratios can be easily detected at video rate. The ultimate in speed, finally, is represented by dual-photometer readout: sub-ms recording is readily achieved, even at very dim excitation, but spatial resolution is sacrificed completely with this method. We detect FRET changes as little as 0.25 % with our setup, but quantification is usually out of the question. Thus, depending on the questions to be answered, it pays to invest in more than just one acquisition setup.

7 Conclusion

In many cell biology labs fluorescence microscopy setups are available that can be employed for measuring FRET-based biosensors. The first priority is to make sure that the right excitation wavelength and correct filters (dichroic and emission filters) are in place to do the FRET measurements. Additional modifications may be necessary for specific requirements (fast imaging or high spatial resolution). To set up the system, it is worthwhile to use FRET biosensors which exhibit high contrast and are easy to manipulate. Good examples are YCam3.6 and the new generation of EPAC-based sensors ($^{T}EPAC^{VV}$). These biosensors are readily expressed in eukaryotic cells, have a high dynamic range and a FRET change that can be conveniently induced by adding small molecule reagents (for more information on measuring cAMP in cells with the EPAC-based sensor see Chapters 4 and 5 in this textbook).

References

1. van der Krogt GN, Ogink J, Ponsioen B, Jalink K (2008) A comparison of donor–acceptor pairs for genetically encoded FRET sensors: application to the Epac cAMP sensor as an example. PLoS One 3:e1916
2. Steyer JA, Almers W (2001) A real-time view of life within 100 nm of the plasma membrane. Nat Rev Mol Cell Biol 2:268–276
3. Nakano A (2002) Spinning-disk confocal microscopy – a cutting-edge tool for imaging of membrane traffic. Cell Struct Funct 27:349–355
4. Gadella Jr TWJ (ed) (2009) FRET and FLIM techniques. Lab Techn Biochem Mol Biol 33: 1–534, Elsevier, Amsterdam, The Netherlands
5. Jares-Erijman EA, Jovin TM (2003) FRET imaging. Nat Biotechnol 21:1387–1395
6. Pietraszewska-Bogiel A, Gadella TW (2011) FRET microscopy: from principle to routine technology in cell biology. J Microsc 241:111–118
7. van Rheenen J, Langeslag M, Jalink K (2004) Correcting confocal acquisition to optimize imaging of fluorescence resonance energy transfer by sensitized emission. Biophys J 86:2517–2529
8. Verveer PJ, Wouters FS, Reynolds AR, Bastiaens PI (2000) Quantitative imaging of lateral ErbB1 receptor signal propagation in the plasma membrane. Science 290:1567–1570
9. Goedhart J et al (2012) Structure-guided evolution of cyan fluorescent proteins towards a quantum yield of 93%. Nat Commun 3:751
10. Ponsioen B et al (2004) Detecting cAMP-induced Epac activation by fluorescence resonance energy transfer: Epac as a novel cAMP indicator. EMBO Rep 5:1176–1180
11. Palmer AE, Giacomello M, Kortemme T, Hires SA, Lev-Ram V, Baker D, Tsien RY (2006) Ca2+ indicators based on computationally redesigned calmodulin-peptide pairs. Chem Biol 13:521–530
12. Goedhart J, Vermeer JE, Adjobo-Hermans MJ, van Weeren L, Gadella TW Jr (2007) Sensitive detection of p65 homodimers using red-shifted and fluorescent protein-based FRET couples. PLoS One 2:e1011
13. Lee SJ, Escobedo-Lozoya Y, Szatmari EM, Yasuda R (2009) Activation of CaMKII in single dendritic spines during long-term potentiation. Nature 458:299–304
14. Waters JC (2009) Accuracy and precision in quantitative fluorescence microscopy. J Cell Biol 185:1135–1148
15. Kardash E, Bandemer J, Raz E (2011) Imaging protein activity in live embryos using fluorescence resonance energy transfer biosensors. Nat Protoc 6:1835–1846

Chapter 3

Quantitative Measurement of Ca^{2+} and Zn^{2+} in Mammalian Cells Using Genetically Encoded Fluorescent Biosensors

J. Genevieve Park and Amy E. Palmer

Abstract

Genetically encoded, ratiometric, fluorescent biosensors can be used to quantitatively measure intracellular ion concentrations in living cells. We describe important factors to consider when selecting a Ca^{2+} or Zn^{2+} biosensor, such as the sensor's dissociation constant (K_d') and its dynamic range. We also discuss the limits of quantitative measurement using these sensors and reasons why a sensor may perform differently in different biological systems or subcellular compartments. We outline protocols for (1) quickly confirming sensor functionality in a new biological system, (2) calibrating a sensor to convert a sensor's FRET ratio to ion concentration, and (3) titrating a sensor in living cells to obtain its K_d' under different experimental conditions.

Key words Genetically encoded sensors, Fluorescence microscopy, Förster resonance energy transfer, Live cell imaging, Zinc, Calcium, Quantitative, Ratiometric

1 Introduction

1.1 Sensor Design

Fluorescent biosensors are valuable tools for observing dynamic changes in intracellular ion concentrations. We will briefly compare and contrast ratiometric to intensiometric sensors, describing how to select and use an appropriate genetically encoded, ratiometric, fluorescent, metal ion sensor in living cells. We will also discuss some important complexities and limitations of quantitative measurements using these sensors.

In practice, there are two major types of fluorescent biosensors for metal ions: "intensiometric" biosensors that change fluorescence intensity when bound to an ion and "ratiometric" biosensors that exhibit a shift in the absorption or emission spectra when bound to an ion. The fluorescence intensity of an intensiometric sensor is dependent on the sensor concentration in each cell and the path length (i.e., the thickness of a cell) in addition to the ion concentration; hence ratiometric biosensors are preferred for quantitative measurements. On the other hand, ratiometric biosensors

Jin Zhang et al. (eds.), *Fluorescent Protein-Based Biosensors: Methods and Protocols*, Methods in Molecular Biology, vol. 1071, DOI 10.1007/978-1-62703-622-1_3,

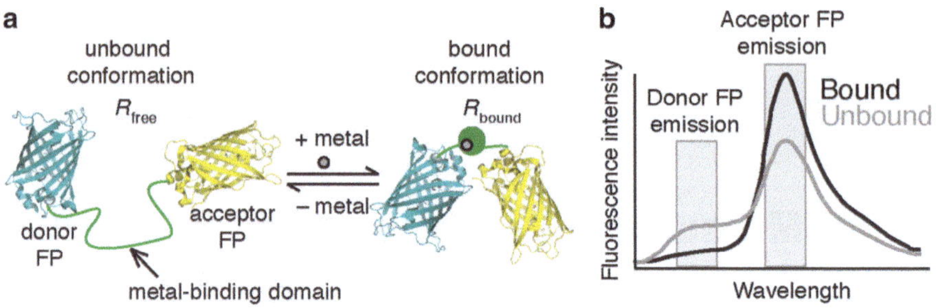

Fig. 1 Design of the Ca^{2+} and Zn^{2+} sensors used in this protocol. (**a**) A genetically encoded, fluorescent, ratiometric sensor contains a Ca^{2+} or Zn^{2+} binding domain fused to a donor FP (usually cyan FP) at is N-terminal end and an acceptor FP (usually yellow FP) at its C-terminal end. When Ca^{2+} or Zn^{2+} reversibly binds to its binding domain, the sensor changes conformation, which leads to a change in FRET efficiency. (**b**) shows how the change in FRET efficiency changes the FRET ratio, which is the ratio of acceptor FP to donor FP emission intensity upon donor FP excitation. Thus, the FRET ratio of the unbound sensor (R_{free}) is distinct from that of the bound sensor (R_{bound})

have several experimental limitations, which include lower sensitivity (i.e., smaller dynamic range), a larger spectral bandwidth, and the need to acquire images with two combinations of fluorescence excitation and emission filters.

In this protocol, we focus on genetically encoded biosensors: proteins encoded by DNA that is introduced into living cells by transient transfection or viral transduction. In contrast to small molecule biosensors, which are chemically synthesized and are introduced to cells immediately before the experiment, genetically encoded biosensors are manufactured by the cell and become functional without further intervention by the investigator. Genetically encoded biosensors are readily targeted to subcellular locations by appending localization sequences to the DNA sequence, a major advantage when the investigator desires the ability to monitor ion concentrations in organelles, such as the ER, Golgi, nucleus, or mitochondria. Subcellular targeting to some locations like vesicles is hindered by the size of some sensors (typical ratiometric FRET sensors are 60–65 kDa).

Almost all available genetically encoded, ratiometric, fluorescent metal ion sensors have the following design (*see* Fig. 1a): a donor fluorescent protein (FP) is attached to an acceptor FP by a linker containing the metal-binding domain. The chromophore of an FP forms auto-catalytically from three amino acid residues inside of a beta barrel. The biosensor undergoes a conformational change upon binding the metal ion, changing the distance and orientation between the two FPs and consequently the efficiency of Förster resonance energy transfer (FRET) (reviewed in refs. 1, 2). The change in FREt alters the emission spectra of the biosensor, decreasing the peak of donor FP emission and increasing the peak of acceptor FP emission (*see* Fig. 1b).

The FRET ratio is defined as the ratio of the acceptor FP emission intensity upon donor excitation to the donor FP emission intensity. The FRET ratio can be converted to an ion concentration when three parameters are known: (1) the sensor affinity in terms of K_d', (2) the FRET ratio in the absence of the ion (R_{free}), and (3) the FRET ratio when the sensor is saturated with the ion (R_{bound}). The sensor affinity can be measured either in vitro or in situ (i.e., in cells) and is usually published in the literature, as discussed below. R_{free} and R_{bound} are measured at the end of each experiment (*see* Subheading 3.2). Typically, R_{free} is the minimum FRET ratio (R_{min}) and R_{bound} is the maximum FRET ratio (R_{max}), but sometimes the sensor response is inverted and the opposite is true [3].

1.2 Understanding Binding Affinity and Dynamic Range

Every biosensor is sensitive to changes within a range of ion concentrations, which spans about two orders of magnitude. This range is largely determined by the sensor's binding affinity for an ion (or ions) and its dynamic range (DR). The binding affinity is reported as the apparent dissociation constant (K_d'), which is equal to the ion concentration when 50 % of the sensor is bound (*see* Fig. 2). Occasionally, when a sensor has multiple binding sites with different binding affinities, multiple K_d' values are reported. A K_d' is determined by fitting a binding curve to experimental data from a sensor titration experiment, in which the FRET ratio of the sensor is measured at different ion concentrations. The simplest binding equation is used to describe the titration data, even though it may not represent actual binding events (*see* Fig. 3). For example, the Ca^{2+} sensor D3cpV is fit to a single-site binding equation, even though each sensor binds to four Ca^{2+} ions [4]. It is important to note that temperature, salt concentration, pH, and other factors can affect the K_d', and that most titrations are performed in vitro with protein purified from a bacterial expression system. However, experiments performed in our group and others indicate that the K_d' in vitro and in cells are often comparable [3, 5]. This is something that each investigator can verify in his or her own experimental system using the protocol outlined in Subheading 3.3.

DR has many definitions in the literature, but it is essentially an indicator of a sensor's measurement sensitivity and its signal-to-noise ratio (SNR). Common definitions of DR include the fold-change in FRET ratio ($DR = R_{max}/R_{min}$); or, the maximum change in FRET ratio ($DR = R_{max} - R_{min}$). SNR is also described as the ratio of R to the standard deviation of R at baseline [6, 7]. Several factors can significantly affect a sensor's DR (or SNR), including the microscopy system used for measurement, the cell type expressing the sensor, and the subcellular location. Accordingly, reported DRs are most useful for relative comparisons and will not always be the same in across experimental systems. Moreover there are numerous examples of sensors exhibiting decreased DR in cells compared to

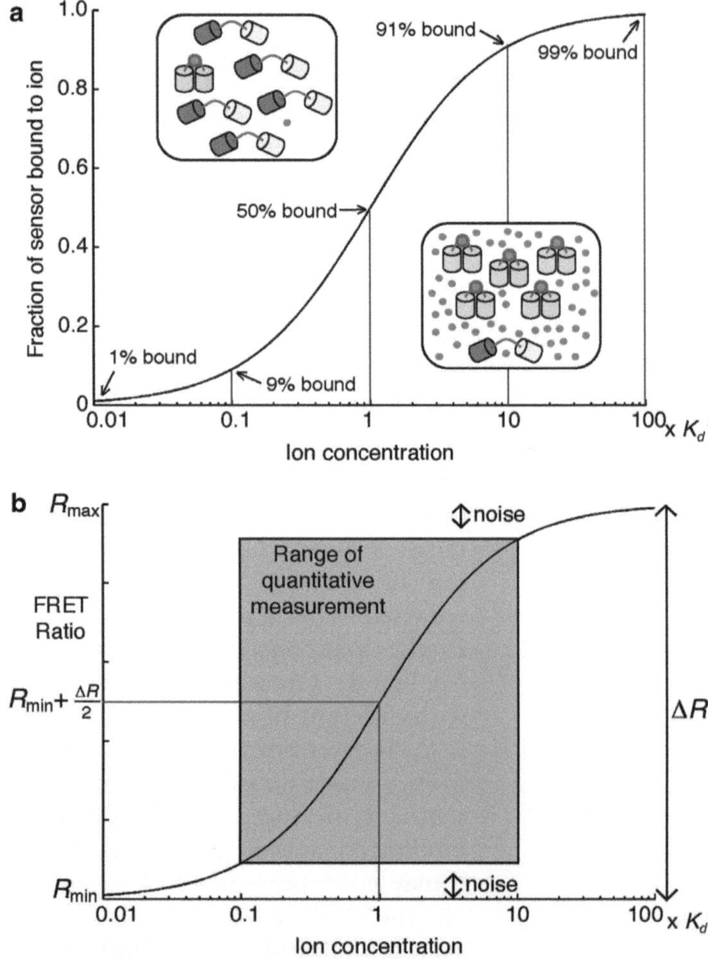

Fig. 2 A sensor's sensitivity to changes in Ca^{2+} or Zn^{2+} concentration is related to its dissociation constant (K_d'). (**a**) When the ion concentration ([ion]) is the same as the sensor's K_d', 50 % of a population of sensors (typically 1–20 μM in cells) is bound to the ion. The fraction of sensor bound changes the most when the [ion] varies within tenfold of the sensor's K_d'. (**b**) The FRET ratio is a linear function of the fraction bound, and so the midpoint of R_{min} and R_{max} corresponds to an [ion] equal to the sensor's K_d'. The SNR and the K_d' limit the range of the [ion] that can be quantified by the sensor

that measured in vitro and hence researchers are encouraged to seek out and compare R_{max} and R_{min} values from in situ experiments, which are often reported graphically in publication figures.

A sensor is most sensitive to changes in ion concentration when it is close to 50 % bound because a change in ion concentration near the K_d' results in a greater change in fraction bound, which is proportional to the change in FRET ratio (*see* Fig. 2b). For example, if a sensor's $K_d' = 1$ μM, a change in ion concentration from 0.1 to 1 μM results in an increase from 9 to 50 % fraction

Fraction bound: $$Y = \frac{R - R_{free}}{R_{bound} - R_{free}}$$

Single-site binding model:

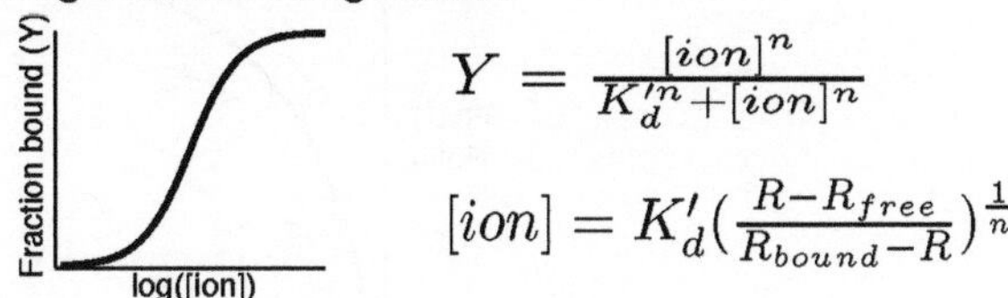

$$Y = \frac{[ion]^n}{K_d'^n + [ion]^n}$$

$$[ion] = K_d'\left(\frac{R - R_{free}}{R_{bound} - R}\right)^{\frac{1}{n}}$$

Two-site binding model:

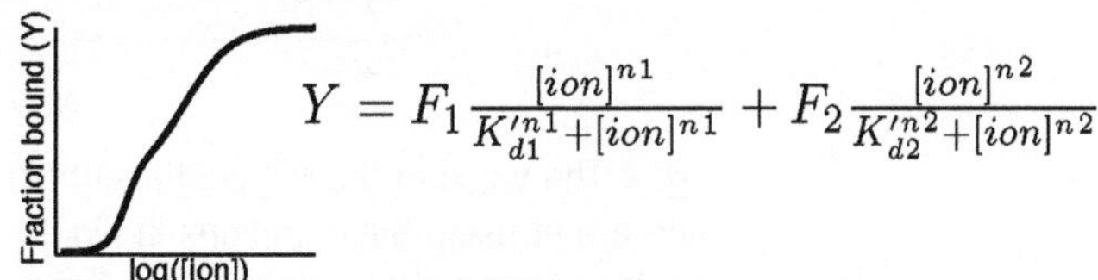

$$Y = F_1 \frac{[ion]^{n1}}{K_{d1}'^{n1} + [ion]^{n1}} + F_2 \frac{[ion]^{n2}}{K_{d2}'^{n2} + [ion]^{n2}}$$

Fig. 3 A sensor's K_d' is calculated by fitting experimental data to a single-site or two-site binding model. An example of each binding curve is shown next to the equation relating the [ion] to the fraction bound (Y), which is calculated from the FRET ratio (R), R_{free}, and R_{bound}. When the single-site binding model is used, the [ion] is readily calculated using the displayed equation

bound, whereas a change from 0.01 to 0.1 μM results an increase from 1 to 9 % fraction bound. Consequently, changes in ion concentration close to the sensor's K_d' are more readily detected. In Fig. 2b, R_{min} and R_{max} correspond to 0 % and 100 % fraction bound, respectively. The variance of R, or noise, relative to the overall change in FRET ratio (ΔR), determines the range of fraction bound that a sensor can reliably report. Thus, a sensor with a larger SNR can be used to measure a greater range of ion concentrations.

Binding cooperativity of multiple ions to one sensor is reported as the Hill coefficient (n or nH). The n value affects the steepness of the binding curve: positive cooperativity results in a Hill coefficient greater than 1 and a steeper curve (*see* Fig. 4). Positive cooperativity also increases the sensitivity of measurement near the K_d' but decreases the range of ion concentration that the sensor can report. In Fig. 4, a change in ion concentration from $1 \times K_d'$ to $2 \times K_d'$ results in a larger change in R when $n = 1.5$ than when $n = 0.5$.

1.3 Choosing a Sensor for a Specific Application

Currently, many biosensors are available for Ca^{2+} or Zn^{2+} imaging, so the investigator must attempt to choose the one best suited to the experimental system. The K_d' and the DR are the most important factors to consider, along with the other factors discussed below, but often the best sensor is revealed by empirical testing of multiple sensors [8].

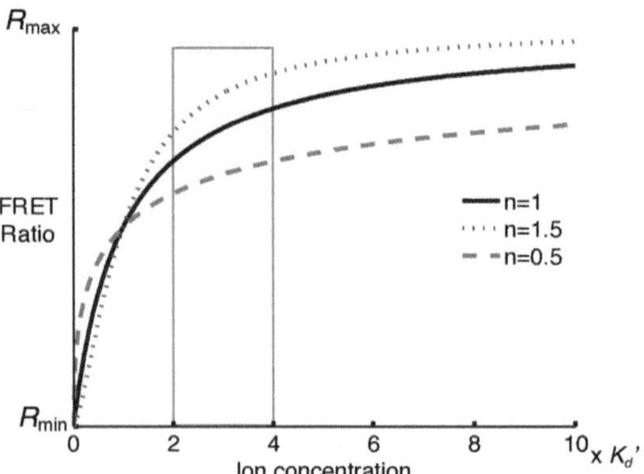

Fig. 4 The value of the Hill coefficient (n) does not affect the sensor's K_d' if the data are fit using the equations in Fig. 3. n affects the change in the FRET ratio resulting from a change in the [ion]. For example, if the [ion] changes from $2 \times K_d'$ to $4 \times K_d'$, the change in FRET ratio is greater when n is larger

First, fluorescence properties of the donor and acceptor FPs affect each sensor. The cyan FP and yellow FP donor-acceptor pair is by far the most common. Single or multiple amino acid mutations in an FP can change its excitation and/or emission wavelengths, brightness, maturation time, or photostability, and so there are several different cyan FPs and yellow FPs with slightly different fluorescence properties [9, 10]. A word of caution regarding yellow FPs is that their fluorescence is quenched by acid and hence it may be challenging to use a sensor containing a yellow FP in an acidic compartment. Circularly permuted (cp) variants of FPs are constructed by changing the N- and C-termini to different loops within the FP. Although circular permutation does not significantly change the fluorescence properties, the incorporation of a cp FP into a sensor can change the DR by altering the relative orientation of the two FPs in the bound and/or unbound conformation of the sensor [4, 11]. In addition, circular permutation can affect the pH sensitivity by changing the pK_a of the chromophore, and rearrangement of cysteine residues can affect its sensitivity to oxidizing organelles. We have found that sensors containing a cpVenus FP have decreased DR and SNR in the ER due to the low signal from the cpVenus FP. Red-green FRET pairs are preferred in acidic compartments, such as the secretory pathway, since sensors containing yellow FPs can be much dimmer at low pH (<6.5). The substitution of different FPs into a sensor can alter its K_d' and DR in unpredictable ways, so the investigator should evaluate each sensor's performance in situ before performing important experiments.

Replacement or mutation of the binding domain also alters the sensor's properties. To monitor increases in ion concentration, it is useful to pick a sensor that is ~20 % saturated at baseline, whereas

Table 1
Selected ratiometric Ca^{2+} biosensors

	Sensor	Ca^{2+}-responsive elements	K_d' for Ca^{2+}	Hill coefficient	Comments	References
Yellow Cameleon series	YC2.60	CaM, M13p	93.5 nM	2.7		[6, 11]
Yellow Cameleon series	YC3.60	CaM E104Q, M13p	215 nM, 779 nM	3.6, 1.2	High dynamic range	[6, 11]
Yellow Cameleon Nano series	YC-Nano50	CaM, M13p	52.5 nM	2.5	Optimized for detecting subtle cytosolic Ca^{2+} transients in living organisms	[6]
D-family Cameleons	D1	mCaM, mM13p	0.8 μM, 60 μM	1.18, 1.67	Does not bind endogenous CaM; optimized for ER	[12]
D-family Cameleons	D3cpV	mCaM, mM13p	0.6 μM	0.74	Does not bind endogenous CaM; optimized for cytosol and mitochondria	[4]
TroponinC family	TN-XXL	mTpC	800 nM	1.5	Optimized for imaging of neurons; fast response	[13]

This table presents the binding properties of some currently used, genetically encoded, ratiometric, Ca^{2+} biosensors. Publications of other sensors within these families (Yellow Cameleon, Yellow Cameleon Nano, Cameleon D, and TroponinC) are referenced in the last column, and Ca^{2+} sensors are more comprehensively reviewed elsewhere [9, 18–20]

CaM calmodulin, *M13p* CaM-binding M13 peptide, *mCaM* and *mM13p* mutant CaM and M13 peptide, *mTpC* modified EF hands I-IV from chicken skeletal muscle troponin C

a sensor that is ~50 % saturated is better for comparing differences in resting ion concentrations in different cells or different environmental conditions. Cameleon-Nano sensors have lower K_d's and are better for quantitative measurement of cytosolic Ca^{2+} in some cell types [6], whereas D1ER is preferred for ER measurement because Ca^{2+} levels are high in the ER and the K_d' of D1ER is much higher than other cameleons [12]. Tables 1 and 2 summarize K_d's and DRs of some current sensors, and Table 3 summarizes sensors used in specific subcellular compartments. In addition to a sensor's K_d' and DR, the investigator may consider properties

Table 2
Selected ratiometric Zn^{2+} biosensors

	Sensor	Zn^{2+}-responsive elements	K_d' for Zn^{2+}	Hill coefficient	Comments	References
eCALWY family	eCALWY4	Atox1 and the WD4 domain of ATP7B	630 pM	1	Optimized for cytosol	[3]
ZinCh family	eZinCh	Zn^{2+}-coordinating residues on CFP and YFP connected by a flexible linker	8.2 μM	1	Targeted to vesicles by fusion to VAMP2	[3, 21]
Zap family	ZapCY1	Zap	2.53 pM	1	Optimized for ER, Golgi, and mitochondria; high dynamic range	[5]
Zap family	ZapCY2	mZap	811 pM	0.44	Optimized for cytosol	[5]

This table presents the binding properties of some currently used, genetically encoded, ratiometric, Zn^{2+} biosensors
Zap first two zinc fingers of *Saccharomyces cerevisiae* Zap1, ***mZap*** mutant Zap with decreased affinity for Zn^{2+}

Table 3
Sensors targeted to subcellular locations

Subcellular location	Ca^{2+} sensors	Zn^{2+} sensors	References
ER	D1ER	ER-ZapCY1	[5, 12]
Golgi	None	Golgi-ZapCY1	[5]
Vesicles	Ycam2	eZinCh	[3, 22]
Mitochondria	4mt-D3cpV	Mito-ZapCY1	[4]
Nucleus	D3cpV	ZapCY2	[5, 19]
Cytosol	D3cpV	eCALWY-4, ZapCY2	[3, 5, 19]

This table summarizes sensors that have been targeted to different subcellular locations by adding a genetically encoded localization tag. Untargeted sensors are typically found in both the cytosol and the nucleus

such as sensitivity to other biologically relevant metal ions, pH, redox balance, salt concentration, or sensor concentration. It is also important to consider the on and off rates of ion binding when attempting to observe dynamic changes in ion concentrations [6–8, 13].

1.4 Outline of Procedures Described in This Protocol

The first step in using a biosensor is to confirm it functions in the biological system used in the experiment, even though most published sensors function to some degree in different cell types and subcellular compartments (*see* Subheading 3.1). Next, calibration of the sensor at the end of an experiment, in single cells, enables the investigator to convert the sensor's FRET ratio to the ion concentration (*see* Subheading 3.2). It is essential to obtain accurate K_d' values, which are used to convert experimental data into ion concentrations. Experimental conditions that significantly differ from those used in a published sensor titration (typically, at pH 7.0–7.4 and 20–25 °C), can change the K_d', and so a protocol for titrating a sensor in living cells (in situ) can be found in Subheading 3.3. This protocol also describes how to use metal-chelate buffer systems to accurately maintain the free Ca^{2+} or Zn^{2+} concentrations at sub-μM concentrations.

2 Materials

2.1 Materials for Preparation of Cells

1. Glass-bottom imaging dishes: available commercially or constructed using 35 mm polystyrene cell culture dishes, SYLGARD 184 silicone elastomer kit (Dow Corning), 18 mm × 18 mm No. 1 glass coverslips, and an industrial-strength 0.375 in. hole punch (Roper Whitney) (*see* **Note 1**).
2. Mammalian expression plasmid encoding the biosensor of choice.
3. Transfection reagent, such as Lipofectamine (Life Technologies) or TransIT-LT1 (Mirus Bio).
4. Cells and cell culture media.

2.2 Reagents for Cellular Imaging

Prepare all solutions using Chelex-treated water and use metal-free, plastic containers to store solutions. Metal-free containers and consumables can be purchased, or metal can be removed from plastic containers and consumables by washing them with dilute acid.

1. Chelex 100 sodium form, 50–100 mesh (Sigma Aldrich).
2. Chelex-treated double deionized H_2O: mix Chelex with autoclaved water in a large plastic container (~3–4 L) on a stir plate for at least 18 h. Let the Chelex settle to the bottom of the container for several hours. Use a bottle-top filter to remove all Chelex from the water, and store the filtered water in a plastic container.
3. For Ca^{2+} measurement, *H*EPES-buffered *H*anks *b*alanced *s*alt *s*olution (HHBSS) with and without Ca^{2+}: dilute 10× HBSS (Life technologies) with water and supplement with 20 mM HEPES and 16.8 mM D-glucose. To prepare Ca^{2+}-free

HHBSS, dilute 10× Ca^{2+}-free, Mg^{2+}-free HBSS (Life Technologies) and supplement with 16.8 mM D-glucose and 20 mM HEPES. For each solution, adjust pH to 7.4 with 1 M sodium hydroxide.

4. For Zn^{2+} measurement, phosphate-free HHBSS with and without Ca^{2+} and Mg^{2+} (*see* **Note 2**): 1.26 mM calcium chloride, 5.4 mM potassium chloride, 1.1 mM magnesium chloride, 137 mM sodium chloride, 16.8 mM D-glucose, 20 mM HEPES, pH 7.4. Mix all components in 900 mL of water. Add 0.3 g of sodium hydroxide to around pH 7, and then adjust with 1 M sodium hydroxide solution to pH 7.4. Add water to 1 L and use a 0.22 μm pore filter to sterilize. Omit calcium chloride and magnesium chloride to make phosphate-, Ca^{2+}-, and Mg^{2+}-free HHBSS.

2.3 Chemicals for Ca^{2+} and Zn^{2+} Perturbations (See Table 4 and Note 3)

1. 0.5 M EGTA stock solution, pH 7.4: dissolve EGTA in 1 M NaOH and adjust pH to 8.0 to solubilize.
2. 1 M calcium chloride stock solution in water.
3. 1 mM ionomycin stock solution in DMSO: Add 141 μL of 100 % DMSO to 1 mg of ionomycin free acid (EMD Millipore, CAS number 56092-81-0) to yield a 10 mM stock solution.

Table 4
Chemicals used for Ca^{2+} and Zn^{2+} calibrations

Chemical	Working concentration	Effect	Comments and references
Ionomycin Bromo-A23187	5–10 μM 0.1–20 μM	Ca^{2+} ionophore	
TPEN	50–150 μM	Zn^{2+} depletion	Membrane-permeable Highest affinity for Zn^{2+}, but also chelates other transition metal ions
EDTA EGTA	2–5 mM 2–5 mM	Zn^{2+} or Ca^{2+} depletion	
S. aureus α toxin	Variable	Membrane permeabilization	A variety of permeabilization conditions have been used for Zn^{2+} calibrations. [3, 5, 23]
Digitonin Alamethicin Saponin	10–20 μM 50 μg/mL 0.01–0.1 %		
Pyrithione	1–500 μM	Zn^{2+} ionophore	Lower concentrations may result in a more stable R_{bound}

Mix well. Make 5 μL aliquots and store at −20 °C. Dilute 10 mM stock solution to 1 mM before use.

4. 3 mM digitonin stock solution in ethanol: dissolve digitonin in 100 % DMSO. Aliquot and store at −20 °C (*see* **Note 4**).
5. 25 mM TPEN (*N*, *N*, *N'*, *N'*-tetrakis-(2-pyridylmethyl)-ethylenediamine) stock solution in 100 % DMSO: Make 50–100 μL aliquots and store at −20 °C.
6. 500 μM pyrithione (2-Mercaptopyridine *N*-oxide sodium salt) stock solution in 100 % DMSO: Make 50–100 μL aliquots and store at −20 °C.
7. 475 μM zinc chloride or zinc sulfate in phosphate-free HHBSS, pH 7.4. Mix well before use (*see* **Note 5**).
8. Metal-chelate buffer stock solutions (*see* Subheading 3.3).

2.4 Equipment for Cellular Imaging

1. Epifluorescence microscope equipped with a mercury or xenon arc lamp and power supply; excitation and emission filter wheels; Lambda SC Smart Shutter controller; cooled CCD camera; 20, 40, 60 and/or 100× plan apochromatic objective (depending on application). Other microscopy systems, such as two-photon confocal, FLIM, and spectral imaging, can also be used to monitor FRET, but they will not be discussed in this protocol.
2. Neutral density filters.
3. Excitation (x) and emission (m) filter sets and dichroic mirrors: CFPx and FRETx 430/24, YFPx 495/10, CFPm 470/24, YFPm and FRETm 535/25, CFP and FRET dichroic 450, YFP dichroic 515. Sputter-coated/ET or brightline filters that provide high transmission will give rise to the brightest images.
4. Microscopy software: MetaFluor (Molecular Devices), NIS Elements (Nikon), Micro-Manager (open-source), or equivalent.
5. Software for image processing: MetaFluor (Molecular Devices), NIS Elements (Nikon), ImageJ (NIH), Fiji (open-source), MATLAB (Mathworks), Excel (Microsoft), or equivalent.
6. Pipets or perfusion system.

3 Methods

3.1 Testing Sensor Functionality

1. Plate cells on several imaging dishes and transfect them with a plasmid encoding the biosensor of choice (*see* **Note 6**).
2. Image cells 24–72 h after transfection, or when sensor expression and cell density are appropriate for the experiment.
3. Gently remove cell culture media from the imaging dish and replace with 2 mL of HHBSS (for Ca^{2+} sensors) or phosphate-free HHBSS (for Zn^{2+} sensors) (*see* **Note 2**). Wash the dish 3–5 times to remove all cell culture media.

4. Place the imaging dish on the microscope stage. Identify and focus on cells.
5. Verify that the sensor is correctly localized to the subcellular compartment. This can be performed using dyes such as MitoTracker, a different fluorescent protein targeted by a different localization tag to the same organelle, or visually if the morphology of the compartment is unique. Alternatively, cells can be fixed, and immunofluorescence can be used to confirm correct localization. It is unnecessary to have 100 % transfection efficiency or 100 % correct localization because analysis will be performed on single cells.
6. Set up acquisition parameters. Limit fluorescent light exposure during acquisition to prevent photobleaching of the sensor. Acquire one set of images that includes a FRET, a donor, an acceptor, and a brightfield/DIC image. Confirm that the fluorescence intensity of the biosensor is about 2.5–8 times the background intensity and does not saturate the camera. Many software programs controlling image acquisition will allow you to select regions of interest and calculate the ratio of two images or regions of interest during acquisition. This function is helpful because it allows you to observe the sensor response during the experiment.
7. Begin the experiment by acquiring images for at least 5 min (if you have added a dye for colocalization purposes, start this step with a different imaging dish and repeat **steps 3** and **4**). Confirm that the sensor responds to increases and decreases of the metal ion of interest by using typical chemical perturbations (*see* Table 4). **Note 7** describes the procedure we use to change imaging solutions manually (i.e., without a perfusion system). You may need to try several different conditions in several imaging dishes. In this protocol, it is sufficient to observe relative changes, whereas quantitative measurement in the next two protocols requires careful and systematic chemical perturbations of metal ion concentration. Remember to wash the cells 3–5 times with HHBSS (Ca^{2+} sensors) or phosphate-free HHBSS (Zn^{2+} sensors) between chemical perturbations and refocus if necessary.
8. After image acquisition, calculate the average FRET ratio of each cell using the full spectrum unprocessed images (usually stored as .tiff files), not images that have been autoscaled and saved as new images (usually .jpg or .png files). This image analysis workflow is illustrated in Fig. 5. Define regions of interest (ROI) and measure the mean intensity of the same ROI in each set of images (CFP, FRET, and YFP) using image analysis software. Subtract the mean background intensity from each ROI's mean intensity in every image.

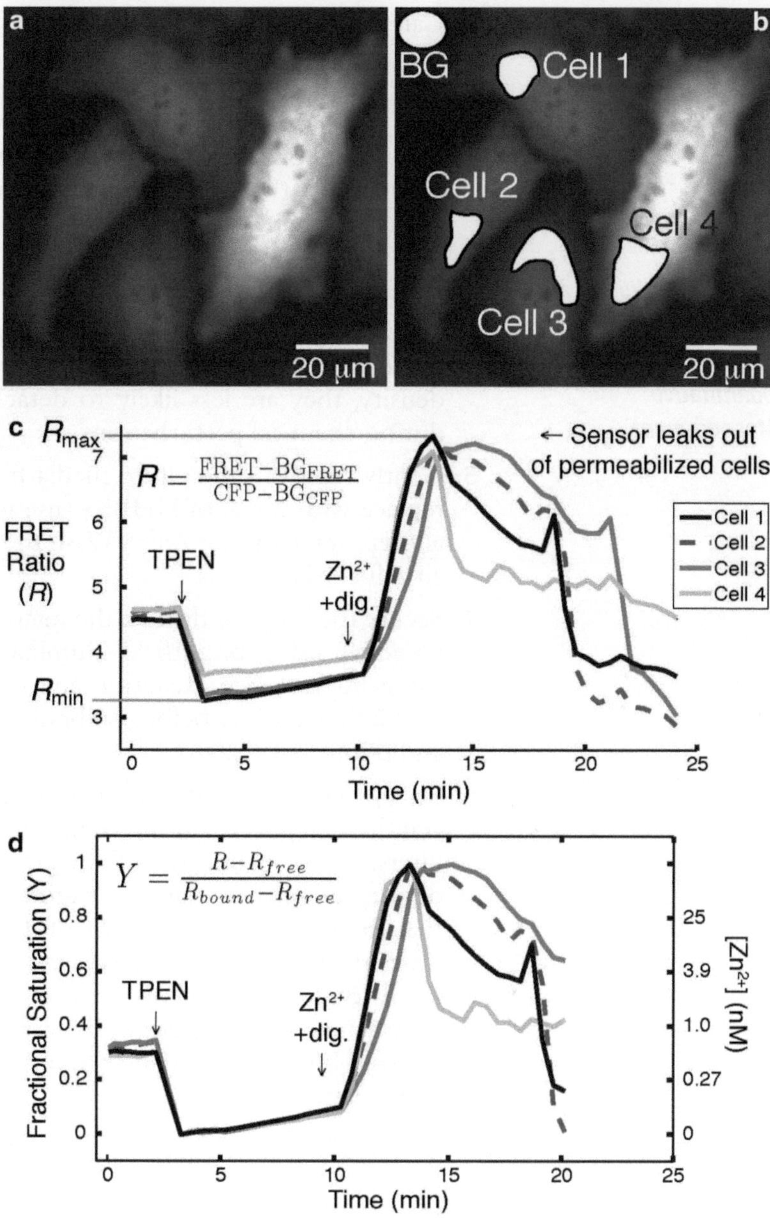

Fig. 5 A typical analysis of a Zn^{2+} sensor calibration is shown. Regions of interest (ROIs), including a background (BG) region, are selected in the acquired FRET and CFP images in (**a**) and (**b**). The mean intensity of each region is exported into a spreadsheet (or equivalent). The mean intensity of the BG region (BG_{FRET} or BG_{CFP}) is subtracted from the mean intensity of each cell's ROI (FRET or CFP), and the equation in (**c**) is used to calculate the FRET ratio (*R*). (**d**) shows a plot of the sensor's fractional saturation (*Y*) over time and indicates the Zn^{2+} concentrations corresponding to *Y*

Calculate the FRET ratio by dividing the ROI's background-subtracted mean intensity in the FRET image by that in the CFP image at each time point. Plot the FRET ratio over time and confirm that the sensor responds to perturbations of metal ion concentration. In addition, plot each ROI's mean intensity in the YFP image over time to observe potential photobleaching.

3.2 Calibration of a Ca^{2+} or Zn^{2+} Biosensor in Living Cells for Quantitative Measurement

1. Plate and transfect cells as in the previous protocol.
2. Image cells 24–72 h after transfection or when appropriate for the experiment. We find that if HeLa cells are at 80–90 % density, they are less likely to detach from the imaging dish during chemical perturbations.
3. Gently remove cell culture media from the imaging dish and replace with 2 mL of HHBSS (use phosphate-free HHBSS in all steps if measuring Zn^{2+}). Wash the dish 3–5 times to remove all cell culture media.
4. Secure the imaging dish on the microscope stage and focus the objective using brightfield illumination. Fine-tune the focus using fluorescence excitation and emission. Limiting fluorescent light exposure before calibration prevents photobleaching and phototoxicity.
5. Set up the image acquisition parameters. Again, limit fluorescent light exposure during acquisition by using neutral density filters. Acquire one set of images that includes a FRET, a donor, an acceptor, and a brightfield/DIC image. Confirm that the fluorescence intensity of the biosensor is about 2.5–8 times the background intensity and does not saturate the camera. If these requirements are not met, change the acquisition parameters and/or select different cells. This step can be tedious, but it is essential for accurate and repeatable quantitative measurements.
6. At this point, make sure that the acquired images are being saved as raw image files and that you are recording the time of each acquisition. Acquire one set of images every 20–60 s for about 5 min. Be sure that the FRET ratio, defined as the intensity of the biosensor in the FRET image divided by its intensity in the donor image, is stable during this time. Most software can plot the FRET ratio of a ROI as images are acquired. If the FRET ratio decreases (or increases), change the acquisition parameters to decrease light exposure (for example, decrease the frequency of image acquisition, decrease the exposure time of each acquisition, etc.). If the FRET ratio does not stabilize after the adjustment of acquisition parameters, change the imaging buffer, select new cells, or go back to **step 3** and start over with a new dish.

7. Begin the experiment by acquiring images for at least 5 min. If the purpose of this experiment is to measure the baseline metal ion concentration, move on to the next step. Otherwise, acquire images while performing any desired environmental or chemical perturbations of the cells. **Note 7** describes the procedure we use to change imaging solutions manually (i.e., without a perfusion system).
8. Measure the FRET ratios of the fully unbound (R_{free}) and saturated (R_{bound}) Ca^{2+} sensor (skip to the next step for Zn^{2+} sensors). Wash cells with Ca^{2+}-free HHBSS before adding EGTA and ionomycin to remove Ca^{2+} from the sensor (final concentrations: 5 μM ionomycin and 3 mM EGTA), and acquire images every 20–30 s until the FRET ratio stabilizes (*see* **Notes 8 and 9**). Wash cells 3–5 times with HHBSS. The FRET ratio should increase as Ca^{2+} from the HHBSS enters cells. Saturate the sensor by adding 5 μM ionomycin and 10 mM $CaCl_2$ to HHBSS. These conditions should be optimized because adding too much Ca^{2+}, too quickly, can lead to cell death before R_{bound} is reached.
9. Measure the FRET ratios of the fully unbound (R_{free}) and saturated (R_{bound}) Zn^{2+} sensor. Add TPEN to a final concentration of 150 μM in phosphate-free HHBSS (*see* **Notes** 7 and **9**). Wash cells 3–5 times with Ca^{2+}-, Mg^{2+}-, and phosphate-free HHBSS (*see* **Note 8**). Acquire 2–3 sets of images to ensure that the cells are in focus and change the acquisition interval to ≤20 s before proceeding because obtaining R_{bound} usually results in rapid cell death. Saturate the sensor with Zn^{2+} appropriately (*see* Table 5). We use 10 μM digitonin/100 μM Zn^{2+} to determine R_{bound} of a cytosolic Zn^{2+} sensor.

Table 5
Metal-chelate buffers for sub-μM Ca^{2+} and Zn^{2+} solutions

Chelator	Ca^{2+} buffering range at pH 7.4	Zn^{2+} buffering range at pH 7.4	Zn^{2+} buffering range at pH 7.4 with competing metal
EDTA	0.002–0.18 μM	0.003–0.3 pM	
EGTA	0.006–0.5 μM	0.15–15 nM	9.7–1,340 nM (with 2 mM Sr^{2+}) 2–134 μM (with 2 mM Ca^{2+})
HEDTA	0.3–25 μM	0.15–15 pM	0.05–7.5 nM (with 2 mM Ca^{2+})
NTA	20–1,670 μM		

The free Ca^{2+} and Zn^{2+} concentrations are accurate at 25 °C and 0.1 M ionic strength. Note the competing metal should have a lower affinity for the chelator than the metal to be buffered. References 14–16 are detailed references and resources

10. Calculate the FRET ratio of individual cells over time (*see* **step 8** in Subheading 3.1 and Fig. 5) and determine R_{min}, R_{max}, and ΔR for each cell (*see* **Note 10**). Convert single-cell FRET ratios to fractional saturation, which is equal to $(R - R_{min})/\Delta R$. Solve for the ion concentration using the appropriate equation (*see* Fig. 3).

3.3 Measurement of Sensor Affinity In Situ with Metal-Chelate Buffers

1. Select the appropriate metal-chelate buffer(s) for the desired free Ca^{2+} or Zn^{2+} concentration(s) (*see* Table 5 and **Note 11**). Some general references for creating different metal-chelate buffers are [14–16]. Detailed descriptions of Zn^{2+}-chelate buffers used in publications are found in the supplementary methods of refs. 3, 5.
2. Prepare two different 0.1 M stock solutions for each buffer. For a single metal-chelate buffer, these solutions are: (1) metal and chelator, and (2) chelator only. For a mixed metal-chelate buffer, these solutions are: (1) Metal #2 *plus* a fixed amount of Metal #1 *plus* the chelator, and (2) a fixed amount of Metal #1 *plus* the chelator. Note that Metal #1 is the competing metal while Metal #2 is the metal to be buffered (e.g., Ca^{2+} or Zn^{2+}). The "pH titration method" [14] should be used to prepare stock solution (1) because it is extremely important to make a 1:1 solution of metal–chelator (*see* **Notes 12 and 13**).
3. Mix the two stock solutions in appropriate proportions to buffer at specific ion concentrations. The stock solution concentration of 0.1 M can be diluted 100× to the 1 mM working concentration in Ca^{2+}- and Mg^{2+}-free HHBSS (for buffered Zn^{2+} solutions, use Ca^{2+}-, Mg^{2+}-, and phosphate-free HHBSS).
4. Measure the FRET ratio of the sensor in living cells when they are permeabilized in the presence of a defined Ca^{2+} or Zn^{2+} concentration. Follow the protocol in Subheading 3.2 to measure R_{free} in single cells. Wash the cells 3–5 times with Ca^{2+}- and Mg^{2+}-free HHBSS, and then replace it with a metal-chelate buffered HHBSS before permeabilizing the cells.
5. Calculate the FRET ratio of individual cells over time (*see* **step 8** in Subheading 3.1 and Fig. 5) and determine R_{min} and R_{final} for each cell. Plot $R_{final} - R_{free}$ vs. the ion concentration and fit data to a binding expression (*see* Fig. 3) using the least squares method.

4 Notes

1. To construct a glass bottomed imaging dish, punch a hole in the middle of the cell culture dish and place it bottom-up. Use an 18-gauge needle and syringe to apply SYLGARD 184 around the rim of the punched hole, and place a coverslip on top. Let the glue cure and then sterilize the dish with ethanol and UV light.

2. Optically clear salt solutions are preferred for cellular imaging because the autofluorescence of Phenol Red and serum in complete media increases background fluorescence. In addition, components of complete media, such as amino acids and proteins, will bind metal ions and affect sensor calibrations. Phosphate-free imaging media is used for Zn^{2+} sensor calibrations because $Zn_3(PO_4)_2$ precipitates and changes the free Zn^{2+} concentration.
3. The appropriate chemical perturbations to reach R_{min} or R_{max} will depend on the cell type, the biosensor, and the organelle to which it is targeted. For example, in HeLa cells transfected with a mitochondrial Zn^{2+} sensor, saturating them with 10 μM $ZnCl_2$ instead of 100 μM $ZnCl_2$ results in a higher R_{max} measurement. Try a few different perturbations and use the one that gives consistent results.
4. For most applications, digitonin is purified by crystallization in ethanol and made freshly before use. We find it sufficient to use commercially purified digitonin for sensor calibrations.
5. While zinc chloride and zinc sulfate are very soluble in water, zinc hydroxide readily precipitates out of solution, decreasing the concentration of free Zn^{2+}. At pH 7.4 and room temperature, up to 475 μM Zn^{2+} is soluble in water. Use the K_{sp} of zinc hydroxide, which is 3.0×10^{-17} at room temperature, to calculate the maximum solubility of Zn^{2+} at the desired pH.
6. Optimal cell plating density and transfection conditions will differ for each cell type and biosensor. For example, we transfect each imaging dish of HeLa cells with 5 μl of Mirus TransIT-LT1 transfection reagent, and 0.5 μg of plasmid DNA encoding a mitochondria-targeted Zn^{2+} sensor or 1.25 μg of plasmid DNA encoding the same Zn^{2+} sensor targeted to the cytosol. In our experience, it is worthwhile to identify conditions that minimize cell toxicity, promote expression levels detectable by the microscope, and maximize correct localization of biosensors targeted to organelles.
7. When working manually (i.e., without a perfusion chamber or a "perfect-focus" system), we find that the following procedure for adding chemicals to the imaging dish minimizes changes in focus. Remove 1 mL of HHBSS from the dish and pipet it into a 1.5 mL tube. Make a 2× solution, containing the chemical perturbation, using this aliquot of HHBSS. Gently pipet the 2X solution into the imaging dish and mix gently.
8. Failure to use Ca^{2+}-free imaging buffer during cell permeabilization will cause rapid cell death.
9. The FRET ratio will initially increase after the addition of ionomycin/EGTA due to the permeabilization of the ER and release of Ca^{2+} stores. It may take up to 10 min to reach R_{min}. High affinity Zn^{2+} sensors, such as ZapCY1, have a very slow

off rate. It can take up to 40 min to reach R_{min}. An alternative is to monitor the FRET ratio for 15–20 min, fit the data to an exponential decay, and calculate R_{min}.

10. Sometimes a cell will have an unusually low dynamic range. This could be due to proteolysis, misfolding, or oxidation of the sensor. These data are difficult to interpret and may have to be discarded.

11. For a sensor titration, these concentrations should vary ~10-fold above and below the expected K_d', and about ten evenly spaced concentrations within this range should be used. Metal-chelate buffers consist of a chelator (e.g., EDTA), the buffered ion (Ca^{2+} or Zn^{2+}), and sometimes a competing ion (e.g., Sr^{2+}). The addition of a competing ion effectively increases the buffering range of the chelator, and other competing metals like Mg^{2+} can affect the metal-chelate buffer. The final concentration of the chelator must be much greater than the free ion concentration, just as in a pH buffer, where the concentration of the weak acid/conjugate base is much greater than the H_3O^+ concentration.

12. The pH affects the protonation state of chelators like EDTA, EGTA, and HEEDTA, and as a result affects the buffering range of the metal-chelate solution. Refer to refs. 14 and [17] to recalculate free ion concentrations at different pH.

13. The pH titration method [14] can be used to make solutions of polyprotic acid chelators (EDTA, EGTA, HEEDTA) and metal ions, where the concentrations of each are verified to be within 0.5 % of each other. For example, when Ca^{2+} binds EGTA, two protons are released, causing a drop in the pH of the solution. Therefore, the change in pH upon Ca^{2+} addition ($\Delta pH/\Delta Ca^{2+}$) decreases when there are equal concentrations of Ca^{2+} and EGTA.

Acknowledgments

Financial support was provided by the Signaling and Cell Cycle Regulation Training Grant (NIH T32 GM08759) to J.G.P. and NIH GM084027 and Alfred P. Sloan Fellowship to A.E.P.

References

1. Lakowicz JR (2006) Principles of fluorescence spectroscopy. Springer, New York
2. Zhang J, Campbell RE, Ting AY et al (2002) Creating new fluorescent probes for cell biology. Nat Rev Mol Cell Biol 3:906–918
3. Vinkenborg JL, Nicolson TJ, Bellomo EA et al (2009) Genetically encoded FRET sensors to monitor intracellular Zn2+ homeostasis. Nat Methods 6:737–740
4. Palmer A, Giacomello M, Kortemme T et al (2006) Ca2+ indicators based on computationally redesigned calmodulin-peptide pairs. Chem Biol 13:521–530
5. Qin Y, Dittmer PJ, Park JG et al (2011) Measuring steady-state and dynamic endoplasmic reticulum and Golgi Zn2+ with genetically encoded sensors. Proc Natl Acad Sci U S A 108:7351–7356

6. Horikawa K, Yamada Y, Matsuda T et al (2010) Spontaneous network activity visualized by ultrasensitive Ca2+ indicators, yellow Cameleon-Nano. Nat Methods 7:729–732
7. Tian L, Hires SA, Mao T et al (2009) Imaging neural activity in worms, flies and mice with improved GCaMP calcium indicators. Nat Methods 6:875–881
8. Yamada Y, Michikawa T, Hashimoto M et al (2011) Quantitative comparison of genetically encoded Ca indicators in cortical pyramidal cells and cerebellar Purkinje cells. Front Cell Neurosci 5:18
9. Newman RH, Fosbrink MD, Zhang J (2011) Genetically encodable fluorescent biosensors for tracking signaling dynamics in living cells. Chem Rev 111:3614–3666
10. Davidson MW, Campbell RE (2009) Engineered fluorescent proteins: innovations and applications. Nat Methods 6:713–717
11. Nagai T, Yamada S, Tominaga T et al (2004) Expanded dynamic range of fluorescent indicators for Ca(2+) by circularly permuted yellow fluorescent proteins. Proc Natl Acad Sci U S A 101:10554–10559
12. Palmer AE, Jin C, Reed JC et al (2004) Bcl-2-mediated alterations in endoplasmic reticulum Ca2+ analyzed with an improved genetically encoded fluorescent sensor. Proc Natl Acad Sci U S A 101:17404–17409
13. Mank M, Santos AF, Direnberger S et al (2008) A genetically encoded calcium indicator for chronic in vivo two-photon imaging. Nat Methods 5:805–811
14. Tsien R, Pozzan T (1989) Measurement of cytosolic free Ca2+ with quin2. Methods Enzymol 172:230–262
15. Cheng KL, Ueno K, Imamura T (1982) CRC handbook of organic analytical reagents. CRC, Boca Raton, FL
16. Smith RM, Martell AE, Motekaitis RJ (2004) NIST critically selected stability constants of metal complexes database. NIST, Gaithersburg, MD
17. Kao JPY, Li G, Auston DA (2010) Chapter 5 – Practical aspects of measuring intracellular calcium signals with fluorescent indicators. Methods Cell Biol 99:113–152
18. Dean KM, Qin Y, Palmer A (2012) Visualizing metal ions in cells: an overview of analytical techniques, approaches, and probes. Biochem Biophys Acta 1823(9): 1406–1415
19. Palmer A, Tsien R (2006) Measuring calcium signaling using genetically targetable fluorescent indicators. Nat Protoc 1: 1057–1065
20. Tian L, Hires SA, Looger L (2012) Imaging neuronal activity with genetically encoded calcium indicators. Cold Spring Harb Protoc 2012(6):647–656
21. Evers TH, Appelhof MA, de Graaf-Heuvelmans PT et al (2007) Ratiometric detection of Zn(II) using chelating fluorescent protein chimeras. J Mol Biol 374:411–425
22. Emmanouilidou E, Teschemacher AG, Pouli AE et al (1999) Imaging Ca2+ concentration changes at the secretory vesicle surface with a recombinant targeted cameleon. Curr Biol 9:915–918
23. Dittmer P, Miranda J, Gorski J et al (2009) Genetically encoded sensors to elucidate spatial distribution of cellular zinc. J Biol Chem 284:16289–16297

Chapter 4

Detecting cAMP with an Epac-Based FRET Sensor in Single Living Cells

J. Klarenbeek and Kees Jalink

Abstract

Cyclic nucleotides such as cGMP and cAMP play pivotal roles as second messengers in many biological processes. Upon stimulation of appropriate signal transduction pathways, the levels of these messengers change rapidly. Such variations in second messenger level may also be spatially restricted within the cell. To detect dynamic and local changes in second messengers, we need to study them in living cells with high spatial and temporal resolution. Focusing on cAMP, here we describe how imaging of an EPAC-based FRET sensor in single cells provides that spatiotemporal resolution.

Key words cAMP, FRET, Biosensor, EPAC, Live cell recording

1 Introduction

Cyclic Adenosine MonoPhosphate (cAMP) is an important second messenger that affects many biological processes. cAMP is generated from ATP by Adenylate Cyclase (AC), an enzyme that is controlled by G-protein coupled receptors through stimulatory ($G\alpha_s$), as well as inhibitory ($G\alpha_i$), G proteins [1]. Cellular effectors of cAMP include Protein Kinase A (PKA) and Exchange Protein directly Activated by cAMP (EPAC) [2]. Traditionally, cAMP was studied by measuring its production from γ-labeled ATP by AC. However, this method it is hard to obtain detailed temporal information and next-to-impossible to get spatial resolution. In addition, isotope assays are expensive and yield radioactive waste. More recently, Förster Resonance Energy Transfer (FRET) sensors for the detection of cAMP (and cyclic Guanidine MonoPhosphate cGMP) were developed. FRET is the radiationless transfer of energy from a donor to a suitable acceptor fluorophore, a process that is highly dependent on fluorophore distance and orientation. FRET sensors are genetically engineered constructs that report on changes in second messenger levels by translating a binding-induced conformational shift into a change in FRET. FRET recordings can be obtained

Jin Zhang et al. (eds.), *Fluorescent Protein-Based Biosensors: Methods and Protocols*, Methods in Molecular Biology, vol. 1071, DOI 10.1007/978-1-62703-622-1_4,

in several ways [3, 4], however for live-cell recordings Fluorescent Lifetime IMaging (FLIM) and imaging of Sensitized Emission (SE) are the most common techniques (for more details, please *see* Chapter 2 of this textbook). SE recordings can be easily implemented on fluorescent microscopes that are readily available in most labs, and will be the subject of this chapter.

Recently, various groups described the development and optimization of FRET sensors for cAMP [4–7]. The first generation of FRET sensors for cAMP were based on PKA [5, 7] , a complex of two regulatory and two catalytic subunits that dissociates upon binding to four molecules of cAMP. The dynamic range of these sensors is considerable, because the donor-labeled regulatory subunit fully dissociates from the acceptor-labeled catalytic subunit, which causes a large increase in the interfluorophore distance. PKA has a high affinity for cAMP, and thus these FRET sensors are very sensitive to small deviations from baseline cAMP levels, but they also have certain disadvantages that prompted the development of alternative strategies [6]. The second generation cAMP sensors are based on EPAC, a GEF for Rap1, and were reported in 2004 by our group as well as others [3, 4, 8, 9]. cAMP binding to EPAC induces a large conformational shift [10] that is read out as a FRET change between donor and acceptor fluorophores fused to the C- and N-terminus. Although a complete overview of EPAC sensors is not within the scope of this chapter, it is important to select a suitable sensor with care. Nikolaev and colleagues [3] used the isolated cAMP-binding region of EPAC to construct a small and compact FRET sensor, whereas DiPilato and colleagues based their design of full-length EPAC1 [8]. Our design is also based on most of the sequence of EPAC1, but we removed a membrane-targeting DEP sequence and rendered the sensor catalytically dead [6]. The advantage of the latter approach is that it prevents spurious activation of the target Rap1 due to expression of the sensor while retaining maximal conformational shift, and, thus, FRET span. Subsequent optimizations with respect to fluorescent moieties have been reported for all of these versions [4, 9, 11, 12]. In this chapter we use our most recently published incarnation of the EPAC1 (ΔDEP) reporter, which consists of a truncated mTurquoise [13] (a CFP variant) fused to EPAC1 (CD, ΔDEP) and a dimer of a circularly permuted and a wild type variant of the YFP derivative Venus (^{cp173}Venus-Venus). This construct will be abbreviated as ^{T}EPAC^{VV}. It has an exceptional signal-to-noise ratio, the fluorophores mature well and it is biochemically inert. mTurquoise is an excellent donor with very high brightness (allowing for dim excitation, so as to minimize phototoxicity) and its strictly mono-exponential decay makes it equally suited for readout by FLIM and by SE. However, other published EPAC sensors work very well too and may have, depending on the application, specific advantages [3, 6].

We here describe the sequential steps to perform successful cAMP recordings from mammalian cells. We cover plating of cells on coverslips, expression of $^{T}EPAC^{VV}$ in these cells, mounting of a coverslip with cells on the microscope, recording of fluorescent ratio data and analysis. Note that this protocol is also directly applicable for recording intracellular cGMP, for which FRET sensors have also been developed [14–16].

2 Materials

All solutions are made with deionized water unless stated otherwise. Diligently follow all waste disposal regulations when disposing of waste materials and/or biological material.

2.1 Cell Culture

For a 70 % solution of ethanol, add 70 ml of ethanol and 30 ml of water together and mix. Store a batch of round 24 mm coverslips in this mix to ensure sterilization. Put one coverslip in each well of a six well plate and allow to dry overnight in the lamellar flow.

2.2 Transfection Reagents

1. For transfection with PEI: Dissolve 3 mg of PolyEthyleneImidine (PEI) in 2 ml of ethanol in a glass container. Store at −20 °C and do not use this solution beyond 3 months (*see* **Note 1**).
2. For transfection with calcium phosphate: prepare 2× HEBS buffer (0.28 M NaCl, 0.05 M HEPES, 0.0015 M Na_2HPO_4, pH = 7.05) and a separate solution of 1 M $CaCl_2$.

2.3 Stock Solutions

1. Carbonate Buffered Saline (CBS): 0.140 M NaCl, 0.005 M KCl, 0.023 M $NaHCO_3$, adjust pH to 7.3 by controlling pCO_2 (~5 %) Store at 4 °C.
2. HEPES-buffer: 1 M HEPES, pH 7.2; sterilize using a 0.22 μm filter and store at 4 °C.
3. Glucose solution: 1 M glucose; sterilize using a 0.22 μm filter and store at 4 °C.
4. HEPES-Buffered-Saline (HBS): 0.140 M NaCl, 0.005 M KCl, 0.01 M HEPES, pH 7.4.

2.4 Work Solutions

1. CBS+/+: Add 0.001 M $MgCl_2$, 0.002 M $CaCl_2$ to the CBS on the day of use (*see* **Note 2**).
2. HBS+/+: On the day of use, add 0.001 M $CaCl_2$, 0.001 M $MgCl_2$, and 0.01 M Glucose to the HBS.
3. HEPES-CBS+/+–Glucose buffer (HCG): for 50 ml, add 0.5 ml of glucose solution and 0.5 ml of HEPES solution to 49 ml of CBS+/+.

2.5 Ratio-FRET Measurement

1. Microscope. Any inverted fluorescence microscope that is capable of (1) exciting CFP (at wavelengths between 430 and 470 nm) and (2) recording donor emission (~470–510 nm) and acceptor emission (~515–600 nm), either simultaneously or consecutively, can be used (*see* **Note 3**).
2. Steel coverslip-holder to mount coverslip with cells on the stage of the inverted microscope, e.g., Attofluor® Cell Chamber, for microscopy A7816 from Invitrogen (Fig. 1a).

3 Methods

3.1 Cell Culture

Culture your cells in your medium of choice, we use Hek293T. Hek293T cells are cultured in DMEM supplemented with 10 % FCS and penicillin/streptomycin. Seed cells on a coverslip in a 6-well plate at a density of 40 % confluency approximately 6 h before transfection (*see* **Note 4**).

3.2 Transfection

1. PEI-transfection: Add 2 μl of PEI solution and 1 μg of ^{T}EPAC^{VV} DNA to 200 μl of DMEM without supplements. Mix gently by tapping and further incubate for 20–60 min. Add the PEI/DNA mixture to the cells and incubate overnight.
2. Calcium phosphate transfection: Add 1 μg of ^{T}EPAC^{VV} DNA to 132 μl of water and 18 μl of Calcium solution and shake gently. Add 150 μl of HEBS drop wise, mix by tapping or gently pipetting the solution up and down. Incubate for 15 min and add the mixture (*see* **Notes 5** and **6**).
3. Other transfection methods like Lipofectamine or Fugene to transfect cells can be substituted when used as described by the manufacturer (*see* **Note 7**).

3.3 Mounting a Coverslip on the Microscope

1. Take a coverslip out of the 6-well plate using sharp tipped forceps and mount it on the bottom ring of the steel coverslip-holder with the cells facing up. Screw on the top ring tight enough to avoid leakage, but not too tight so as to avoid breaking the coverslip.
2. Immediately add 2 ml of HCG or HBS to the cells (*see* **Note 8**) to make sure they do not dry out.
3. Dry the bottom side of the coverslip with tissue paper. To make sure that no leakage occurs, press a fresh (dry) part of the tissue paper to the ring and hold it there for a few seconds. Leakage will immediately be apparent from dark coloration of the tissue paper. Repeat this at least one more time to make sure the steel ring doesn't leak (*see* **Note 9**).
4. Make sure that the microscope and associated hardware and software are turned on (*see* **Note 10**).

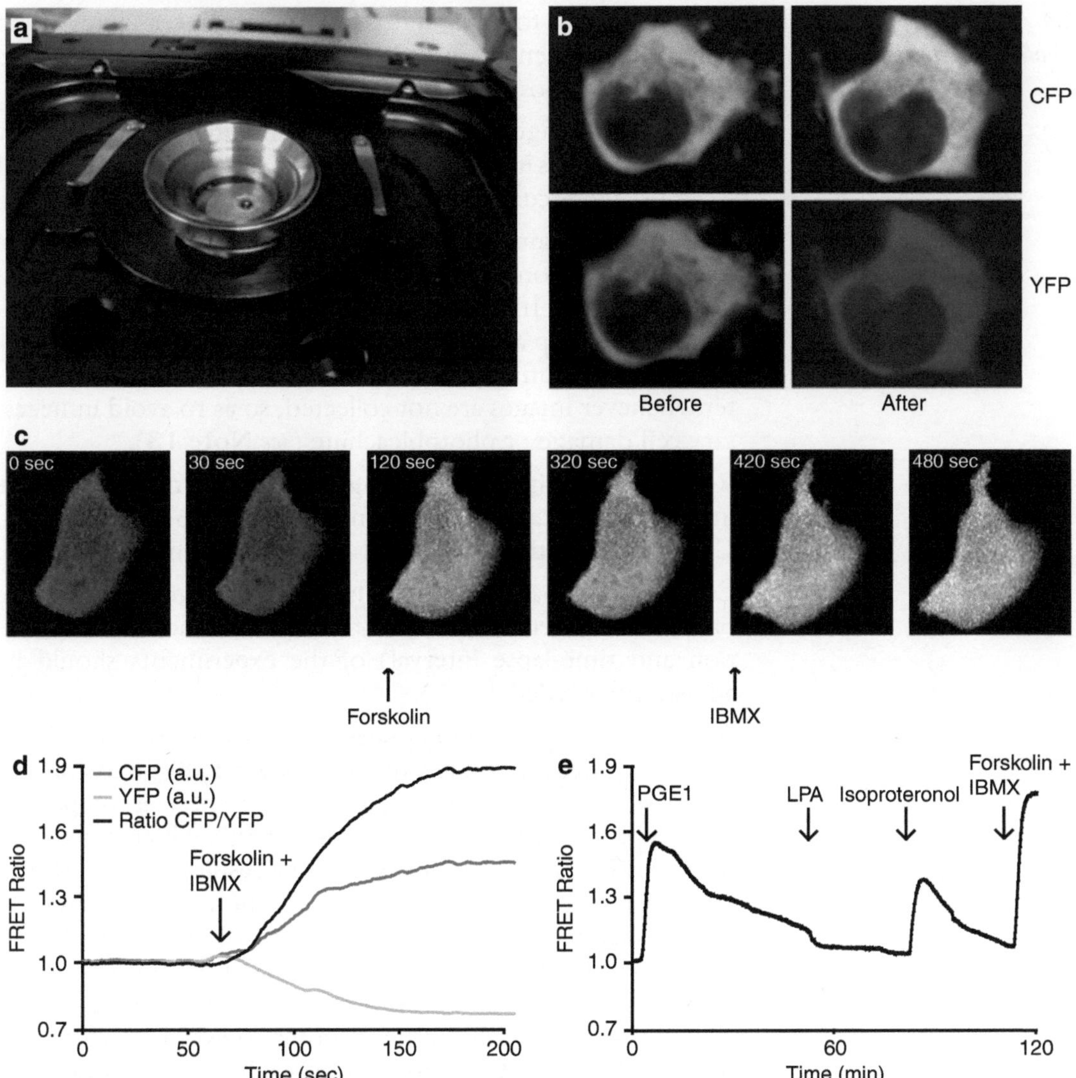

Fig. 1 Time-lapse imaging of cAMP. (**a**) Steel coverslip holder with coverslip and medium mounted on a confocal microscope. (**b**) Images of HEK293T cells transfected with $^{T}EPAC^{VV}$. The intensity of cells in the CFP- and YFP-channels, excited with a 458 nm laser, before and after stimulation with Forskolin (25 μM) and IBMX (100 μM) is shown. (**c**) Ratio CFP/YFP intensity of a representative cell stimulated with Forskolin (25 μM) and IBMX (100 μM). See also http://www.youtube.com/watch?v=jHPj6GOIvOY for a full color movie. (**d**) Ratio imaging on a dual-PMT setup using a wide-field system with CFP excitation. CFP- and YFP intensity (a.u.) are shown in *dark* and *light grey*, respectively, and CFP/YFP ratio is shown in *black*. Cells were stimulated with Forskolin (25 μM) and IBMX (100 μM). (**e**) Experiments showing prolonged ratiometric imaging of a single cell, stimulated with PGE1 (5 μM), LPA (1 μM), Isoproteronol (10 μM) and Forskolin (25 μM) and IBMX (100 μM), as indicated

5. Put a droplet of immersion oil in the middle at the bottom of the coverslip. Put the assembled steel ring on the microscope (*see* **Note 11**) (Fig. 1a).

3.4 Detection of cAMP-Levels

1. Identifying cells for imaging: Using the eyepieces, quickly find a (few) fluorescent and healthy cell(s) (*see* **Note 12**). Avoid unnecessary exposure to bright excitation light.
2. Viewing the image of the cell on the computer screen: Properly position the cells in the field of view and use the fine-adjustment to adjust the focus.
3. Setting-up imaging: If using a confocal microscope, choose 458 nm excitation and collect CFP and YFP images simultaneously (Fig. 1b). If using a wide-field system, choose CFP excitation and set up a sequence so that CFP- and YFP emission is collected alternatingly. Make sure to close the excitation shutter whenever images are not collected, so as to avoid unnecessary cell damage or photobleaching (*see* **Note 13**).
4. Real-time monitoring of imaging experiments: Using the imaging software, draw a region of interest around the cell(s). Calculate the ratio of CFP over YFP in real-time (Fig. 1c).
5. Establishing a baseline: Start time lapse imaging and record a baseline for several minutes (*see* **Note 14**). The timing (duration and time-lapse interval) of the experiments should be adjusted as needed.
6. Addition of stimuli from concentrated stocks: Stimuli will be added by pipetting in a few microliters of a 1,000-fold concentrated stock solution and mixing. This is best accomplished using a procedure that we call "turbo mix", which consists of the following: using a P20 PIPETMAN, aspirate 2 μl of stimulus in a pipette tip. Carefully remove the tip containing the stimulus from the P20 and transfer it to a P200 PIPETMAN, while keeping the piston slightly depressed. As soon as the tip containing 2 μl of stimulus is put on the P200 pipette, gradually release the piston a little in order to avoid expelling the stimulus from the tip. This should result in the vehicle solution with the stimulus being at the end of the pipette tip that is now tightly fitted on the p200 with the piston pressed in. Insert the pipette tip into the medium in the dish and release the piston further, so the stimulus is premixed with ~200 μl of medium. Then squeeze the piston to transfer the mix to the dish and further mix by pipetting up and down with the P200 8–10 times. With some training, this is a fast and easy way to add stimuli evenly to the cells from concentrated stocks. A red LED flashlamp may be used to guide pipetting in the dark.
7. Negative control (mock) stimulus: To test whether turbo-mixing by itself may induce a signal change, we add 2 μl of vehicle and mix just as if the stimulus were included (*see* **Notes 15** and **16**).
8. If cells are not affected by turbo-mixing (i.e., the negative control at the onset of the experiment), continue the experiment by adding vehicle plus stimulus.

9. Measuring the response and calibrating the response: Wait until the signal has reached maximum and/or has recovered. Add additional stimuli if necessary (*see* **Notes 17** and **18**). Calibration of the response may be achieved by adding a positive control. We use a combination of the PDE inhibitor IBMX (100 μM) and forskolin (25 μM), which directly activates AC (Fig. 1d, e).
10. Finishing the imaging experiment: When the experiment is finished, stop the time lapse and save your data (*see* **Notes 19** and **20**).
11. Clean-up: Throw away the coverslip with the cells and wash the steel ring with three repetitive steps of first deionized water followed by 70 % ethanol (*see* **Note 21**). After the ethanol wash, wash once with deionized water. Store the steel ring in 0.5 M NaOH.

4 Notes

1. After 3 months, transfection efficiency drops dramatically. Storage at 4 °C is possible but −20 °C is recommended to prevent evaporation. It is strongly recommended to use glass containers, since ethanol in an eppendorf tube by itself sometimes affects cells.
2. $CaCO_3$ precipitates slowly at 4 °C. This process is reversible by heating the solution but this will take a long period of time. Adding the $CaCO_3$ on the day of use is therefore preferred.
3. Choice of filters is not very critical, as long as donor and acceptor emission are well-covered. High-NA immersion objectives are preferred because of their high light-collecting efficiency. We have had good results using:
   (a) A confocal microscope (any vendor) capable of exciting CFP with, for example, a 458 nm Argon laser line or a 442 nm HeCd laser line, and recording at ~470–505 and 515–590 nm simultaneously with two PMT detectors. Software should be able to quantify the ratio of YFP to CFP intensity in a region of interest (ROI).
   (b) A wide-field fluorescence microscope, featuring controlled excitation such as a Xenon arc lamp with shutter, filtered at 436 ± 10 nm, or a power LED at ~450 nm. A main dichroic beam splitter at ~455 nm is used to separate the emission from the excitation. The two emission channels, e.g., 475 ± 15 nm filter for CFP emission and 540 ± 20 nm filter for YFP, can be recorded sequentially by placing the emission filters in a fast filter wheel under control of the recording software. The fluorescent images are recorded

with a sensitive CCD camera. Alternatively, an emission beam splitting device ("dual-view") can be used to record CFP and YFP images simultaneously. Camera, shutter and filter wheel are under the control of a computer running any of a wide variety of software capable of gathering image sequences and quantifying the ratio signals.

(c) A wide-field system as in b, but the filter wheel and CCD camera are replaced by a beamsplitter hooked up to two light detectors (PMTs). In this configuration, spatial information is lost, but millisecond acquisition speed is easily obtained for detailed kinetic analysis.

4. To obtain a homogenous layer of cells, move the plate twice in the south/north direction followed by twice in the west/east direction. This will prevent cells from piling up in the middle of the well.
5. Do not exceed 20 min of incubation since large precipitates are formed which will decrease transfection efficiency.
6. Longer incubation times often lead to higher expression levels, however it is recommended to refresh the medium 18–24 h after transfection to prevent side-effects of the PEI or calcium phosphate.
7. Transfection efficiency differs for every cell line with every method. Therefore multiple methods of transfection should be compared to determine the condition with the highest transfection efficiency. Keep in mind that commercial transfection agents often are very expensive.
8. The choice of buffer depends on the setup used. When a setup has a supply of 5 % CO_2, then HCG is the preferred buffer; otherwise, HBS works fine. However, other buffers appropriate to the cell type under study should also work. Note that fluorescence of for example pH indicators (phenol red) may deteriorate the signal-to-noise levels.
9. Leakage of a steal ring will influence recordings during the course of the experiment by interfering with the immersion fluids. This could lead to out of focus recordings that exhibit suboptimal or even artificial data. Furthermore, uncleaned spills may severely damage the objective.
10. Allow the lamp to warm up for ~30 min for best stability.
11. Depending on the objective, other immersion fluids may be applicable, i.e., water or glycerol.
12. Ensure that the fluorescent cell looks like the untransfected cells as an indication that it is healthy. Very bright cells often have different morphology (e.g., are more rounded) and these are to be avoided. Fluorescent cells that exhibit intermediate fluorescence intensity often show a higher signal to noise ratio.

13. Make sure that the target cells are not too bright or too dim. The signal of cells that are too bright might reach the maximal signal too fast, ending in false measurements. On the other hand, cells that are too dim might show a low signal to noise ratio. Adjust the gain of the PMT's or intensity of the laser/LED/Arc lamp to reach a suitable intensity.
14. The sampling time may vary and is dependent on both the response time of the signaling pathway as well as the fluorophores used in the sensor. A high sampling frequency gives an excellent time resolution; however, due to bleaching effects, especially of the donor fluorophore, a lower sampling frequency might be required for stable long-term experiments.
15. Mix the medium gently with a P200 pipette. Some cell lines show activation by sheer stress and/or solvent of the stimulus, effects from stimuli can thereby be misinterpreted.
16. Don't aim the pipette tip to the cell as this might wash the cell away.
17. Usually the cell will respond to several stimuli when given one after another. However beware that after a few stimuli the cell might show "fatigue" and is not able to respond anymore. Change the order of addition of stimuli between experiments to assess this effect.
18. When multiple cell lines, constructs or stimuli are used, it is recommendable to randomize conditions on a single day, e.g., first do cell line A and then cell line B, then A again followed by B, etc. This "interlacing" of conditions avoid systematic bias due to, for example, differences in time after transfection.
19. Saving all the individual images could easily result in very large data files, while saving only the data from the ROI intensity in most cases will be enough. If spatial information is to be retained, consider recording a lower-resolution time-lapse series (e.g., by binning the camera).
20. Normally SE is calculated as the ratio of YFP over CFP, but calculating the inverse (CFP over YFP) gives the same information and has the advantage that an increase in cAMP results in an upward deflection of the curve.
21. Some stimuli appear to bind to the stainless steel of the ring; by these repetitive washing steps with deionized water and ethanol, most of these stimuli are cleaned away. When the first experiments on a day are better than the subsequent ones, test whether proper cleaning may improve things. Some stimuli actually can only be removed by cleaning with concentrated NaOH solutions.

References

1. Gilman AG (1995) Nobel lecture. G proteins and regulation of adenylyl cyclase. Biosci Rep 15:65–97
2. de Rooij J, Zwartkruis FJ, Verheijen MH, Cool RH, Nijman SM et al (1998) Epac is a Rap1 guanine-nucleotide-exchange factor directly activated by cyclic AMP. Nature 396:474–477
3. Nikolaev VO, Bunemann M, Hein L, Hannawacker A, Lohse MJ (2004) Novel single chain cAMP sensors for receptor-induced signal propagation. J Biol Chem 279:37215–37218
4. van der Krogt GN, Ogink J, Ponsioen B, Jalink K (2008) A comparison of donor–acceptor pairs for genetically encoded FRET sensors: application to the Epac cAMP sensor as an example. PLoS One 3:e1916
5. Adams SR, Harootunian AT, Buechler YJ, Taylor SS, Tsien RY (1991) Fluorescence ratio imaging of cyclic AMP in single cells. Nature 349:694–697
6. Ponsioen B, Zhao J, Riedl J, Zwartkruis F, van der Krogt G et al (2004) Detecting cAMP-induced Epac activation by fluorescence resonance energy transfer: Epac as a novel cAMP indicator. EMBO Rep 5:1176–1180
7. Zaccolo M, De Giorgi F, Cho CY, Feng L, Knapp T et al (2000) A genetically encoded, fluorescent indicator for cyclic AMP in living cells. Nat Cell Biol 2:25–29
8. DiPilato LM, Cheng X, Zhang J (2004) Fluorescent indicators of cAMP and Epac activation reveal differential dynamics of cAMP signaling within discrete subcellular compartments. Proc Natl Acad Sci U S A 101:16513–16518
9. Klarenbeek JB, Goedhart J, Hink MA, Gadella TW, Jalink K (2011) A mTurquoise-based cAMP sensor for both FLIM and ratiometric read-out has improved dynamic range. PLoS One 6:e19170
10. Rehmann H, Wittinghofer A, Bos JL (2007) Capturing cyclic nucleotides in action: snapshots from crystallographic studies. Nat Rev Mol Cell Biol 8:63–73
11. Salonikidis PS, Zeug A, Kobe F, Ponimaskin E, Richter DW (2008) Quantitative measurement of cAMP concentration using an exchange protein directly activated by a cAMP-based FRET-sensor. Biophys J 95:5412–5423
12. Salonikidis PS, Niebert M, Ullrich T, Bao G, Zeug A et al (2011) An ion-insensitive cAMP biosensor for long term quantitative ratiometric fluorescence resonance energy transfer (FRET) measurements under variable physiological conditions. J Biol Chem 286: 23419–23431
13. Goedhart J, van Weeren L, Hink MA, Vischer NO, Jalink K et al (2010) Bright cyan fluorescent protein variants identified by fluorescence lifetime screening. Nat Methods 7:137–139
14. Honda A, Sawyer CL, Cawley SM, Dostmann WR (2005) Cygnets: in vivo characterization of novel cGMP indicators and in vivo imaging of intracellular cGMP. Methods Mol Biol 307: 27–43
15. Niino Y, Hotta K, Oka K (2010) Blue fluorescent cGMP sensor for multiparameter fluorescence imaging. PLoS One 5:e9164
16. Russwurm M, Mullershausen F, Friebe A, Jager R, Russwurm C et al (2007) Design of fluorescence resonance energy transfer (FRET)-based cGMP indicators: a systematic approach. Biochem J 407:69–77

Chapter 5

Analysis of Compartmentalized cAMP: A Method to Compare Signals from Differently Targeted FRET Reporters

Alessandra Stangherlin, Andreas Koschinski, Anna Terrin, Anna Zoccarato, He Jiang, Laura Ashley Fields, and Manuela Zaccolo

Abstract

Förster resonance energy transfer (FRET)-based reporters are important tools to study the spatiotemporal compartmentalization of cyclic adenosine monophosphate (cAMP) in living cells. To increase the spatial resolution of cAMP detection, new reporters with specific intracellular targeting have been developed. Therefore it has become critical to be able to appropriately compare the signals revealed by the different sensors. Here we illustrate a protocol to calibrate the response detected by different targeted FRET reporters involving the generation of a dose–response curve to the cAMP raising agent forskolin. This method represents a general tool for the accurate analysis and interpretation of intracellular cAMP changes detected at the level of different subcellular compartments.

Key words Imaging, Förster resonance energy transfer (FRET), Targeted sensors, cAMP compartmentalization

1 Introduction

Förster resonance energy transfer (FRET), often also referred to as fluorescence resonance energy transfer, is a non-radiative process of energy transfer from an excited donor fluorophore to an acceptor fluorophore [1]. FRET efficiency is inversely proportional to the sixth power of the donor–acceptor distance and this phenomenon has been extensively used to monitor protein–DNA interactions, protein–protein interactions, and protein conformational changes. Among many applications, this technique has been exploited to generate genetically encoded sensors to visualize cyclic adenosine monophosphate (cAMP) dynamics in intact living cells [2]. A FRET-based indicator of cAMP is generally composed

---

Alessandra Stangherlin and Andreas Koschinski contributed equally to this work.

Jin Zhang et al. (eds.), *Fluorescent Protein-Based Biosensors: Methods and Protocols*, Methods in Molecular Biology, vol. 1071, DOI 10.1007/978-1-62703-622-1_5,

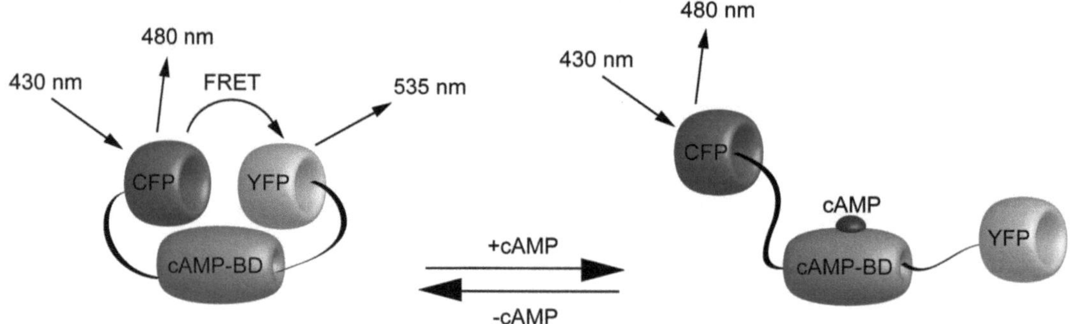

Fig. 1 Representation of a FRET-based sensor for cAMP and its mechanism. In the cAMP-free conformation of the sensor, CFP and YFP are close enough for FRET to occur. Upon excitation at 430 nm, part of the excited state energy of CFP is transferred by resonance to YFP, which, in turn, becomes excited and both emissions from CFP at 480 and YFP at 535 nm can be recorded. Binding of cAMP to the probe induces a conformational change which moves the two fluorophores apart, abolishing FRET. *cAMP*, cyclic adenosine monophosphate; *CFP*, cyan fluorescent protein; *YFP*, yellow fluorescent protein; *cAMP-BD*, cAMP-binding domain; *FRET*, Förster resonance energy transfer

of a cAMP-binding domain and the cyan and yellow variants of the green fluorescent protein, CFP and YFP, respectively. Binding of cAMP to the sensor induces a conformational change which alters the distance between the two fluorophores, thereby affecting the efficiency of energy transfer (Fig. 1). Because such sensors are genetically encoded, they can be expressed in living cells upon transfection or infection and used in real-time imaging experiments to monitor changes in intracellular cAMP levels based on changes in FRET.

The high spatial and temporal resolution of FRET-based real-time imaging has provided an initial understanding of the molecular mechanisms regulating the specificity of cAMP-PKA signaling and its compartmentalization. Thanks to this approach, it was possible to establish that cAMP is generated in discrete microdomains and to unravel some of the mechanisms responsible for shaping AMP gradients within the cell [3–6].

To increase the spatial resolution of cAMP detection and to better investigate the subcellular dynamics of this second messenger, cAMP indicators have been genetically engineered to achieve targeting to different subcellular compartments, including the nucleus [4], the plasma membrane [4], mitochondria [7], different A-kinase anchoring proteins (AKAPs) [5], or a variety of protein complexes [8, 9]. These are very useful tools as for example they allow one to estimate the amplitude and the dynamics of the cAMP signal elicited by a specific stimulus at different subcellular sites. The comparison of the responses detected with differently targeted sensors can subsequently be used to determine whether the cAMP signal is regulated in a compartment-specific manner [10–12]. However, differences in the physicochemical properties of the targeted sensors might be a source of errors and artifacts.

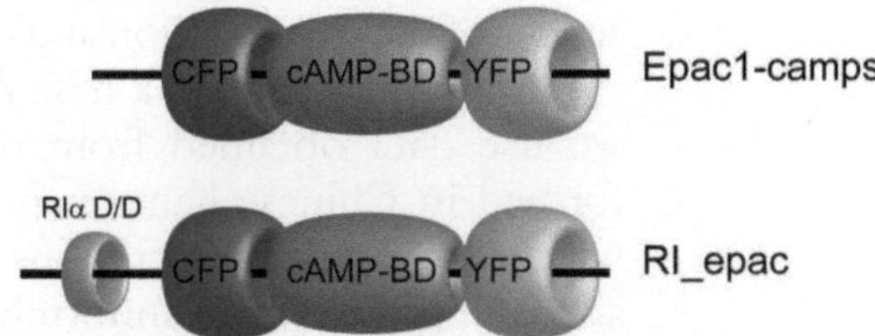

Fig. 2 Schematic representation of Epac1-camps and RI_epac sensors. *CFP*, cyan fluorescent protein; *YFP*, yellow fluorescent protein; *cAMP-BD*, cAMP-binding domain; *RIα DD*, dimerization and docking domain of the PKA regulatory subunit RIα

For instance, the fusion of a targeting sequence to the cAMP sensor may affect the FRET signal via a number of mechanisms, including altered interaction between the fluorophores, interference with the conformational change induced by cAMP binding, changes in the affinity of the sensor for cAMP or exposure of the fluorophores to different environmental conditions (e.g., local variations in pH, Cl^- concentration, etc.). These mechanisms may affect fluorescence emission independently of cAMP signals. As a consequence, FRET efficiency might be increased, reduced, or abolished [9] and absolute FRET change values detected by different sensors cannot be directly compared. When dealing with reporters that show different FRET change to a stimulus of fixed intensity, some investigators have resolved to normalize the FRET changes recorded to the maximal FRET change for that specific sensor at saturation [4, 5, 13]. This procedure is correct when the sensors that are compared show superimposable dose–response curves [4, 5]. However, such information is not always available and in some cases the comparison is based on the assumption that the relationship between signal intensity and FRET change is linear, with a single measured calibration point (FRET change at saturation) as a basis for the calculation. If the compared sensors have different affinities for the signal molecule or different slopes of the dose–response curves, the results obtained using this approach are misleading.

The following protocol illustrates how to compare the signal generated by two differently targeted FRET reporters by using a forskolin-to-FRET change dose–response curve as a reference and by translating the FRET changes in “forskolin-equivalent” values. In the first part of the protocol, we describe the general steps to generate dose–response curves for FRET sensors. As an example, we use the sensors Epac1-camps [14] and RI_epac [5] (Fig. 2). When expressed in cells, Epac1-camps is uniformly distributed throughout the cytosol, whereas RI_epac is targeted to RI-selective AKAPs (RI compartment) via fusion to the amino-terminus of the sensor of the dimerization and docking domain from the regulatory subunit RIα of protein kinase A (PKA). To illustrate this approach we use data obtained from real-time imaging experiments performed in Chinese hamster ovary (CHO) cells upon stimulation with increasing concentrations of forskolin (FRSK), an activator of adenylyl cyclases that uniformly increases cAMP within the cell. In the protocol below, we provide the relevant general formulas and specific examples in which the formulas are applied to real primary data. The second part of the protocol illustrates how to use the forskolin-to-FRET change dose–response curves to calibrate FRET changes detected by the two different sensors, Epac1-camps and RI_epac.

This approach provides a tool for the analysis of cAMP dynamics in different compartments and can be applied to any cell type, experimental condition and to any reporter. Importantly, it can be easily implemented in any laboratory.

2 Materials

2.1 Cell Culture and Transfection

1. CHO cells.
2. CHO growth medium: HAM'S-F12, 10 % fetal bovine serum, 100 U/mL penicillin, 100 μg/mL streptomycin, and 2 mM glutamine.
3. 0.05 % trypsin.
4. Phosphate buffer solution (PBS): 137 mM NaCl, 2.7 mM KCl, 100 mM Na_2HPO_4, 2 mM KH_2PO_4.
5. Transfection reagent: TransIT®-LT1 Transfection Reagent (MIR 2300, Mirus).

2.2 FRET Experiments and Analysis

1. Imaging buffer (modified Tyrode solution): 125 mM NaCl, 5 mM KCl, 1 mM Na_3PO_4, 1 mM $MgSO_4$, 20 mM HEPES, 1 mM $CaCl_2$, 5.5 mM Glucose, adjusted to pH 7.4 with NaOH.
2. Forskolin (FRSK).
3. FRET setup [15].
4. Graph Pad Prism™ software.

3 Methods

3.1 Generation of Forskolin Dose–Response Curves for Epac1-Camps and RI_epac

1. Seed CHO cells on high-quality glass coverslips and transfect them with the vector constructs encoding the appropriate sensor. In our example experiment, we used Epac1-camps and RI_epac.

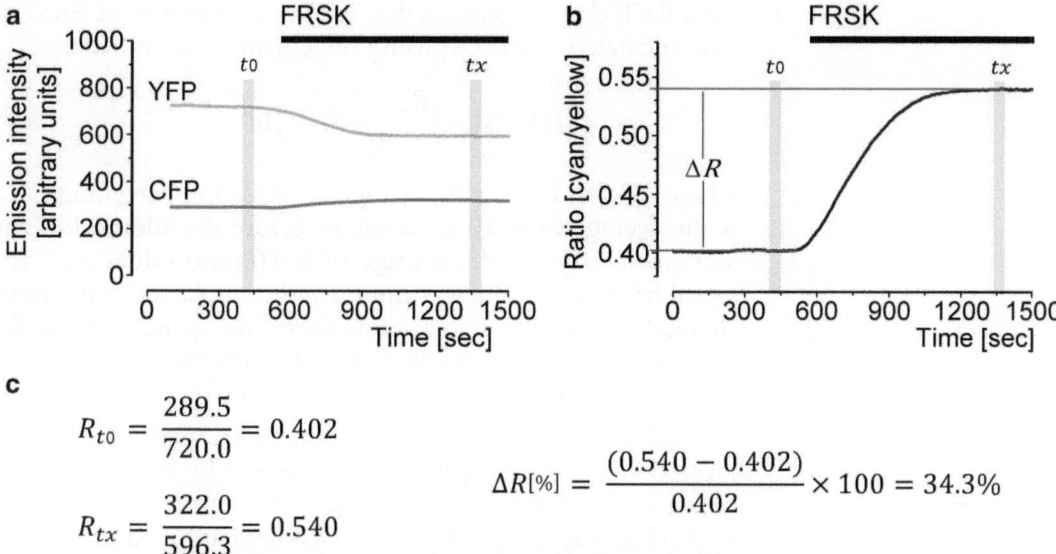

Fig. 3 Generation of the FRET change curve and calculation of $\Delta R_{[\%]}$. Representative example performed in CHO cells transfected with Epac1-camps and treated with 5 μM FRSK. (**a**) Plot of background-subtracted CFP and YFP emission intensity values over time. (**b**) Plot of the ratio values (*R*) calculated from (**a**) over time. The *black bar* at the top of panels (**a**) and (**b**) indicates application of the stimulus. The *gray vertical bars* in (**a**) and (**b**) indicate the time interval within which FRET ratio values have been averaged to calculate R_{t0} and R_{tx}. (**c**) Calculation of the ratio change. *R*, $Em_{480\,nm}/Em_{535\,nm}$; R_{t0}, ratio value calculated before the stimulus is applied; R_{tx}, ratio value calculated at the plateau of the response; ΔR, change in ratio value

2. For both sensors perform a series of FRET change measurements, using cells bathed in imaging buffer and treated with decreasing concentrations of FRSK (e.g., 25, 10, 5, and 2.5 μM). The aim is to reach a concentration of stimulus at which a FRET change is no longer detected (*see* **Note 1**). As the scope of this chapter is not to describe how to perform real-time FRET imaging, for a more detailed description of microscope settings and the acquisition technique *see* introductory Chapter 1 and [15, 16].
3. For each FRSK concentration, cells are excited at 430 nm and the emission intensities at 480 nm (CFP emission) and 535 nm (YFP emission) are recorded over time (Fig. 3a) (*see* **Note 2**). A time-course of FRET change is then calculated by computing the ratio value (*R*) for each acquisition time point using the following equation (*see* **Note 3**):

$$R = \frac{\text{Emission}_{480\text{nm}}}{\text{Emission}_{535\text{nm}}}. \quad (1)$$

4. The ratio values over time are plotted in order to obtain a FRET change curve (Fig. 3b).

5. The FRET change generated at each concentration of FRSK is then calculated as ΔR [%] using the equation (*see* **Note 4**):

$$\Delta R\,[\%] = \frac{(R_{tx} - R_{t0})}{R_{t0}} \times 100, \quad (2)$$

where ΔR [%] is the ratio change expressed in percentage, R_{t0} is the average of 5–10 ratio values before the addition of the stimulus and R_{tx} is the average of 5–10 ratio values after the signal has reached a new, stimulus-induced plateau. The signal should be reasonably stable within these time-spans (*see* **Note 5**). An example of values obtained in CHO cells expressing Epac1-camps and treated with 5 μM FRSK is shown in Fig. 3.

6. The ΔR [%] values obtained at each FRSK concentration are plotted and a curve is fitted to the points by applying the fit function for a sigmoidal dose–response curve with variable slope (Fig. 4a). This can be done using Graph Pad Prism™ or a similar software program (*see* **Note 6**). For each curve the specific parameters EC_{50}, Hill-coefficient (n), R_0 (the value of FRET change in the absence of stimulus), and R_{max} (the value of FRET change at saturating concentration of the stimulus) are automatically calculated (*see* **Note 7**). The values obtained for Epac1-camps and RI_epac in our example experiment are given in Table 1.

7. To facilitate the comparison of dose–response curves generated by different sensors, the curves can be normalized to the maximal FRET change of the respective sensor (*see* **Note 8**). The normalized dose–response curves for Epac1-camps and RI_epac are shown in Fig. 4b (*see* **Note 9**).

8. Compare EC_{50}, Hill coefficients and maximal FRET change values of the sensors. If they are the same, the FRET changes detected by the sensors can be directly compared. If the values differ, as in the case of Epac1-camps and RI_epac in our example (Fig. 4 and Table 1), the responses detected with these sensors should be transformed in FRSK-equivalents before they can be compared.

3.2 Calculation of FRSK-Equivalent Values

1. Perform the experiment of choice. In our example, CHO cells were treated with a fixed concentration of a cAMP raising stimulus and Epac1-camps generated a FRET change of 6.3 ± 1.3 % (SEM, n = 9), whereas the FRET change generated by RI_epac was 4.0 ± 0.8 % (SEM, n = 9) (Fig. 5a).

2. For each sensor, calculate the FRET change detected upon treatment with the stimulus as a percentage of the maximal FRET change for that sensor (Fig. 5b). The normalized responses in our example are 15.3 ± 3.2 % (SEM, n = 9) of the

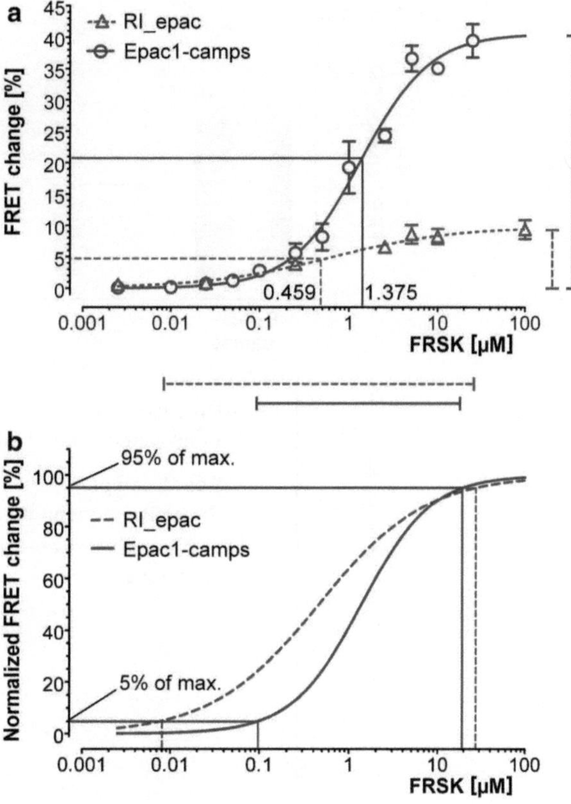

Fig. 4 Dose–response curves for Epac1-camps and RI_epac. (**a**) Dose–response curves for Epac1-camps (*solid*) and RI_epac (*dotted*). EC_{50} values for each sensor are indicated on the *X*-axis. *Bars* below the *X*-axis show the sensitivity ranges (5–95 % of maximal FRET change) for each sensor. *Bars* on the *right* indicate the dynamic range of the sensors. All values represent mean ± SEM, $n = 3$–14. (**b**) Normalized dose–response curves for Epac1-camps (*solid*) and RI_epac (*dotted*). The useful sensitivity range (5–95 %) is indicated for both sensors

Table 1
Parameters of the dose–response curves for Epac1-camps and RI_epac

	Epac1-camps	RI_epac
R_0 [%]	0.00	0.00
R_{max} [%]	41.17	9.54
n (Hill-coefficient)	1.121	0.7286
EC_{50} [μM]	1.375	0.4588
Sensitivity range	99 nM–19 μM	8 nM–26 μM

Parameters for the curves shown in Fig. 4 as calculated by Graph Pad Prism™. R_0 is the minimal FRET change, R_{max} is the maximal FRET change. EC_{50} is the concentration of a stimulus that generates a half-maximal FRET change, n is the Hill-coefficient. The sensitivity range is shown in the bottom line

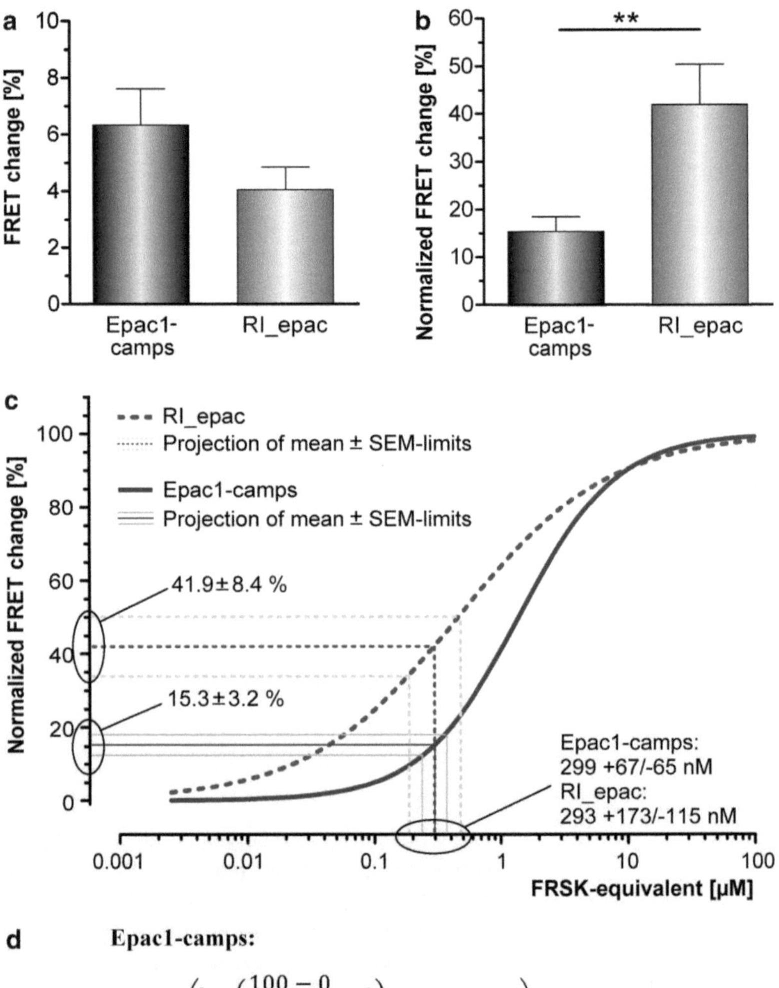

d

Epac1-camps:

$$x = -\left(\frac{\log\left(\frac{100-0}{15.3-0}-1\right)}{1.121} - \log\ 1.375\right) = -0.5247$$

$$\text{FRSK-equivalent} = 10^{x} \Rightarrow 10^{-0.5247} = 299\ \text{nM}$$

RI_epac:

$$x = -\left(\frac{\log\left(\frac{100-0}{41.9-0}-1\right)}{0.7286} - \log\ 0.4588\right) = -0.5332$$

$$\text{FRSK-equivalent} = 10^{x} \Rightarrow 10^{-0.5332} = 293\ \text{nM}$$

Fig. 5 Calculation of FRSK-equivalent values. (**a**) Average FRET change recorded in CHO cells expressing Epac1-camps or RI_epac upon treatment with a cAMP-raising stimulus. (**b**) Normalized FRET changes for Epac1-camps and RI_epac, calculated as percentage of the maximal FRET change of the respective sensors at saturation. Unpaired student's *t*-test reveals a significant difference in the cAMP responses in the two compartments ($p = 0.0098$). (**c**) Normalized dose–response curves for Epac1-camps (*solid*) and RI_epac (*dotted*), and calculation

maximal response for Epac1-camps and 41.9 ± 8.4 % (SEM, $n=9$) of the maximal response for the RI_epac sensor. Note that, based on this calculation, it could be concluded that the specific stimulus we have used in this experiment generates a significantly ($p=0.0098$) higher response in the RI compartment (where RI_Epac is targeted) as compared to the bulk cytosol (the signal detected by the untargeted sensor, Epac1-camps).

3. For each sensor, convert the normalized response to "FRSK-equivalents". Calculate the FRSK concentration that would lead to the measured FRET change using the graphical approach illustrated in Fig. 5c or Eq. 3:

$$x = -\left(\frac{\log\left(\frac{R_{max} - R_0}{\Delta R - R_0} - 1\right)}{n} - \log EC_{50}\right), \tag{3}$$

where n is the Hill-coefficient, ΔR is the FRET change value at which the FRSK-equivalent (the concentration of FRSK that would generate the same FRET change) is calculated, x is the logarithm of the FRSK concentration, R_0 is the minimal and R_{max} the maximal FRET change. Note that, because of the nonlinear relationship between FRSK concentration and FRET response, the upper and lower limits of the respective errors (SD or SEM) will differ. In our example, the FRSK-equivalent value calculated for RI_epac is 293 +173/−115 nM, whereas the FRSK-equivalent value calculated for Epac1-camps is 299 +67/−65 nM (Fig. 5c, d).

4. Compare the responses detected by the different sensors in terms of FRSK-equivalents. In our example, this analysis shows that the cAMP response in the two compartments is identical. As noted above, if the FRET changes detected by the two sensors were simply expressed as the percentage of maximal FRET change, the wrong conclusion would be drawn (i.e., that the cAMP response generated in the RI compartment is significantly higher than the cAMP response in the bulk cytosol) (*see* **Note 10**).

Fig. 5 (continued) of FRSK-equivalent values using the graphical approach. Error bars indicate SEM, $n=9$ for Epac1-camps and $n=9$ for RI_epac. (**d**) The FRET changes detected by Epac1-camps and RI_epac in our example experiment are converted in FRSK-equivalents following Eq. 3. Note that when normalization is applied R_{max} is set to 100 and R_0 to 0 %

4 Notes

1. If even at very low concentrations of FRSK a FRET change can still be detected, control experiments with 0 μM FRSK should be performed. If 0 μM FRSK generates a response, vehicle effects and addition artifacts should be investigated. Changes in osmolarity, pH, and temperature are common causes for such experimental artifacts. If the vehicle produces such artificial responses, the measurements should be corrected to subtract the effect of vehicle.
2. It is critical to correct for background signal and drifting baseline. A useful guide to correction procedures is discussed in ref. [16]. Some acquisition programs, e.g., Metafluor™, provide some built in background-correction functions. However, these functions change the original data irrevocably. In fact, if such a function is applied during acquisition, all the recorded data will already be background-subtracted and it will not be possible to go back to the original data. Therefore, it is advisable to perform all experiments without any a priori correction. Raw data should be exported into any adequate spreadsheet program (e.g., MS Excel™ or Open Office Calc™) and analyzed off-line.

   In our experience, correction for bleed-through and cross-excitation is normally not necessary [17, 18]. Without this correction, the FRET ratio signal is altered by a certain factor (the value of which depends on the specific setup used for the measurements), which, however, always will be the same if the same equipment is used.
3. Binding of cAMP to Epac1-camps and RI_epac increases the distance between the fluorophores (Fig. 1), leading to reduced FRET at increasing cAMP concentrations. For convenience, the ratio is calculated as CFP emission at 480 nm over YFP emission at 545 nm so that an increase in ratio correlates with an increase in cAMP. However, the ratio should be reversed when using sensors with inverse features [19, 20].
4. Equation 2 directly expresses the FRET change as a percentage. The equation $\Delta R = (R_{tx} - R_{t0}) / R_{t0}$, which expresses the ratio change as a numeric value, can also be used.
5. If the signal never reaches a stable plateau, but shows a constant increase or decrease (drift), the reasons should be investigated. In our experience the composition of the extracellular buffer, its pH, temperature changes or bleaching of the fluorophores can be reasons for such a drift. In general all experimental conditions should be optimized in order to minimize drifting ratios.

If the drift cannot be eliminated, it may also be possible to apply a mathematical correction to compensate for a constant drift, e.g., by linear interpolation of the baseline and, depending on direction of the drift, subtraction or addition of the respective interpolated values before calculating the ratio change. In this case control experiments should be performed to verify that this drift typically is constant over the complete time of a standard experiment. However, every experiment will still have to be judged critically before applying the adequate correction.

6. Graph Pad Prism™ calculates and fits the curve according to the least squares fit method, using an iterative approach on the basis of the following equation, which has been partly adapted to the terms used in the protocol:

$$\Delta R = R_0 + \frac{R_{max} - R_0}{1+10^{((\log EC_{50} - x)\times n)}},$$

where x is the logarithm of the concentration of the agonist and ΔR is the FRET change. ΔR starts at a minimum (R_0 or "bottom") and goes to the maximum (R_{max} or "top") with a sigmoidal shape.

7. EC_{50} is the concentration of a stimulus needed to generate a half-maximal FRET change. It is a commonly used parameter to characterize the sensitivity of a sensor. The lower this value, the higher the sensitivity of the sensor to its ligand. The Hill-coefficient (or Hillslope) (n) provides a quantification of cooperativity effects in the reaction and correlates with the steepness of the curve. A higher numeric value of the Hill-coefficient usually corresponds to a steeper curve, leading typically to a narrower useful sensitivity range.

8. For normalization, each value is expressed as the percentage (or fraction) of the maximal FRET change achievable by the respective sensor. Note that normalization is not absolutely necessary and FRSK-equivalent values can also be calculated directly on the FRET change dose–response curve. However, normalization allows one to compare the shapes of the curves, which is necessary when comparing responses in different cell systems (*see* **Note 9**).

9. Sensors give more accurate and resolved readings when they are not used at the extreme ends of their sensitivity range. Minimal (ΔR_0) and maximal (ΔR_{max}) FRET changes define the absolute sensitivity range of the sensor. The higher the maximal change, the better is the signal-to-noise ratio, typically leading to a better resolution of responses. However, because most biological sensors possess a sigmoidal concentration–response curve, the resolution of these sensors at very low and

very high concentrations is poor. Therefore the useful sensitivity range is smaller than the absolute range and depends strongly on the shape of the curve. If a signal is expected to be at the borders or beyond the useful sensitivity range, another sensor with appropriate sensitivity should be used, if available. For the sensors presented here we defined the useful sensitivity range to be between 5 and 95 % of the maximal response of the sensor. The sensitivity range can be calculated either graphically, using the curve shown in Fig. 4b, or mathematically, using Eq. 3. In our example, the sensitivity range for the Epac1-camps sensor is within 99 nM–19 μM, whereas the sensitivity range for RI-epac is within 8 nM–26 μM.

10. A final note of caution: it should not be assumed that the dose–response curve of a given sensor will be the same in all cell types, nor that different sensors, when expressed in different cell types, would present similar changes in the shape of the dose–response curve. Therefore, when using the sensors in cell systems different from the one in which the dose–response curve has been generated, it is important to verify that this reference curve is still valid.

Acknowledgments

We thank Alsbetha Hulikova and Thomas Fritz for valuable discussions. The work described in this paper was supported by the British Heart Foundation (PG/07/091/23698) and the NSF–NIH CRCNS program (NIH R01 AA18060).

References

1. Clegg R (1996) Fluorescence resonance energy transfer. In: Wang XF, Herman B (eds) Fluorescence imaging spectroscopy and microscopy, vol 137. Wilely, New York, pp 179–251
2. Zaccolo M, De Giorgi F, Cho CY et al (2000) A genetically encoded, fluorescent indicator for cyclic AMP in living cells. Nat Cell Biol 2: 25–29
3. Zaccolo M, Pozzan T (2002) Discrete microdomains with high concentration of cAMP in stimulated rat neonatal cardiac myocytes. Science 295:1711–1715
4. Terrin A, Di Benedetto G, Pertegato V et al (2006) PGE(1) stimulation of HEK293 cells generates multiple contiguous domains with different [cAMP]: role of compartmentalized phosphodiesterases. J Cell Biol 175:441–451
5. Di Benedetto G, Zoccarato A, Lissandron V et al (2008) Protein kinase A type I and type II define distinct intracellular signaling compartments. Circ Res 103:836–844
6. Monterisi S, Favia M, Guerra L et al (2012) CFTR regulation in human airway epithelial cells requires integrity of the actin cytoskeleton and compartmentalized cAMP and PKA activity. J Cell Sci 125(Pt 5):1106–1117
7. DiPilato LM, Cheng X, Zhang J (2004) Fluorescent indicators of cAMP and Epac activation reveal differential dynamics of cAMP signalling within discrete subcellular compartments. Proc Natl Acad Sci U S A 101: 16513–16518
8. Sin YY, Edwards HV, Li X et al (2011) Disruption of the cyclic AMP phosphodiesterase-4 (PDE4)-HSP20 complex attenuates the beta-agonist induced hypertrophic response in cardiac myocytes. J Mol Cell Cardiol 50: 872–883

9. Herget S, Lohse MJ, Nikolaev VO (2008) Real-time monitoring of phosphodiesterase inhibition in intact cells. Cell Signal 20: 1423–1431
10. Mohamed TM, Oceandy D, Zi M et al (2011) Plasma membrane calcium pump (PMCA4)-neuronal nitric-oxide synthase complex regulates cardiac contractility through modulation of a compartmentalized cyclic nucleotide microdomain. J Biol Chem 286:41520–41529
11. Matthiesen K, Nielsen J (2011) Cyclic AMP control measured in two compartments in HEK293 cells: phosphodiesterase K(M) is more important than phosphodiesterase localization. PLoS One 6:e24392
12. Wachten S, Masada N, Ayling LJ et al (2010) Distinct pools of cAMP centre on different isoforms of adenylyl cyclase in pituitary-derived GH3B6 cells. J Cell Sci 123:95–106
13. Castro LR, Gervasi N, Guiot E et al (2010) Type 4 phosphodiesterase plays different integrating roles in different cellular domains in pyramidal cortical neurons. J Neurosci 30: 6143–6151
14. Nikolaev VO, Bunemann M, Hein L et al (2004) Novel single chain cAMP sensors for receptor-induced signal propagation. J Biol Chem 279:37215–37218
15. Gesellchen F, Stangherlin A, Surdo N et al (2011) Measuring spatiotemporal dynamics of cyclic AMP signaling in real-time using FRET-based biosensors. Methods Mol Biol 746: 297–316
16. Borner S, Schwede F, Schlipp A et al (2011) FRET measurements of intracellular cAMP concentrations and cAMP analog permeability in intact cells. Nat Protoc 6:427–438
17. Berrera M, Dodoni G, Monterisi S et al (2008) A toolkit for real-time detection of cAMP: insights into compartmentalized signaling. Handb Exp Pharmacol 186:285–298
18. Evellin S, Mongillo M, Terrin A et al (2004) Measuring dynamic changes in cAMP using fluorescence resonance energy transfer. Methods Mol Biol 284:259–270
19. Zhang J, Ma Y, Taylor SS et al (2001) Genetically encoded reporters of protein kinase A activity reveal impact of substrate tethering. Proc Natl Acad Sci U S A 98:14997–15002
20. Nikolaev VO, Gambaryan S, Lohse MJ (2006) Fluorescent sensors for rapid monitoring of intracellular cGMP. Nat Methods 3:23–25

Chapter 6

Genetically Encoded Fluorescent Biosensors for Live Cell Imaging of Lipid Dynamics

Moritoshi Sato

Abstract

Fluorescence imaging provides a powerful technique to visualize spatiotemporal dynamics of biomolecules in living cells, if fluorescent biosensors for the relevant biomolecules become available. Here I describe a fluorescent biosensor for a lipid second messenger, phosphatidylinositol 3,4,5-triphosphate (PI(3,4,5)P_3). The biosensor overcomes limitations of existing methods for the lipid analysis and allows us to pinpoint that the PI(3,4,5)P_3 concentrations are increasing and/or decreasing not only at the plasma membrane but also at organelle membranes, such as the Golgi apparatus membranes and endoplasmic reticulum membranes. The present biosensor has also been shown to be applicable to a variety of lipid second messengers, including diacylglycerol and phosphatidylinositol 3,4-bisphosphate.

Key words Fluorescent biosensors, Fluorescence resonance energy transfer (FRET), Live cell imaging, Lipid second messengers, Phosphatidylinositol 3,4,5-triphosphate (PI(3,4,5)P_3), Subcellular dynamics

1 Introduction

PI(3,4,5)P_3 is a lipid second messenger that regulates diverse cellular functions, including cell proliferation and apoptosis, and has a role in the progression of diabetes and cancer (Fig. 1a) [1–3]. PI(3,4,5)P_3 is generated by phosphoinositide 3-kinase (PI3K) [4]. When generated at cellular membranes, PI(3,4,5)P_3 recruits and activates its binding proteins, including Akt, PDK1, Btk, and GRP1, to regulate various cellular functions. To analyze the PI(3,4,5)P_3 levels, labeling cells with [^{32}P]orthophosphate has been widely used. However, because millions of cells must be smashed and analyzed to obtain sufficient radiochemical signals this method has several limitations with respect to its spatial and temporal resolution. Recently, several PI(3,4,5)P_3 biosensors have been constructed by fusing *Aequorea victoria* green fluorescent protein (GFP) to a PI(3,4,5)P_3-binding domain derived from Btk, GRP1, ARNO, or Akt. These sensors, which translocate from the

Jin Zhang et al. (eds.), *Fluorescent Protein-Based Biosensors: Methods and Protocols*, Methods in Molecular Biology, vol. 1071, DOI 10.1007/978-1-62703-622-1_6,

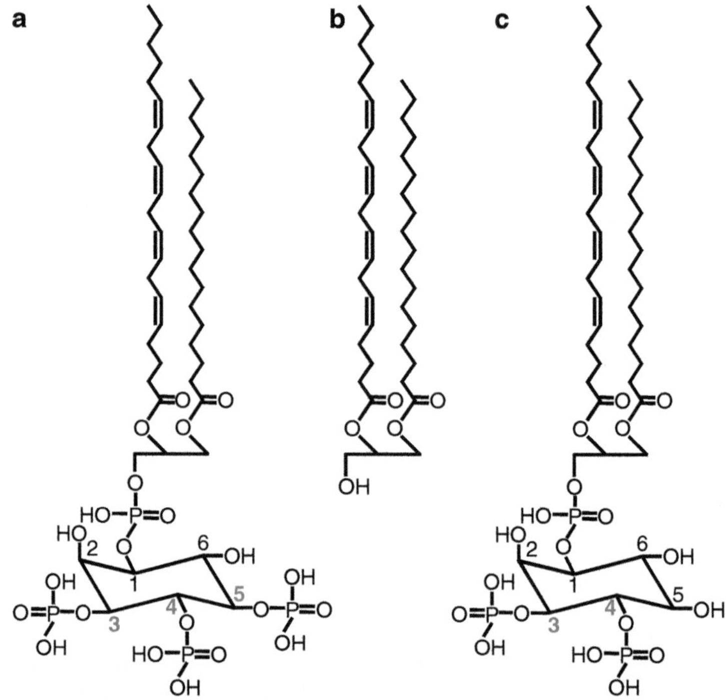

Fig. 1 Structure of lipid second messengers. (**a**) Phosphatidylinositol 3,4,5-triphosphate (PI(3,4,5)P_3). (**b**) Diacylglycerol (DAG). (**c**) Phosphatidylinositol 3,4-bisphosphate (PI(3,4)P_2)

cytosol to the membrane following PI(3,4,5)P_3 generation, have been used to visualize changes in the PI(3,4,5)P_3 levels in the cellular membrane[5–8]. However, several factors frequently observed during fluorescence imaging experiments, such as membrane ruffles and changes in the cell shape, affect the fluorescence intensity change in the region of interest in a PI(3,4,5)P_3-independent manner, causing serious artifacts. Moreover, it is difficult with these types of fluorescent fusion proteins to distinguish to which membranes the fusion proteins are translocated in the cell.

To overcome these limitations of existing methods, we have developed a fluorescent biosensor for PI(3,4,5)P_3 based on fluorescence resonance energy transfer (FRET) (Fig. 2) [9]. The biosensor, named Fllip, is composed of two different-colored mutants of GFP and a PI(3,4,5)P_3-binding domain. The PI(3,4,5)P_3 level was observed by dual-emission ratio imaging, thereby allowing stable observation of PI(3,4,5)P_3 without suffering from the artifacts described above. Fllip is further connected with the membrane localization sequence (MLS) for the localized analysis of the PI(3,4,5)P_3 concentration in subcellular membranes. Fllip has allowed us to visualize the spatiotemporal regulation of the PI(3,4,5)P_3 generation in single living cells.

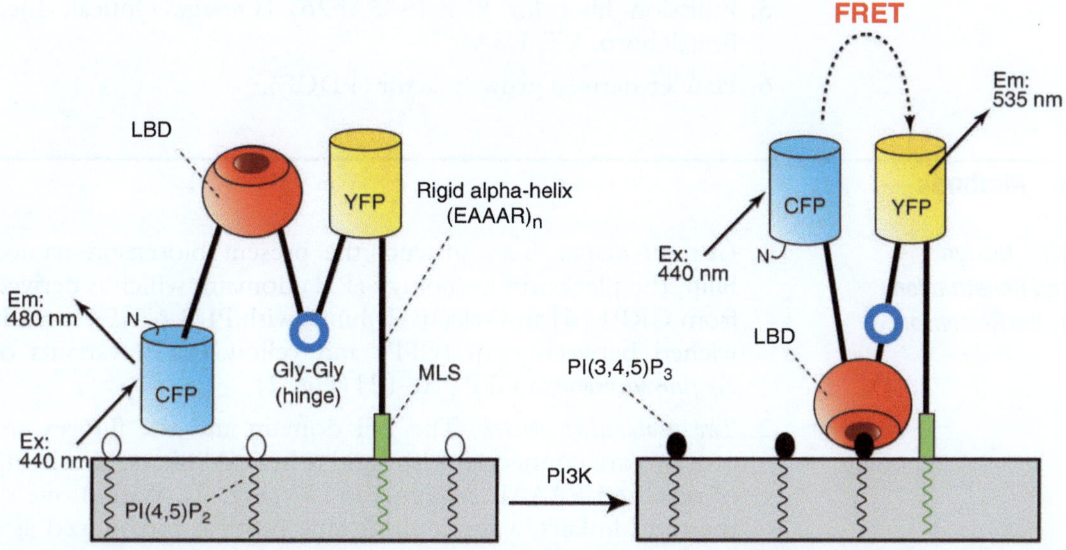

Fig. 2 Principle of the present fluorescent biosensor, Fllip. Abbreviations: *CFP* cyan fluorescent protein, *YFP* yellow fluorescent protein, *LBD* lipid binding domain, *MLS* membrane localization domain, *PI(4,5)P*$_2$ phosphatidylinositol 4,5-bisphosphate, *PI(3,4,5)P*$_3$ phosphatidylinositol 3,4,5-triphosphate, *PI3K* phosphatidylinositol 3-kinase, *Ex* excitation, *Em* emission, *FRET* fluorescence resonance energy transfer

2 Materials

2.1 Materials for Cell Culture and Transfection

1. Chinese hamster ovary cells expressing platelet-derived growth factor receptor (CHO-PDGFR cells).
2. Ham's F-12 medium.
3. Fetal bovine serum.
4. Penicillin and Streptomycin.
5. Trypsin–EDTA.
6. PBS tablet.
7. LipofectAMINE 2000.
8. OptiMEM.
9. Glass-bottomed dish.

2.2 Materials for Live Cell Imaging

1. Hanks balanced salt solution.
2. Excitation filter for CFP (440AF21) (Omega Optical, Inc., Brattleboro, VT, USA).
3. Dichroic mirror (455DRLP) (Omega Optical, Inc., Brattleboro, VT, USA).
4. Emission filter for CFP (480AF30) (Omega Optical, Inc., Brattleboro, VT, USA).

5. Emission filter for YFP (535AF26) (Omega Optical, Inc., Brattleboro, VT, USA).
6. Platelet-derived growth factor (PDGF).

3 Methods

3.1 Design and Construction of the Biosensor

1. *General design.* To construct the present biosensor, named Fllip, the pleckstrin homology (PH) domain, which is derived from GRP1 [6] and selectively binds with PI(3,4,5)P_3, is sandwiched between cyan (CFP) and yellow (YFP) variants of *Aequorea victoria* GFP [10–12] (Fig. 2).
2. *The molecular switch.* The PH domain and the fluorescent proteins are connected with rigid α-helical linkers, consisting of repeated EAAAR sequences [13] (Fig. 2). Within one of the rigid linkers, a single di-glycine motif is introduced as a hinge. When PI(3,4,5)P_3 is produced at the membrane upon activation of phosphatidylinositol 3-kinase (PI3K), the PH domain of Fllip binds with PI(3,4,5)P_3 (Fig. 2) and induces a significant conformational change through the flexible di-glycine motif introduced in the rigid α-helical linker. The flip-flop type of conformational change exhibited by Fllip results in intramolecular FRET from CFP to YFP, allowing detection of the PI(3,4,5)P_3 dynamics at the cellular membrane.
3. *Membrane targeting.* Fllip can be tethered to the membrane by fusing it with an MLS via the rigid α-helical linker (Fig. 2). Three different sequences, termed MLS_{pm}, MLS_{em}, and MLS_{mit}, have been tested. These sequences tether Fllip to the plasma membrane (pm), endomembranes (em), and mitochondrial outer membranes (mit), respectively (Fig. 3). The endomembranes include the Golgi apparatus membranes and the endoplasmic reticulum membranes. The amino acid sequence of MLS_{pm} (QGCMGLPCVVM) is derived from the CAAX box motif of N-Ras [14]. The N-terminal and C-terminal cysteines are palmitoylated and farnesylated, respectively. MLS_{em} (QGSMGLPCVVM) is derived from a mutated CAAX box motif of N-Ras, in which the original N-terminal cysteine is substituted with a serine not to be palmitoylated [14]. The amino acid sequence of MLS_{mit} (FNRWFLTGMTVAGVVLLGSLFSRK) is derived from the C-terminal sequence of Bcl-xL [15]. Fllip biosensors connected with MLS_{pm}, MLS_{em}, and MLS_{mit} are named Fllip-pm, Fllip-em, and Fllip-mit, respectively.

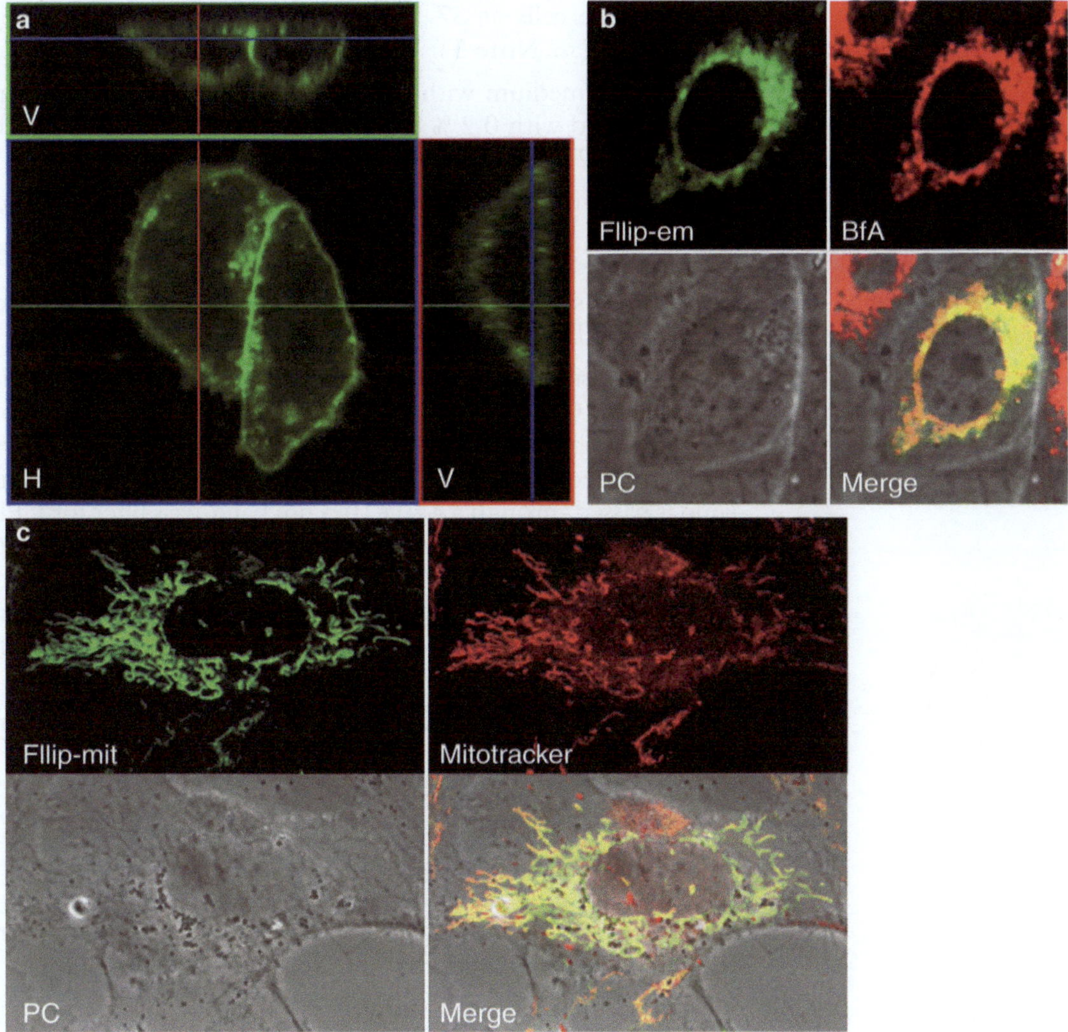

Fig. 3 Subcellular localization of Fllip-pm (**a**), Fllip-em (**b**), and Fllip-mit (**c**). *V* vertical sections, *H* horizontal section, *BfA* BODIPY-brefeldin A, *PC* phase contrast images

3.2 Cell Culture and Transfection

1. Plate CHO-PDGFR cells on 35-diameter glass-bottomed dishes with Ham's F-12 medium supplemented with 10 % fetal calf serum.
2. Incubate the cells at 37 °C in 5 % CO_2 until the cells are 50–80 % confluent.
3. Transfect the cells with 0.8 μg per dish of cDNA encoding Fllip using LipofectAMINE 2000 according to the manufacturer's instructions.
4. The transfection mixture may be replaced with fresh 1.5 mL of Ham's F-12 medium supplemented with 10 % fetal calf serum after 5–12 h of transfection, if desired.

5. Incubate the cells at 37 °C in 5 % CO_2 for 1–3 days after transfection (*see* **Note 1**).
6. Replace the medium with a serum-free Ham's F-12 medium supplemented with 0.2 % bovine serum albumin and incubate the cells for 2–6 h for serum starvation (*see* **Note 2**).
7. Wash with 1 mL of Hank's balanced salt solution (HBSS) and add 1 mL of HBSS for fluorescence imaging of PI(3,4,5)P_3.

3.3 Fluorescence Imaging

1. Place the cells on the stage of an inverted microscope equipped with a cooled CCD camera (*see* **Note 3**).
2. Image acquisition and processing are controlled by a PC connected to the CCD camera and a filter wheel using the MetaFluor software. A shutter for excitation in front of the xenon lamp is also controlled by the PC. Excitation light from a 75-mW xenon lamp is passed through a 440 ± 10.5 nm band-pass filter for CFP excitation (440AF21). The light is reflected onto the cells using a dichroic mirror (455DRLP). The emitted light is collected with a 40× or 63× objective and passed through a 480 ± 15 nm band-pass filter (480AF30) for CFP emission and a 535 ± 13 nm band-pass filter (535AF26) for YFP emission.
3. Define several factors for image acquisition, such as excitation power, time of exposure to the light, image acquisition interval, and binning (*see* **Note 4**).
4. By browsing the cells on the dish, choose moderately bright cells, in which Fllip is well localized in the subcellular membranes of interest, for example, the plasma membrane and endomembranes (Golgi membranes and ER membranes) in the case of Fllip-pm and Fllip-em, respectively (*see* **Note 5**).
5. Draw a region of interest within the cell on the field of view of the CCD camera (*see* **Note 6**).
6. Start to acquire images every 2–10 s with the 440 ± 10.5 nm excitation. The software MetaFluor records fluorescence intensities of the CFP and YFP emissions of Fllip, together with their emission ratios in the selected regions of the cells over time.
7. When the emission ratio of CFP to YFP reaches a steady state, add ligands, such as PDGF, to the dish to stimulate the cells and monitor the PI(3,4,5)P_3 levels through the change in the emission ratio of Fllip (Fig. 4).
8. After finishing all the experiments, wipe the residual immersion oil out of the objective.

3.4 Customizing the Biosensor

1. It should be mentioned that the modular design of Fllip allows it to be easily adapted to detect a variety of other lipid second messengers. Fllip has two key components: the MLS and the lipid binding domain (LBD). These two components can be tailored to detect the lipid-of-interest.

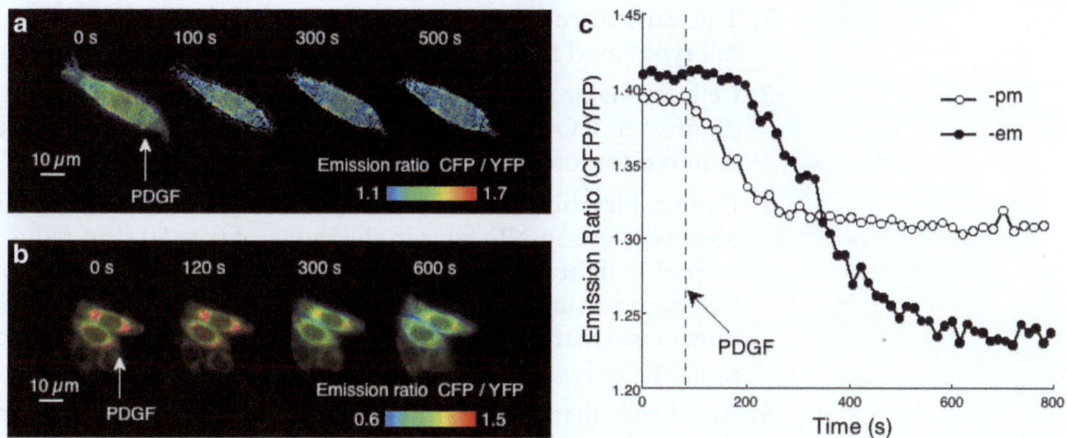

Fig. 4 (**a**, **b**) Pseudocolor images of the CFP/YFP emission ratio before (0 s) and 100, 300, or 500 s after addition of platelet-derived growth factor (PDGF). Results were obtained with CHO-PDGFR cells expressing Fllip-pm (**a**) or Fllip-em (**b**). (**c**) Time course of the CFP/YFP emission ratio of Fllip-pm and Fllip-em before and after stimulation with PDGF

Table 1
Changing the selectivity of the present fluorescent biosensor

Lipids	Lipid binding domains (LBDs)	References
$PI(3,4,5)P_3$	Pleckstrin homology domain derived from Grp1	[9]
DAG	Cysteine rich domain derived from PKC β	[16]
$PI(3,4)P_2$	Pleckstrin homology domain derived from TAPP1	[17]

2. By connecting each specific MLS, Fllip can be directed to various subcellular membranes, such as the plasma membrane and organelle membranes, including membranes of the Golgi apparatus, endoplasmic reticulum, mitochondrion, and the nucleus.

3. Lipid second messengers, such as diacylglycerol (DAG) [16] (Fig. 1b) and phosphatidylinositol 3,4-bisphosphate ($PI(3,4)P_2$) [17] (Fig. 1c) can be detected through use of appropriate LBDs, instead of the $PI(3,4,5)P_3$-binding domain, as performed previously (Table 1).

4 Notes

1. Incubation at 28 °C often improves subcellular localizations of Fllip-pm and Fllip-em in the plasma membrane and endomembranes, respectively.

2. This experimental step may not be necessary, depending upon cell types used for imaging.
3. Cells are viable and healthy for at least 60 min at ambient conditions. A CO_2 chamber and a temperature controller are required for long-term experiments.
4. If photobleaching of fluorescent proteins, in particular YFP, is observed, use an ND filter in the range of 0.1–10 % to reduce excitation light intensity and/or decrease exposure time. Also, binning can sum the signal from multiple pixels on the CCD camera so that less light is required while keeping a good signal-to-noise ratio.
5. Avoid cells that are too bright in order to reproducibly obtain quantitative FRET signals. Also avoid cells that are too dim, because these cells often exhibit noisy images.
6. The intensity in the region should be more than five times the background intensity.

Acknowledgments

This work was supported by grants from the Ministry of Education, Science and Culture, Japan and Japan Science and Technology Agency.

References

1. Cantley LC (2002) The phosphoinositide 3-kinase pathway. Science 296:1655–1657
2. Czech MP (2000) PIP_2 and PIP_3: complex roles at the cell surface. Cell 100:603–606
3. Marte BM, Downward J (1997) PKB/Akt:connecting phosphoinositide 3-kinase to cell survival and beyond. Trends Biochem Sci 22:355–358
4. Wymann MP, Pirola L (1998) Structure and function of phosphoinositide 3-kinases. Biochim Biophys Acta 1436:127–150
5. Varnai P, Rother KI, Balla T (1999) Phosphatidylinositol 3-kinase-dependent membrane association of the Bruton's tyrosine kinase pleckstrin homology domain visualized in single living cells. J Biol Chem 274: 10983–10989
6. Venkateswarlu K, Gunn-Moore F, Tavare JM, Cullen PJ (1998) Nerve growth factor- and epidermal growth factor-stimulated translocation of the ADP ribosylation factor-exchange factor GRP1 to the plasma membrane of PC12 cells requires phosphatidylinositol 3-kinase and the GRP1 pleckstrin homology domain. Biochem J 335:139–146
7. Venkateswarlu K, Oatey PB, Tavare JM, Cullen PJ (1998) Insulin-dependent translocation of ARNO to the plasma membrane of adipocytes requires phosphatidylinositol 3-kinase. Curr Biol 8:463–466
8. Watton SJ, Downward J (1999) Akt/PKB localisation and 3′ phosphoinositide generation at sites of epithelial cell-matrix and cell–cell interaction. Curr Biol 9:433–436
9. Sato M, Ueda Y, Takagi T, Umezawa Y (2003) Production of $PtdInsP_3$ at endomembranes is triggered by receptor endocytosis. Nat Cell Biol 5:1016–1022
10. Miyawaki A, Tsien RY (2000) Monitoring protein conformations and interactions by fluorescence resonance energy transfer between mutants of green fluorescent protein. Methods Enzymol 327:472–500
11. Vogel SS, Thaler C, Koushik SV (2006) Fanciful FRET. Sci STKE 2006:re2
12. Sato M, Ozawa T, Inukai K, Asano T, Umezawa Y (2002) Fluorescent indicators for imaging protein phosphorylation in single living cells. Nat Biotechnol 20: 287–294

13. Merutka G, Shalongo W, Stellwagen E (1991) A model peptide with enhanced helicity. Biochemistry 30:4245–4248
14. Choy E et al (1999) Endomembrane trafficking of Ras: the CAAX motif targets proteins to the ER and Golgi. Cell 98:69–80
15. Garcia-Saez AJ, Mingarro I, Perez-Paya E, Salgado J (2004) Membrane-insertion fragments of Bcl-xL, Bax, and Bid. Biochemistry 43:10930–10943
16. Sato M, Ueda Y, Umezawa Y (2006) Imaging diacylglycerol dynamics at organelle membranes. Nat Methods 3:797–799
17. Aoki K, Nakamura T, Inoue T, Meyer T, Matsuda M (2007) An essential role for the SHIP2-dependent negative feedback loop in neuritogenesis of nerve growth factor-stimulated PC12 cells. J Cell Biol 177:817–827

Chapter 7

Live-Cell Imaging of Cytosolic NADH–NAD⁺ Redox State Using a Genetically Encoded Fluorescent Biosensor

Yin Pun Hung and Gary Yellen

Abstract

NADH is an essential redox cofactor in numerous metabolic reactions, and the cytosolic NADH–NAD^+ redox state is a key parameter in glycolysis. Conventional NADH measurements rely on chemical determination or autofluorescence imaging, which cannot assess NADH specifically in the cytosol of individual live cells. By combining a bacterial NADH-binding protein and a fluorescent protein variant, we have created a genetically encoded fluorescent biosensor of the cytosolic NADH–NAD^+ redox state, named Peredox (Hung et al., Cell Metab 14:545–554, 2011). Here, we elaborate on imaging methods and technical considerations of using Peredox to measure cytosolic NADH:NAD^+ ratios in individual live cells.

Key words NADH, Glycolysis, Lactate dehydrogenase, Sensor calibration, Single cell imaging

1 Introduction

Nicotinamide adenine dinucleotide (oxidized: NAD^+; reduced, NADH) is an essential redox cofactor in numerous metabolic reactions. In mammalian cells, the cytosolic NADH–NAD^+ redox state is a key parameter in glucose metabolism by glycolysis [1]. Conventional NADH measurements rely on chemical determination or autofluorescence imaging; however, these approaches cannot assess NADH specifically in the cytosol of individual live cells. Chemical methods cannot measure NADH in different cell types within a population or in single cells, while autofluorescence, which reflects the combined signal of both NADH and NADPH, is not specific for NADH alone [2, 3].

We have therefore created a genetically encoded fluorescent biosensor to report on the cytosolic NADH–NAD^+ redox state [4]. This biosensor, named Peredox, is constructed by linking a green fluorescent protein (GFP) variant, circularly permuted T-Sapphire [5], to a bacterial NADH-binding protein Rex [6–8]. Upon binding NADH, Rex undergoes a conformational change that is coupled to an increase in the GFP fluorescence. Through competition between

Jin Zhang et al. (eds.), *Fluorescent Protein-Based Biosensors: Methods and Protocols*, Methods in Molecular Biology, vol. 1071,
DOI 10.1007/978-1-62703-622-1_7, © Springer Science+Business Media, LLC 2014

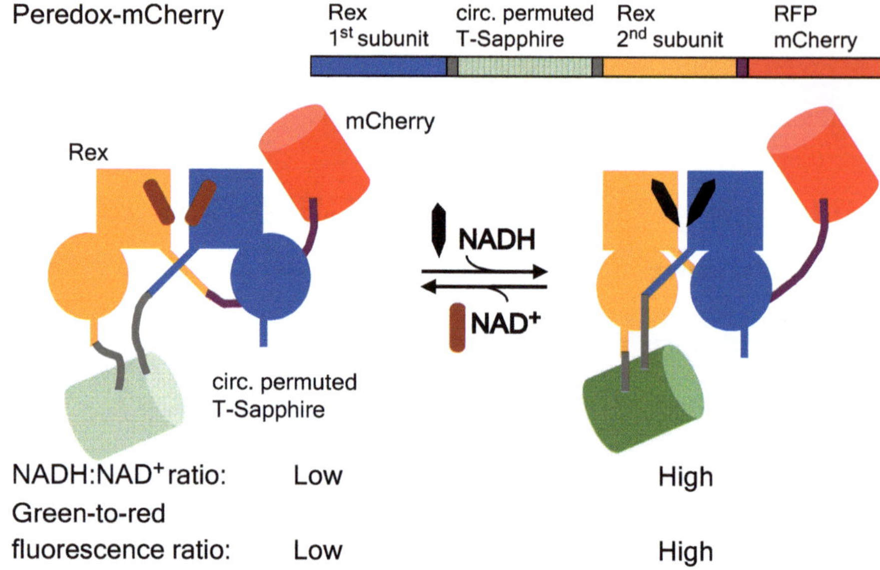

Fig. 1 Schematic showing the design of the fluorescent sensor of the cytosolic NADH–NAD$^+$ redox state, Peredox-mCherry: A circularly permuted GFP T-Sapphire (*green*) is interposed between the two Rex subunits (*blue* and *orange*), with a C-terminal RFP mCherry to normalize for the green fluorescence. Upon binding to NADH (*black*) but not NAD$^+$ (*brown*), Rex undergoes a conformational change that is coupled to an increase in the GFP fluorescence. Thus, the green-to-red fluorescence ratio increases with NADH:NAD$^+$ ratio. Adapted from Hung et al. [4]

NADH and NAD$^+$ for binding, Peredox reports the cytosolic NADH:NAD$^+$ ratios in mammalian cells (*see* Fig. 1). Peredox has been optimized to be pH-resistant to facilitate imaging experiments. We calibrate the sensor response by varying the amounts of exogenous lactate and pyruvate, and we have monitored the cytosolic NADH–NAD$^+$ redox dynamics in various cultured and primary cell types during energetic challenges. Together with other fluorescent biosensors [9–11], Peredox can reveal how glucose metabolism and the cytosolic NADH–NAD$^+$ redox state are regulated in the context of intact cells. Here, we elaborate on imaging methods, biosensor calibration using lactate and pyruvate, and other technical considerations of using Peredox to measure cytosolic NADH:NAD$^+$ ratios in individual live cells.

2 Materials

2.1 Cell Culture

1. Plasmid DNA encoding Peredox-mCherry or other versions of this reporter in a mammalian expression vector, such as pcDNA3.1 or GW1 (*see* **Notes 1–4**; available from www.addgene.org).

2. Mouse neuroblastoma cell line Neuro-2a or other adherent mammalian cell lines (American Type Culture Collection; *see* **Note 5**).
3. Culture medium, serum, and supplements appropriate for the chosen cell line (*see* **Note 5**).
4. Phosphate-buffered saline (PBS) and trypsin/ethylenediamine tetraacetic acid (EDTA).
5. Standard 10-cm tissue culture dishes and multi-well tissue culture plates.
6. Transfection reagents (for instance, QIAGEN Effectene) or viral transduction reagents (*see* **Notes 6** and **10**).
7. Flame-sterilized glass coverslips coated with protamine (Sigma) at a concentration of 1 mg/ml in water (*see* **Note 7**).

2.2 Cell Microscopy

1. Extracellular solution: 121.5 mM NaCl, 25 mM $NaHCO_3$, 2.5 mM KCl, 2 mM $CaCl_2$, 1.25 mM NaH_2PO_4, and 1 mM $MgCl_2$ (*see* **Note 8**).
2. Sodium lactate (Sigma) at a stock concentration of 500 mM in water (*see* **Note 9**).
3. Sodium pyruvate (Sigma) at a stock concentration of 500 mM in water (*see* **Note 9**).
4. Widefield fluorescence microscope, equipped with appropriate optical filters, and environmental control (*see* **Notes 11–13**).

3 Methods

3.1 Cell Culture

1. Neuro-2a or other adherent cell lines are passaged before confluence and plated for transient transfection: Cells are rinsed with PBS, dissociated using trypsin/EDTA, and plated sparsely onto sterilized protamine-coated coverslips in a 24-well plate.
2. After several hours or the next day, the adherent cells are transfected with plasmid DNA encoding Peredox-mCherry (or other reporters of interest) using the transfection reagent according to the manufacturer's protocol. The following day, Neuro-2a cells are maintained in low-serum culture medium supplemented with 10–20 μM retinoic acid for differentiation.
3. Cells are imaged 1–5 days after transient transfection.

3.2 Imaging Peredox in Various Lactate: Pyruvate Ratios

Beyond qualitative assessment, Peredox can report quantitatively the cytosolic NADH:NAD^+ ratios in individual live cells (*see* **Note 14**). To do so, we first need to calibrate the biosensor to establish the relationship between its fluorescence response and the cytosolic NADH:NAD^+ ratio. The quantitative details of the biosensor response can vary depending on the fluorescence optics and the light source, as well as the intracellular environment, making it important to calibrate the sensor in the target cells. As detailed in this section, we accomplish this by varying the extracellular concentrations of lactate and pyruvate in order to poise the cytosolic NADH–NAD$^+$ redox state in individual cells [12]. Presumably, lactate and pyruvate enter cells via monocarboxylate transporters, equilibrate readily between extracellular and intracellular compartments, and lead to exchange between NADH and NAD$^+$ catalyzed by endogenous lactate dehydrogenases (LDH):

$$\text{Lactate} + \text{NAD}^+ \leftrightarrow \text{Pyruvate} + \text{NADH} + \text{H}^+.$$

By setting the extracellular lactate:pyruvate ratios, we poise the cytosolic NADH:NAD$^+$ ratios and measure the corresponding biosensor readouts (*see* **Notes 15** and **16**).

1. Prior to imaging, prepare the following lactate and pyruvate solutions by diluting the lactate and/or pyruvate stocks in the extracellular solution (*see* Table 1 and **item 1** of Subheading 2.2).
2. All solutions are maintained at a near physiological temperature of ~35 °C and bubbled with 95 % air and 5 % CO_2. Continuous perfusion of solutions to the imaging chamber is performed using a peristaltic pump at a flow rate of ~3 ml/min. We begin with Solution A (the extracellular solution containing 10 mM lactate).
3. Place the coverslip containing transfected Neuro-2a cells onto the imaging chamber. Immobilize the coverslip using wax or other adhesives.
4. To monitor Peredox-mCherry, acquire green (T-Sapphire) and red fluorescence images at constant time intervals, for instance, 20 s. The exposure time depends on the optics (magnification and

Table 1
Extracellular solution with various lactate:pyruvate ratios

[Lactate] (mM)	[Pyruvate] (mM)	Lac:Pyr ratio	Solution
10	0.00	∞	A
10	0.02	500	B
10	0.07	160	C
10	0.20	50	D
10	0.50	20	E
10	1.67	6	F
0	10	0	G

numerical aperture of the objective, density filters), pixel binning, and biosensor expression. To reduce sensor photobleaching and cell phototoxicity, choose the minimum exposure time that yields a good signal-to-noise ratio for fluorescence detection. For both green and red images, we typically used a 20× objective and an exposure time of 20 ms.

5. Measure a baseline sensor response in Solution A for ~15 min. Switch sequentially to Solutions B, C, D, E, F, and G (the order can be varied), and measure the sensor response in each condition for ~15 min. In the case of real-time monitoring by image acquisition software, one can select a region of interest per cell and monitor its green-to-red fluorescence ratio throughout the experiment; one can switch the solutions once the steady state is achieved for the previous solution condition.
6. To verify the lactate:pyruvate ratio sensing of Peredox, repeat the experiment with an extracellular solution containing different total concentrations of lactate and pyruvate. For instance, double the lactate and pyruvate concentrations in Table 1 of Subheading 3.2 while keeping the indicated lactate:pyruvate ratios constant.

3.3 Data Analysis and Sensor Calibration

1. Using image analysis software such as ImageJ (http://rsb.info.nih.gov/ij/) or TILLvisION 4.0.1 (TILL photonics), subtract background from fluorescence images.
2. For each time point, divide the green image by the red image to generate a pixel-by-pixel green-to-red fluorescence ratio image. Set a signal threshold for the red image to avoid rationing artifacts.
3. For each cell, select a region of interest, and plot the time course of its green-to-red fluorescence ratio using Excel (Microsoft), MATLAB (MathWorks), Origin 6.0 (MicroCal), or other graphic programs.
4. To normalize the data in each cell, divide its fluorescence ratio time course by the minimum ratio, which is achieved by perfusion with pyruvate alone. The normalized response of Peredox-mCherry should range from 1 to ~2.5 (*see* Fig. 2; **Notes 10** and **17**).
5. From the normalized fluorescence ratios of multiple cells, plot the steady-state response against the corresponding lactate:pyruvate ratio in the extracellular solution. The sensor response versus the lactate:pyruvate ratio is fitted using a logistic function (*see* **Note 18**).
6. By assuming a constant physiological pH of 7.4 and that the LDH reaction is at equilibrium, we convert the lactate:pyruvate ratios into predicted NADH:NAD+ ratios. Therefore, by calibrating the biosensor using lactate and pyruvate, we can determine the cytosolic NADH:NAD+ ratio from its response (*see* Fig. 3; **Notes 16** and **19**).

Peredox-mCherry data analysis

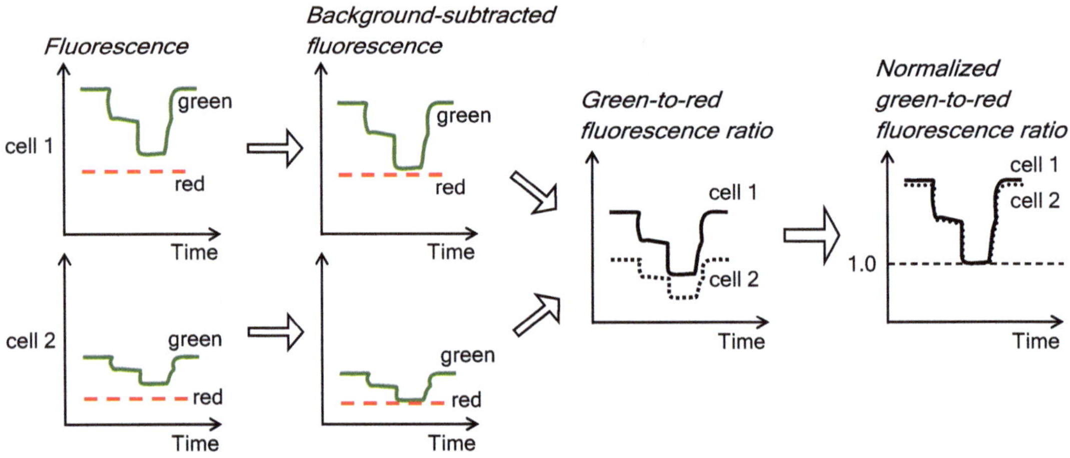

Fig. 2 Flow chart depicting data analysis of Peredox-mCherry: After plotting the time course of both green and red fluorescence for each cell, we first subtract the background. We then generate the green-to-red fluorescence ratio at each time point for each cell. Finally, we normalize the data by the minimal green-to-red ratio signal in pyruvate alone (or by the maximal ratio signal in lactate); the normalized green-to-red ratio data for each cell facilitates comparison of biosensor responses among cells. Data analysis is described in Subheading 3.3

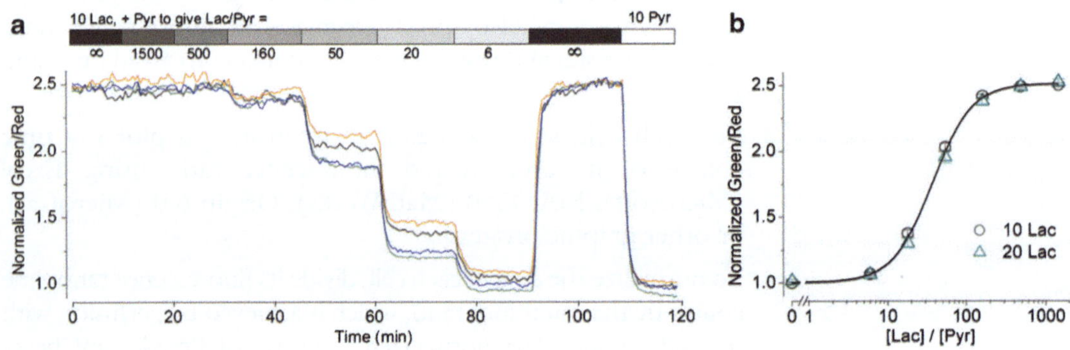

Fig. 3 Example of Peredox-mCherry calibration data by varying lactate:pyruvate ratios in the extracellular solution. (**a**) Time course of normalized green-to-red fluorescence ratios of four individual cultured mouse neuroblastoma Neuro-2a cells transfected with Peredox-mCherry; cells were placed in extracellular solutions with indicated lactate:pyruvate ratios. (**b**) Steady-state green-to-red fluorescence ratios of Neuro-2a cells plotted against extracellular lactate:pyruvate ratios, with lactate of 10 or 20 mM (mean ± SEM, *n* = 12–15 cells). Adapted from Hung et al. [4]

3.4 Measuring Cytosolic NADH:NAD+ Ratios in Individual Live Cells

1. Acquire the fluorescence responses of cells expressing Peredox-mCherry in the condition of interest (*see* **steps 2** and **3** of Subheading 3.2, **Notes 19–21**).
2. Perfuse cells with the extracellular solution containing lactate (10 mM), and measure the maximum green-to-red ratio signal. Then, perfuse cells with the extracellular solution containing pyruvate (10 mM), and measure the minimum ratio signal (*see* **Notes 15** and **16**).
3. On the acquired imaging data, perform background subtraction, ratio determination, and data normalization (*see* **steps 1–4** of Subheading 3.3).
4. Using the calibration curve determined in **step 6** of Subheading 3.3, we convert the normalized fluorescence ratio response into measurements of cytosolic NADH:NAD+ ratios.

4 Notes

1. Both pcDNA3.1 (Invitrogen) and GW1 (from the former British Biotech) are mammalian expression vectors that are driven by the cytomegalovirus (CMV) promoter. For most cell lines, the pcDNA3.1 expression vector has been widely used, whereas the GW1 expression vector has been used in neuronal cell lines.
2. Based on the GFP T-Sapphire, Peredox is not spectrally ratiometric, as its fluorescence excitation or emission spectrum contains only a single major peak. To normalize its readouts for sensor concentration, Peredox is attached in tandem to the RFP mCherry (or the YFP mCitrine); the green-to-red (or green-to-yellow) fluorescence ratio then indicates the sensor readout. As the two fluorescent proteins in Peredox-mCherry (or Peredox-mCitrine) differ in photostability, the fluorescence ratio may drift over time; however, depending on the illumination condition, such drifts appear minor in most imaging setups.
3. Many RFPs, including mCherry, can aggregate in lysosomes [15]. Indeed, we observe bright red puncta in certain cell types expressing Peredox-mCherry. These fluorescent puncta render the green-to-red fluorescence ratios heterogeneous within a cell and difficult to compare among distinct cells. Nonetheless, we minimize puncta by using stable low biosensor expression rather than transient transfection with high expression. We can also preclude these puncta by using Peredox-mCitrine or nuclear-targeted Peredox-mCherry.
4. Peredox-mCitrine can be used together with a red biosensor (such as pHRed [13]) or a red small-molecule indicator (such as the tetramethylrhodamine methyl ester, TMRM) for simultaneous imaging of two metabolic readouts.

5. Consult the American Type Culture Collection (http://www.atcc.org) for protocols on cell culture. We purchase all cell culture media, serum, and supplements from Invitrogen.
6. Consult [14] for protocols on generation of stable cell lines using retroviral transduction.
7. For coating the growth surface for cell adhesion, other cell types may prefer poly-D-lysine or collagen.
8. When bubbled with 95 % air and 5 % CO_2, this extracellular solution should yield an osmolarity of ~280 Osm and a pH of 7.4. For other cell types, the salt concentration can be varied to optimize the solution osmolarity.
9. Once prepared, stock solutions can be stored at −20 °C for several months. Repeated freezing and thawing are not recommended [16].
10. The NADH biosensor can be expressed in mammalian cells transiently via transfection or stably via viral transduction of the encoding plasmid DNA. To decide between transient and stable expression, the two main considerations are time and uniformity of sensor expression. On one hand, the major advantage of transient over stable expression is time: Given an optimized protocol, cells can be imaged as soon as 1–2 days after transient transfection, whereas it takes several weeks to generate and expand individual stable cell clones. On the other hand, the major advantage of stable expression is that both the sensor concentration and the fluorescence ratio response appear more uniform, rendering signal comparison among individual cells more reliable. In contrast, early on after transient transfection, the baseline green-to-red (or green-to-yellow) fluorescence ratios can vary widely across cells, as the two fluorescent proteins in Peredox-mCherry (or Peredox-mCitrine) differ in their maturation periods, and it could take days for their fluorescence in cells to arrive at steady states. Here, while the baseline green-to-red fluorescence ratios can be nonuniform, fluorescence ratios from each cell can be normalized after sensor calibration. We acquire the sensor signal of each cell in well-defined conditions in terms of cytosolic NADH–NAD^+ redox state: We obtain the minimum green-to-red fluorescence ratio in each cell by perfusion with pyruvate alone, and we use that signal to normalize the data series of that cell accordingly. Once normalized, fluorescence responses can be compared among cells (*see* Fig. 2).
11. The basic requirements for live-cell fluorescence imaging include a fluorescence microscope with appropriate filters, image acquisition software, and environmental control. For long-term imaging, we recommend a widefield microscope over a confocal microscope. Despite a lack of optical sectioning,

widefield microscopy requires lower illumination intensity, thus reducing sensor photobleaching and cell phototoxicity. We use either an upright microscope (such as Olympus BX51) or an inverted microscope (such as Nikon Eclipse Ti), although the use of a dipping objective will slow solution exchange.

12. Sensor fluorescence can be acquired using appropriate filter sets controlled by automated filter wheels (*see* Table 2). When using a light source with an integrated monochromator (such as TILL Photonics Polychrome IV), a filter set can be used with selective excitation to acquire both T-Sapphire and mCherry fluorescence (*see* Table 3) or both T-Sapphire and mCitrine fluorescence (*see* Table 4).

13. Proper environmental control is a prerequisite for cell viability and long-term imaging. Cells can be continuously perfused with

Table 2
Chroma filters for imaging Peredox-mCherry or -mCitrine without a monochromator

FP	Filter set	Excitation filter	Dichroic	Emission filter
T-Sapphire	–	ET405/20×	T425LPXR	ET525/50 m
RFP mCherry	41043	HQ575/50×	Q610LP	HQ640/50 m
YFP mCitrine	41028	HQ500/20×	Q515LP	HQ535/30 m

Table 3
Filters for imaging Peredox-mCherry with a monochromator

FP	Excitation (nm)	Excitation filter	Dichroic	Emission filter
T-Sapphire	405	Chroma 69002×	Chroma 69002bs	Semrock FF01-524/628-25
RFP mCherry	575			

Table 4
Chroma filters for imaging Peredox-mCitrine with a monochromator

FP	Excitation (nm)	Excitation filter	Dichroic	Emission filter
T-Sapphire	405	–	515DCXR	D535/25 m
YFP mCitrine	485			

defined extracellular solutions (*see* **item 1** of Subheading 2.2), which have been bubbled with 95 % air and 5 % CO_2 at a near physiological temperature of ~35 °C. To avoid outgassing, solutions are first warmed to physiological temperature before bubbling. Temperature can be maintained using an in-line heater or chamber heater, and perfusion rate can be regulated by a peristaltic pump or gravity flow. As an alternative to continuous perfusion, cells can be placed within an environmental chamber with temperature and CO_2 regulation but no solution exchange, for instance, in a multi-well microplate format.

14. In qualitative assessment, we monitor the *relative* changes in the cytosolic NADH:NAD^+ ratio, whether it increases, decreases, or remains unchanged. In contrast, with quantitative measurements, we determine the *actual* cytosolic NADH:NAD^+ ratio. Unlike qualitative assessment, quantitative measurements require proper calibration of the biosensor and are thus experimentally more demanding.
15. We calibrate the biosensor response by setting cytosolic NADH:NAD^+ ratios in individual live cells using exogenous lactate and pyruvate without cell permeabilization. As this calibration method relies on the equilibration between intracellular and extracellular lactate and pyruvate concentrations, cells lacking monocarboxylate transporters or LDH may necessitate other calibration methods: For instance, one can permeabilize cells using α-toxin or saponin [17] and perfuse solutions containing NADH and NAD^+ at various NADH:NAD^+ ratios.
16. For effective live-cell calibration using lactate and pyruvate, we need to establish the following two conditions: First, we eliminate the effect of glycolysis on the cytosolic NADH–NAD^+ redox state; otherwise, lactate and pyruvate alone would not be sufficient to set the NADH–NAD^+ redox state, and the calibration curve would be shifted upward [4]. Thus, in the calibration procedure, we do not supply cells with glucose. Alternatively, in conditions where glucose cannot be easily removed, cells can be calibrated in the presence of a glycolytic inhibitor, such as iodoacetate; after application of 20 mM pyruvate with 0.5 mM iodoacetate, we can obtain the minimal ratio response of Peredox [4]. Second, we effectively control the extracellular concentrations of lactate and pyruvate by using continuous perfusion of fresh solutions; only then can we set intracellular lactate:pyruvate ratios and cytosolic NADH:NAD^+ ratios. Notably, for multi-well microplates and other formats that are incompatible with continuous perfusion, extracellular concentrations of lactate and pyruvate can vary dramatically due to cell metabolism. To ensure constant extracellular concentrations of lactate and pyruvate for calibration, one can plate a small number of cells in a large volume of

solution, for instance, 1,000 cells in 3 ml of solution in the well of a 24-well plate.

17. With optimal optics, Peredox-mCherry has a dynamic range of roughly 2.5-fold, whereas Peredox-mCitrine has a smaller dynamic range of about twofold, due to the greater spectral overlap between T-Sapphire and mCitrine. The dynamic range may appear smaller with nonoptimal optics.

18. We find that a lactate:pyruvate ratio of 37 corresponds to a half-maximal response in Peredox-mCherry. For data fitting by a logistic function, we use a Hill coefficient of 1.7. From the calibration data, we derive the following equation:

$$F = 1 + \frac{1.5}{1 + \left(\frac{37}{X}\right)^{1.7}},$$

where F is the normalized fluorescence response and X is the extracellular lactate:pyruvate ratio.

Assuming a constant physiological pH of 7.4 and that the LDH reaction is at equilibrium,

$$[\mathrm{H}^+] = 10^{-7.4}.$$

$$K = \frac{[\mathrm{Pyr}] \times [\mathrm{NADH}] \times [\mathrm{H}^+]}{[\mathrm{Lac}] \times [\mathrm{NAD}^+]} = 1.1 \times 10^{-11}.$$

We derive the following equation:

$$\frac{[\mathrm{NADH}]}{[\mathrm{NAD}^+]} = 2.8 \times 10^{-4} \times \frac{[\mathrm{Lac}]}{[\mathrm{Pyr}]}.$$

Combining the equations, we get the following equation:

$$F = 1 + \frac{1.5}{1 + \left(\frac{0.01}{R}\right)^{1.7}},$$

or

$$R = \frac{0.01}{\left[\left(\frac{1.5}{F-1}\right) - 1\right]^{0.588}}.$$

Where F is the normalized fluorescence response and R is the cytosolic NADH:NAD+ ratio.

19. Common cell culture medium such as Dulbecco's Modified Eagle's Medium (DMEM) contains millimolar pyruvate, which can be catalyzed by LDH to lower the cytosolic NADH:NAD$^+$ ratios in cells. To elevate the baseline response of Peredox, cells can be placed in a pyruvate-deficient culture medium such as Roswell Park Memorial Institute (RPMI)-1640.
20. While tuned to sensing the cytosolic NADH:NAD$^+$ ratio, Peredox cannot report physiological changes in the mitochondrial NADH:NAD$^+$ ratio, which has been estimated to be 100- to 1,000-fold higher than the cytosolic NADH:NAD$^+$ ratio [18]. Although a fluorescent sensor has been engineered to monitor the mitochondrial NADH pool [19], it does not report the mitochondrial NADH:NAD$^+$ ratio.
21. Beyond live-cell microscopy, Peredox may be monitored by fluorescence-activated cell sorting (FACS) or a fluorescence platereader. For adherent cells, we do not recommend using Peredox with FACS, as metabolic state may be altered by cell detachment required by this technique [20].

Acknowledgments

We thank Mathew Tantama for careful reading of this manuscript. This work was supported by the Albert J. Ryan fellowship, the Stuart H.Q. and Victoria Quan predoctoral fellowship in neurobiology (both to Y.P.H.), and the U.S. National Institutes of Health (R01 NS055031 to G.Y.).

References

1. Nicholls DG, Ferguson SJ (2002) Bioenergetics, 3rd edn. Academic, London
2. Avi-Dor Y, Olson JM, Doherty MD, Kaplan NO (1962) Fluorescence of pyridine nucleotides in mitochondria. J Biol Chem 237: 2377–2383
3. Rocheleau JV, Head WS, Piston DW (2004) Quantitative NAD(P)H/flavoprotein autofluorescence imaging reveals metabolic mechanisms of pancreatic islet pyruvate response. J Biol Chem 279:31780–31787
4. Hung YP, Albeck JG, Tantama M, Yellen G (2011) Imaging cytosolic NADH–NAD$^+$ redox state with a genetically encoded fluorescent biosensor. Cell Metab 14:545–554
5. Zapata-Hommer O, Griesbeck O (2003) Efficiently folding and circularly permuted variants of the Sapphire mutant of GFP. BMC Biotechnol 3:5
6. Sickmier EA, Brekasis D, Paranawithana S, Bonanno JB, Paget MSB, Burley SK, Kielkopf CL (2005) X-ray structure of a Rex-family repressor/NADH complex insights into the mechanism of redox sensing. Structure 13: 43–54
7. Wang E, Bauer MC, Rogstam A, Linse S, Logan DT, von Wachenfeldt C (2008) Structure and functional properties of the Bacillus subtilis transcriptional repressor Rex. Mol Microbiol 69:466–478
8. McLaughlin KJ, Strain-Damerell CM, Xie K, Brekasis D, Soares AS, Paget MSB, Kielkopf CL (2010) Structural basis for NADH/NAD$^+$ redox sensing by a Rex family repressor. Mol Cell 38:563–575
9. Berg J, Hung YP, Yellen G (2009) A genetically encoded fluorescent reporter of ATP:ADP ratio. Nat Methods 6:161–166
10. Frommer WB, Davidson MW, Campbell RE (2009) Genetically encoded biosensors based on engineered fluorescent proteins. Chem Soc Rev 38:2833–2841
11. Tantama M, Hung YP, Yellen G (2012) Optogenetic reporters: Fluorescent protein-based

genetically encoded indicators of signaling and metabolism in the brain. Prog Brain Res 196: 235–263

12. Bücher T, Brauser B, Conze A, Klein F, Langguth O, Sies H (1972) State of oxidation-reduction and state of binding in the cytosolic NADH-system as disclosed by equilibration with extracellular lactate-pyruvate in hemoglobin-free perfused rat liver. Eur J Biochem 27:301–317
13. Tantama M, Hung YP, Yellen G (2011) Imaging intracellular pH in live cells with a genetically encoded red fluorescent protein sensor. J Am Chem Soc 133:10034–10037
14. Debnath J, Muthuswamy SK, Brugge JS (2003) Morphogenesis and oncogenesis of MCF-10A mammary epithelial acini grown in three-dimensional basement membrane cultures. Methods 30:256–268
15. Katayama H, Yamamoto A, Mizushima N, Yoshimori T, Miyawaki A (2008) GFP-like proteins stably accumulate in lysosomes. Cell Struct Funct 33:1–12
16. Passonneau JV, Lowry OH (1993) Enzymatic analysis: a practical guide. Humana, Totowa
17. Schulz I (1990) Permeabilizing cells: some methods and applications for the study of intracellular processes. Methods Enzymol 192:280–300
18. Williamson DH, Lund P, Krebs HA (1967) The redox state of free nicotinamide-adenine dinucleotide in the cytoplasm and mitochondria of rat liver. Biochem J 103:514–527
19. Zhao Y, Jin J, Hu Q, Zhou H-M, Yi J, Yu Z, Xu L, Wang X, Yang Y, Loscalzo J (2011) Genetically encoded fluorescent sensors for intracellular NADH detection. Cell Metab 14:555–566
20. Schafer ZT, Grassian AR, Song L, Jiang Z, Gerhart-Hines Z, Irie HY, Gao S, Puigserver P, Brugge JS (2009) Antioxidant and oncogene rescue of metabolic defects caused by loss of matrix attachment. Nature 461:109–113

Chapter 8

Measuring Membrane Voltage with Microbial Rhodopsins

Adam E. Cohen and Daniel R. Hochbaum

Abstract

Membrane voltage (V_m) is a fundamental biological parameter that is essential for neuronal communication, cardiac activity, transmembrane transport, regulation of signaling, and bacterial motility. Optical measurements of V_m promise new insights into how voltage propagates within and between cells, but effective optical contrast agents have been lacking. Microbial rhodopsin-based fluorescent voltage indicators are exquisitely sensitive and fast, but very dim, necessitating careful attention to experimental procedures. This chapter describes how to make optical voltage measurements with microbial rhodopsins.

Key words Voltage indicator, Microbial rhodopsin, Fluorescent protein, Neurons, Action potential, Optogenetics

1 Introduction

Membrane voltage mediates long-range coupling of all transmembrane proteins: a change in charge distribution or ionic conductivity in one part of a cell produces a voltage change which propagates, typically within milliseconds, throughout the cell. The voltage biases the energy landscapes of all membrane process that involve ions or charged residues. Thus, as with temperature or pH, V_m is a global regulator of cellular activity.

Temperature and pH can both be measured with electrode-based probes, but in cells or tissues it is often convenient to map spatial distributions with molecular indicators. The same applies to V_m: direct electrode- or pipette-based recordings are possible, but these techniques are slow and laborious, lack spatial information, and frequently damage the cell. Targets for electrodes must be sufficiently large, have a readily accessible membrane, and be stationary. Many interesting samples do not meet these criteria, such as bacteria, mitochondria, plant and fungus cells, and motile microorganisms. Electrodes are also not able to perform long-term measurements of V_m in large populations of cells, such as one might like to measure in a developing embryo.

Jin Zhang et al. (eds.), *Fluorescent Protein-Based Biosensors: Methods and Protocols*, Methods in Molecular Biology, vol. 1071, DOI 10.1007/978-1-62703-622-1_8, © Springer Science+Business Media, LLC 2014

Starting in the late 1960s, scientists sought to develop voltage-sensitive organic dyes [1, 2]. These molecules have been used effectively in neuronal [3] and cardiac recordings [4]. However, phototoxicity from dyes typically limits recording times to <1 min, and small signals, gradual internalization of the dye, and the inability to target genetically defined subpopulations of cells limit widespread use of this technology.

Genetically encoded voltage indicators are traditionally based on a fusion of one or more GFP homologues to a voltage sensing domain of a transmembrane protein [5–7]. A conformational change in the voltage sensing domain changes the fluorescence properties of the reporters. These probes have been used in neuronal [6] and cardiac [8] preparations, in vitro and in vivo. Traditionally protein-based indicators have had low sensitivity and slow response. Recently proteins with good sensitivity were developed, though slow response remains a challenge. Thus there is a need for genetically encoded voltage indicators with improved speed, sensitivity, and stability.

We recently introduced a new class of fluorescent voltage indicating proteins based on microbial rhodopsins [9, 10]. In the wild, microbial rhodopsin proton pumps convert solar energy into transmembrane voltage. Upon expression in neurons, these proteins convert light into a hyperpolarizing potential, suppressing neuronal activity [11]. We found that one could run these proteins "in reverse," converting changes in membrane potential into detectable changes in fluorescence.

Voltage imaging with microbial rhodopsins has been highly effective in cultured cells, but has not yet been demonstrated in vivo, except in larval zebrafish heart [12]. The low brightness of these indicators is a challenge for imaging in mouse brain slices or in vivo. Imaging in tissue culture has only been performed with 1-photon epifluorescence. For reasons described below, imaging via confocal or 2-photon microscopies has been challenging.

Here we describe the state of the art in voltage imaging with microbial rhodopsins. This field is evolving rapidly, with new indicators being developed monthly. This paper provides a snapshot of the best protocols as of June 2012.

2 Materials

2.1 Choosing a Voltage Indicator

Several voltage indicators are available, and the choice of indicator depends on the details of the experiment. One may wish to vary the promoter, fluorescent protein fusion, cell-type, or gene delivery method. Plasmids and viruses are available from several sources (Table 1), and new constructs may be created through straightforward cloning. Below we describe some considerations in selecting or designing a voltage indicating construct and expressing it in a host.

Table 1
Microbial rhodopsin-based voltage indicators

Construct name	Available From	Description
GPR(D97N)	Addgene 33780	Functions only in bacteria. No photocurrent. Available with Ara or IPTG-inducible promoters
GPR(D97N)-pHluorin	Cohen Lab, Harvard	Simultaneously reports V_m and cytoplasmic pH in bacteria
Arch	Addgene 22217	Fast ($\tau \sim 0.5$ ms). Generates a photocurrent. Brightness increases nonlinearly with illumination intensity
Arch(D95N)	Addgene 34616	Slow ($\tau \sim 40$ ms). No photocurrent
eArch 3.0	Addgene 35514 or 35516	Improved trafficking in neurons
eArch 3.0 (D95N)	Cohen Lab, Harvard	
CaViar: Arch(D95N)-GCaMP5G	Cohen Lab, Harvard	Simultaneously reports cytoplasmic [Ca^{2+}] and V_m
pHlaVor: Arch(D95N)-pHluorin	Cohen Lab, Harvard	Simultaneously reports cytoplasmic pH and V_m
Transgenic animals		
Floxed Arch mouse	Jackson Labs 012735	Ai35D, ref. [15]
Zebrafish expressing pan-neuronal Arch	Florian Engert, Harvard	*Hu*C promoter
Zebrafish expressing cardiac Arch(D95N)	Florian Engert, Harvard	cmlc2 promoter

1. Choice of Microbial rhodopsin.
   The first voltage indicator we developed was the Proteorhodopsin Optical Proton Sensor (PROPS) [9], based on the D97N mutant of green-absorbing proteorhodopsin (GPR; Uniprot Q9F7P4). The D97N mutation blocked the photocurrent, so this gene served as a non-perturbative indicator. PROPS indicated V_m in *E. coli* and, anecdotally in *B. subtilis* and *P. aeruginosa*. Despite significant effort, we failed to create a version of PROPS that trafficked to the plasma membrane in mammalian cells.

   Archaerhodopsin 3 (Arch; Uniprot P96787) [11], from the Dead Sea microorganism *Halorubrum sodomense*, trafficked well in mammalian cells and resolved single action potentials in rat, mouse, and human iPSC-derived neurons [10].

The wild-type protein responded to a voltage step extremely quickly (< 0.5 ms response time), but generated a slight hyperpolarizing photocurrent upon illumination. Red-shifted illumination (640 nm) minimizes photocurrent while preserving voltage-sensitive fluorescence. Currently Arch WT is the best choice for imaging action potentials in mammalian neurons.

The mutant Arch(D95N) did not perturb membrane potential and was 50 % more sensitive than Arch WT, but had a slower response time (41 ms at room temperature). At 35 °C Arch(D95N) showed a biphasic step response, with roughly half of the response occurring in <1 ms, and the other half taking ~30 ms. At room temperature Arch(D95N) was too slow to report neuronal action potentials with high accuracy, but was effective in reporting cardiac action potential waveforms. Arch(D95N) is the best choice for voltage imaging in cardiomyocytes and in any sample where the voltage dynamics are slower than ~10 ms.

We have anecdotally observed voltage-sensitive fluorescence in other microbial rhodopsin-based proton pumps. Blue-absorbing proteorhodopsin (BPR; Uniprot Q9AFF7) reported V_m in *E. coli.* Cruxrhodopsin (CR-1; Uniprot Q57101) and ArchT (GenBank HM367071) also reported voltage in mammalian cells. Approximately 5,000 microbial rhodopsin genes have been identified, and we speculate that voltage-sensitive fluorescence is widespread among proton pumping members of this class. Only GPR and Arch have been characterized in this regard in detail.

2. Choice of promoter.
   Microbial rhodopsin-based voltage indicators can be used with any highly expressing promoter. One can select either a cell type-specific promoter or a ubiquitous one. Common promoters in neurons are α – CaMKII and human synapsin 1. Common ubiquitous promoters are CMV, CAG, EF-1α, and ubiquitin. In bacteria, we have successfully used the IPTG- and arabinose-inducible promoters.

3. Choice of fluorescent protein fusion.
   Fusion to a fluorescent protein can add significant new capabilities to a voltage indicator. C-terminal fusion to eGFP facilitates determination of expression level and membrane localization using a conventional epifluorescence microscope. In principle, the ratio of rhodopsin:eGFP fluorescence can yield absolute membrane voltage, independent of variations in expression level. However, differential rates of photobleaching or imperfect membrane trafficking can introduce errors in this determination.

   One can also fuse a microbial rhodopsin to a spectrally distinct indicator of a second modality. For instance, we created

Arch(D95N)-gCaMP5G, where gCaMP5G is a blue-excited Ca^{2+} indicator [13]. This construct reports Ca^{2+} and V_m, and thus we call it CaViar [12]. Similarly, we created Arch(D95N)-pHluorin, where pHluorin is a blue-excited pH indicator [14]. This construct reports pH and V_m, and thus we call it pHlaVor.

Fusion of a microbial rhodopsin to a GFP homologue dramatically improves trafficking in neurons, e.g. Arch-eGFP traffics far better than does Arch without eGFP. The mechanism for this effect is not understood, though we speculate that the GFP improves protein stability. When spectral interference from the GFP is undesirable, the GFP can be mutated to a nonfluorescent state by mutating the chromophore TYG → GGG. This procedure preserves the salubrious trafficking properties.

2.2 Cell Culture

1. Cells types: rat, mouse, or human iPS-derived neurons, human embryonic kidney (HEK) cells.
2. Glass-bottomed culture dishes (e.g. Mattek P35G-1.5-14-C).
3. Fetal bovine serum (FBS).
4. Cytarabine.
5. Puromycin.

2.3 Transfection/ Gene Delivery

1. Calcium phosphate.
2. Viral delivery system (e.g. lentivirus or adeno-associated virus).

2.4 Imaging

1. Inverted fluorescence microscope equipped with ~100 mW laser emitting between 594 and 640 nm (*see* **Note 1**). We use a Coherent Obis 637LX laser, 140 mW or a Cobolt Mambo 594 nm laser, 100 mW (*see* **Note 2**).
2. High numerical aperture (NA) objective. We use 60×, either NA 1.45 oil or NA 1.2 water.
3. Filter set with an emission band between 660 and 760 nm. Any filter set designed for imaging Cy5, such as Chroma set 49006, is appropriate. Filter sets with a long-pass emission filter rather than bandpass, such as Semrock LF635/LP-B-000 collect a larger fraction of the fluorescence.
4. High-speed and high-sensitivity EMCCD camera. The camera should have an exposure time commensurate with the duration of the events of interest. To image neuronal action potentials, the frame rate should be at least 1,000 frames/s. Small-format frame-transfer EMCCDs such as the Andor iXon DU-860 can achieve this speed when used with a small region of interest (ROI). CMOS cameras, while fast, appear to lack adequate sensitivity.

5. Neuronal stimulation system. For the purpose of signal averaging, one typically wishes to initiate action potentials at defined times during the recording. This can be done by field stimulation, patch clamp, or optogenetic means.
6. Imaging medium (in mM): 125 NaCl, 2 KCl, 3 $CaCl_2$, 1 $MgCl_2$, 10 HEPES, 30 glucose, pH 7.3 (NaOH adjusted), 305–310 mOsm (sucrose adjusted).

3 Methods

3.1 Cell Culture

Rat, mouse, or human iPS-derived neurons can be cultured on coverslips or in glass-bottomed dishes (e.g. Mattek P35G-1.5-14-C) using standard procedures [16]. The key requirement for voltage imaging is that the neurons be healthy, with a resting membrane potential between −60 and −70 mV.

Culture cells at a density of 10,000–30,000 cm^{-2}. A glial monolayer improves cell health. Initially culture cells in the presence of 10 % fetal bovine serum (FBS) to promote glial growth. After the glia have reached 50–70 % confluency (typically 2–3 days), add cytarabine to inhibit further cell growth. From a stock of culture medium containing 4 μM cytarabine and 5 % serum, add an equal volume to the medium already in the culture dish.

Cells can be patterned to form autapses via microisland culture [17–19], or in a variety of other shapes and connectivities via microfluidic chambers [20, 21] or patterning of adhesion molecules.

3.2 Gene Delivery

Any of the commonly used gene delivery methods are appropriate for voltage imaging. In neurons, we have successfully used calcium phosphate and viral delivery.

One can also make stably expressing cell lines or organisms. We have made lines of HEK cells stably expressing Arch WT and Arch(D95N) via puromycin selection, as well as transgenic zebrafish expressing Arch WT under a pan-neuronal promoter *Hu*C or Arch(D95N) under a cardiac promoter, cmlc2. A mouse line expressing Arch WT under control of a Cre recombinase is also available [15].

Below we describe the protocol for calcium phosphate transfection, which yields sparse expression while preserving cell health, followed by protocols for both high-titer and low-titer viral infections.

1. Calcium phosphate transfection.
   Our method for calcium phosphate transfection closely follows ref. [22], with the modification that we use exchanges of medium rather than changes in CO_2 concentration to induce the pH changes needed to dissolve calcium phosphate precipitates

after transfection. The transfection medium is at pH 7.1. After transfection, replace the transfection medium with identical medium at pH 6.0–6.5. Incubate for 5 min. at 37 °C (*see* **Note 3**).

2. High-titer viral infection.

   Cultures can be infected with adeno-associated virus (AAV; high-titer) or lentivirus (high-titer or low-titer). High-titer lentivirus or AAV may be acquired from the University of North Carolina Gene Therapy Center vector core (http://genetherapy.unc.edu/services.htm). Protocols for in-house production of lentivirus can be found on Prof. Ed Boyden's lab Web site [23]. We have successfully expressed Arch in rat hippocampal cultures using lentivirus (α-CaMKII promoter) and in mouse hippocampal or cortical culture using calcium phosphate, lentivirus, or AAV pseudotypes AAV2/1, AAV2/8, or AAV2/9 (α-CaMKII or synapsin promoters). In human iPS-derived cardiac cells we did not have success with lentivirus, but we did have success with lipofection (TransIT-LT1, Mirus Bio), AAV2/1 or AAV2/8.

   Cultures may be infected with high-titer AAV or lentivirus 2–7 days after plating. For maximum lentiviral infection efficiency, the virus should be added at least 24 h. before addition of cytarabine. Add 1 μL of high-titer virus, in a volume of 2 mL culture medium. Assay expression level 7 days after infection and adjust quantity of virus accordingly in subsequent experiments (*see* **Note 4**).

3. Low-titer lentiviral infection.

   Cultured cells may be infected efficiently using low-titer lentivirus, which is more economical and easier to produce than high-titer virus. One begins virus production as for high-titer virus, but directly uses the supernatant from lentivirus producing HEK293T cells. Add lentiviral supernatant 2–3 days post plating in a 1:1 ratio with culture medium. After a 24 h incubation at 37 °C, aspirate the medium and replace with culture medium (with cytarabine).

3.3 Imaging

Imaging experiments are typically carried out 12–16 days post plating.

1. Replace culture medium with imaging medium.
2. Add drugs or ion channel blockers as desired.
3. Locate a suitable neuron. It is best to find neurons expressing Arch-eGFP by imaging GFP fluorescence at low magnification. Healthy neurons will have GFP fluorescence uniformly distributed along the cell membrane, and will lack large blebs or bright inclusions. Often neurons that overexpress the protein appear sick or dead.

4. Switch to high magnification and red or orange illumination. Record movies of electrical activity at high frame rate under high-intensity illumination. During intervals when data is not being acquired the laser should be shuttered to minimize photobleaching and phototoxicity. One may wish to combine illumination and imaging at two wavelengths to monitor a second modality, such as Ca^{2+} or pH (*see* **Note 5**).
5. For important considerations when imaging the prototypical voltage indicator, Arch, and other microbial rhodopsin-based indicators, please *see* **Note 6**.

3.4 Data Analysis

Raw movies of action potentials typically appear quite noisy. To determine whether action potentials are being detected, one should average the signal from a subset of pixels corresponding to the neuron of interest. This selection and averaging is easily done in the free software imageJ (Fig. 1a). Alternatively, we developed an algorithm to combine pixel intensity values with weights adjusted to maximize the signal-to-noise ratio in the output. A Matlab implementation of this algorithm is given in the supplementary material to ref. [10].

To make a movie showing propagation of an action potential, one must typically average movies of many (~100) action potentials. This averaging is easily done in Matlab by identifying the time corresponding to the peak of each action potential, and then averaging corresponding movie frames referenced to each action potential (Fig. 1b).

4 Notes

1. Microbial rhodopsins are spectrally dissimilar to GFP, and thus successful voltage imaging requires careful attention to the microscope design. There is a description of the microscope in the supplement of ref. [10].
2. LED or lamp illumination is not sufficiently intense to image fluorescence of microbial rhodopsins.
3. Precipitates should dissolve during incubation. If precipitates remain visible after the 5 min incubation, aspirate and repeat the incubation with low-pH medium.
4. Notes on membrane trafficking.
   Efficient trafficking to the plasma membrane is essential for a voltage indicator to work well. Internalized protein contributes background fluorescence and noise, but no signal. The determinants of good trafficking in eukaryotic cells are not well understood. Addition of export and trafficking motifs has been reported to improve trafficking in neurons [24]. In our

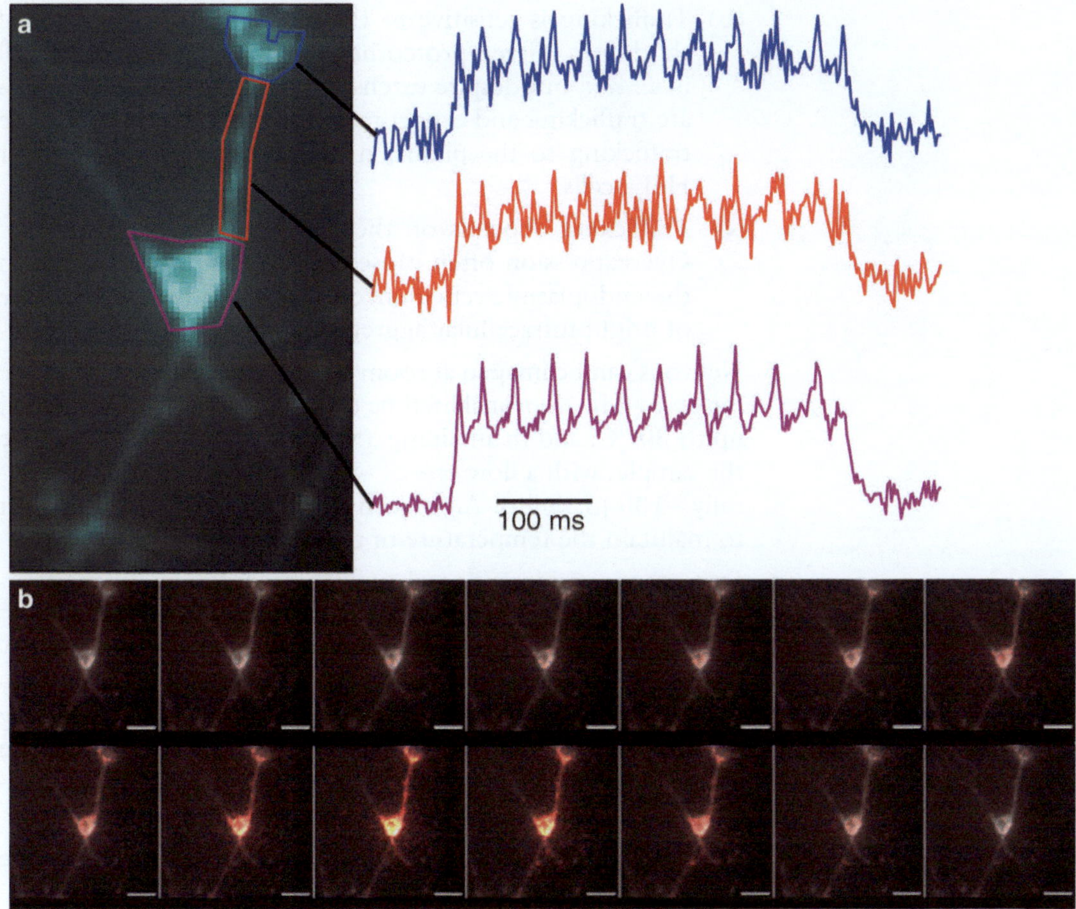

Fig. 1 Imaging action potentials with Arch. A rat hippocampal neuron was infected with lentivirus encoding *FCK-Arch-eGFP* and imaged through a 60× NA 1.45 oil immersion objective with illumination at 1,800 W/cm^2 at 640 nm. Trains of action potentials were induced via injection of 200 pA through a patch pipette into the cell body. (**a**) Single-trial fluorescence intensity traces from subcellular regions during a train of action potentials. (**b**) Filmstrip showing propagation of an action potential generated by averaging $n=98$ movies of individual action potentials. Frames separated by 1 ms, scale bar 25 μm

experience, trafficking is highly dependent on cell type, microbial rhodopsin gene, and timecourse of expression. With these caveats, we have observed the following trends.

(a) Trafficking differs dramatically depending on cell type and expression mechanism. We observe excellent trafficking of Arch 3.0 and of Arch in mouse neurons, and less good trafficking in HEK cells or iPSC-derived cardiomyocytes. We observed moderate trafficking of Arch in zebrafish neurons and heart, though many cells showed intracellular aggregates. We observed weak, but detectable, voltage sensitivity of Arch in *C. elegans*, probably due to low expression and/or poor membrane trafficking.

(b) Trafficking is sensitive to the structure of the microbial rhodopsin. Green proteorhodopsin (GPR) trafficked well in *E. coli*, but despite extensive efforts to attach appropriate trafficking and export motifs, never showed detectable trafficking to the plasma membrane in HEK, CHO, or HeLa cells.

(c) Trafficking depends on the method of gene expression. Overexpression often causes an apparent "traffic jam" in the endoplasmic reticulum and Golgi and the appearance of bright intracellular aggregates.

5. Neurons can be imaged at room temperature and will be stable for up to 2 h. Neuronal lifetime can be increased by imaging at up to 30 °C, and maintaining a very thin liquid volume above the sample, with a flow rate of ~1 chamber volume/min (typically ~100 μL/min). An objective heater is typically sufficient to maintain the temperature of the culture.

6. Important considerations when imaging Arch and other microbial rhodospins
   Microbial rhodopsins have complex photophysics, with many (typically >7) spectroscopically distinct states, connected by optical and thermal transitions. In Arch, the state exhibiting voltage-sensitive fluorescence is not the ground state; it is a photogenerated intermediate. This fact has important implications for imaging.

(a) The fluorescence excitation spectrum differs from the ground-state absorption spectrum. Arch absorbs maximally at 558 nm, but voltage-sensitive fluorescence peaks with excitation around 600 nm. To minimize photocurrent and voltage-insensitive background, image with illumination at 640 or 594 nm. The microbial rhodopsins studied to date emit fluorescence in a broad band in the near infrared, typically extending from ~660 to 760 nm, with a peak around 710 nm.

(b) The brightness of Arch increases superlinearly with illumination intensity. The ground state of Arch is virtually nonfluorescent. Illumination increases the population in the fluorescent intermediate state, which then must be optically excited to emit light. In fluorescence imaging of GFP one can compensate for lower illumination intensity with longer exposure time. This is not the case for Arch: intense illumination is essential. To image neuronal action potentials, one should illuminate at ~1,000 W/cm^2 at 640 nm. This intensity can be achieved only with laser illumination. LED or lamp illumination is too dim.

(c) After initial photoexcitation, transition from the ground state to the fluorescent state takes ~3 ms, and the fluorescent state has a lifetime of ~30 ms. In point-scanning microscopies, such as confocal or 2-photon, the light source does not dwell at each point long enough for the molecules to reach the fluorescent state. Thus to apply these microscopies one must carefully design the illumination scan trajectory to sample molecules while they are in the fluorescent state. To achieve optical sectioning in 3-dimensional tissue, Selective Plane Illumination Microscopy (SPIM) is a promising alternative [25, 26].

Acknowledgments

This work was supported by the Harvard Center for Brain Science, ONR grant N000141110-549, NIH grants 1-R01-EB012498-01 and New Innovator grant 1-DP2-OD007428, the Harvard/MIT Joint Research Grants Program in Basic Neuroscience, a Sloan Foundation Fellowship, and a Dreyfus Teacher Scholar Award. D.R.H is supported by an NSF graduate research fellowship.

References

1. Tasaki I, Watanabe A, Sandlin R, Carnay L (1968) Changes in fluorescence, turbidity, and birefringence associated with nerve excitation. Proc Natl Acad Sci U S A 61:883–888
2. Cohen L et al (1974) Changes in axon fluorescence during activity: molecular probes of membrane potential. J Membr Biol 19:1–36
3. Peterka DS, Takahashi H, Yuste R (2011) Imaging voltage in neurons. Neuron 69:9–21
4. Herron TJ, Lee P, Jalife J (2012) Optical imaging of voltage and calcium in cardiac cells & tissues. Circ Res 110:609–623
5. Mutoh H, Perron A, Akemann W, Iwamoto Y, Knöpfel T (2011) Optogenetic monitoring of membrane potentials. Exp Physiol 96:13–18
6. Akemann W, Mutoh H, Perron A, Rossier J, Knopfel T (2010) Imaging brain electric signals with genetically targeted voltage-sensitive fluorescent proteins. Nat Methods 7:643–649
7. Jin L et al (2011) Random insertion of split-cans of the fluorescent protein venus into shaker channels yields voltage sensitive probes with improved membrane localization in mammalian cells. J Neurosci Methods 199:1–9
8. Tsutsui H, Higashijima S, Miyawaki A, Okamura Y (2010) Visualizing voltage dynamics in zebrafish heart. J Physiol 588:2017–2021
9. Kralj JM, Hochbaum DR, Douglass AD, Cohen AE (2011) Electrical spiking in escherichia coli probed with a fluorescent voltage indicating protein. Science 333:345–348
10. Kralj JM, Douglass AD, Hochbaum DR, Maclaurin D, Cohen AE (2012) Optical recording of action potentials in mammalian neurons using a microbial rhodopsin. Nat Methods 9:90–95
11. Chow BY et al (2010) High-performance genetically targetable optical neural silencing by light-driven proton pumps. Nature 463:98–102
12. Kralj JM, Hou, JH, Douglass AD, Wortzman J, Engert F, Cohen AE (2012) Simultaneous measurement of membrane voltage and calcium with a genetically encoded reporter. Submitted
13. Akerboom J et al (2012) Optimization of a GCaMP calcium indicator for neural activity imaging. J Neurosci 32(40):13819–13840
14. Miesenbock G, De Angelis DA, Rothman JE (1998) Visualizing secretion and synaptic transmission with pH-sensitive green fluorescent proteins. Nature 394:192–195
15. Madisen L et al (2012) A toolbox of Cre-dependent optogenetic transgenic mice for light-induced activation and silencing. Nat Neurosci 15:793–802
16. Goslin K (1998) Culturing nerve cells. The MIT Press, Cambridge, MA

17. Burgalossi A et al (2012) Analysis of neurotransmitter release mechanisms by photolysis of caged Ca^{2+} in an autaptic neuron culture system. Nat Protoc 7:1351–1365
18. Ricoult SG, Goldman JS, Stellwagen D, Juncker D, Kennedy TE (2012) Generation of microisland cultures using microcontact printing to pattern protein substrates. J Neurosci Methods 208(1):10–7
19. Sgro AE et al (2011) A high-throughput method for generating uniform microislands for autaptic neuronal cultures. J Neurosci Methods 198(2):230–235
20. Wheeler BC, Brewer GJ (2010) Designing neural networks in culture. Proc IEEE 98:398–406
21. Feinerman O, Rotem A, Moses E (2008) Reliable neuronal logic devices from patterned hippocampal cultures. Nat Phys 4:967–973
22. Jiang M, Chen G (2006) High Ca^{2+} -phosphate transfection efficiency in low-density neuronal cultures. Nat Protoc 1:695–700
23. Lentivirus production for high-titer, cell-specific, in vivo neural labeling, http://syntheticneurobiology.org/protocols/protocoldetail/31/12
24. Mattis J et al (2011) Principles for applying optogenetic tools derived from direct comparative analysis of microbial opsins. Nat Methods 9:159–172
25. Weber M, Huisken J (2011) Light sheet microscopy for real-time developmental biology. Curr Opin Genet Dev 21(5):566–572
26. Tomer R, Khairy K, Amat F, Keller PJ (2012) Quantitative high-speed imaging of entire developing embryos with simultaneous multiview light-sheet microscopy. Nat Methods 9:755–763

Chapter 9

Imaging the Activity of Ras Superfamily GTPase Proteins in Small Subcellular Compartments in Neurons

Ana F. Oliveira and Ryohei Yasuda

Abstract

Resolving the spatiotemporal dynamics of intracellular signaling is important for understanding the molecular mechanisms of various cellular processes induced by extracellular signals. Two-photon fluorescence lifetime imaging microscopy (2pFLIM) in combination with a fluorescence resonance energy transfer (FRET)-based signaling sensors allows one to image signaling within small subcellular compartments, such as dendritic spines of neurons, with high sensitivity and spatiotemporal resolution. In this protocol, we describe the procedures and equipment required for imaging intracellular signaling activity, with a particular focus on signaling mediated by the Ras superfamily of small GTPase proteins.

Key words FLIM, FRET, Lifetime, Ras GTPase, Ras, RhoA, Cdc42

1 Introduction

Small GTPase proteins in the Ras superfamily (Ras GTPase proteins) include ~150 members in five major subfamilies (Ras, Rho, Rab, Arf, and Ran) and are important for many cellular processes, such as cell division, motility, and differentiation [1]. Ras GTPases, which are generally monomeric, membrane-associated and ubiquitously expressed, bind to and hydrolyze guanosine triphosphate (GTP) into guanosine diphosphate (GDP) [2]. Ras GTPase proteins act like a signaling switch; they are active when bound to GTP and inactive when bound to GDP [2]. Guanosine nucleotide exchange factors (GEFs) activate Ras GTPase proteins by exchanging GDP with GTP, while GTPase-activating proteins (GAPs) inactivate them by promoting hydrolysis of GTP into GDP. Active Ras GTPases recruit and activate effector molecules. Signaling dynamics mediated by Ras GTPases have been extensively studied by biochemical assays, which have limited temporal resolution and essentially no subcellular spatial resolution.

The spatiotemporal dynamics of Ras GTPase signaling have been mostly unknown until the recent advent of imaging

Jin Zhang et al. (eds.), *Fluorescent Protein-Based Biosensors: Methods and Protocols*, Methods in Molecular Biology, vol. 1071, DOI 10.1007/978-1-62703-622-1_9, © Springer Science+Business Media, LLC 2014

techniques based on Förster resonance energy transfer (FRET) [3]. FRET is the process of non-radiative energy transfer from an excited donor fluorophore to an acceptor fluorophore. Because FRET strongly depends on the distance between the donor and acceptor, FRET can be used as a readout of protein–protein interactions for proteins fused to fluorophores or to detect conformational changes of a protein tagged with two fluorophores [4]. Typically, the relative change (i.e., before and after stimulation) in FRET is measured by calculating the ratio between the fluorescence emission of the donor fluorophore and that of the acceptor fluorophore (ratiometric imaging). However, this measurement is sensitive to the local fluorophore concentration ratio between donor and acceptor. To avoid this problem, a ratiometric FRET sensor usually contains both donor and acceptor in one polypeptide and, thus, the stoichiometry is fixed. Based on the ratiometric FRET technique, a couple of pioneering studies enabled the measurement of the spatiotemporal dynamics of Ras GTPase signaling activity in live cells using a series of sensors called "Ras and interacting protein chimeric unit (Raichu)" [5, 6]. Raichu consists of the fusion of enhanced cyan fluorescent protein (ECFP), Ras binding domain (RBD), Ras-GTPase proteins (without the CAAX membrane targeting domain) and enhanced yellow fluorescent protein (EYFP), with ECFP anchored in the plasma membrane using the CAAX tail [5]. In the inactive GDP-bound form, ECFP and EYFP are located away from each other, thereby resulting in low FRET. Following Ras GTPase stimulation, the RBD associates with the active Ras GTPase. This brings ECFP and EYFP into close proximity, thereby increasing FRET [5, 7]. However, Raichu is not suitable for imaging in small compartments due to limited sensitivity. Furthermore, since Raichu is extensively modified from the original Ras GTPases, localization of Raichu may not be the same as that of endogenous Ras GTPase proteins.

The recent development of fluorescence lifetime imaging microscopy (FLIM) in combination with a FRET signaling sensor extensively optimized for FLIM addressed some of the issues in ratiometric FRET imaging, permitting the quantification of signaling activity in small subcellular compartments [8–13]. In this technique, changes in the fluorescence lifetime, the time elapsed from the excitation of the fluorophore until the emission of a photon, are used as a readout of FRET. When FRET occurs, FRET, as well as photon emission, contributes to the decay of the excited state of the donor, thereby decreasing the fluorescence lifetime of the donor. Because only the donor fluorescence is typically used, fluorescence lifetime is independent of the relative concentration of donor to acceptor and wavelength-dependent light scattering (Fig. 1) [4]. When multiple populations with different FRET efficiency coexist, the fluorescence decay curve becomes multi-exponential (Fig. 1).

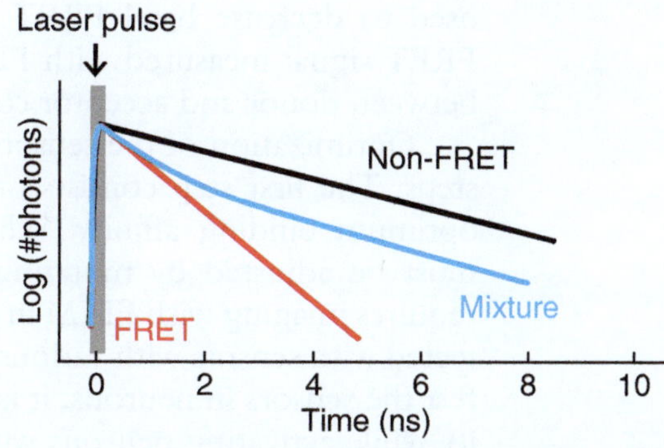

Fig. 1 Fluorescence lifetime decay curves after excitation of the donor fluorophore with a short laser pulse. FRET decreases the fluorescence lifetime (*red*). When free donors and donors bound to acceptors coexist, the fluorescence decay curve contains two exponentials (*blue*). Adapted from ref. 8 with permission

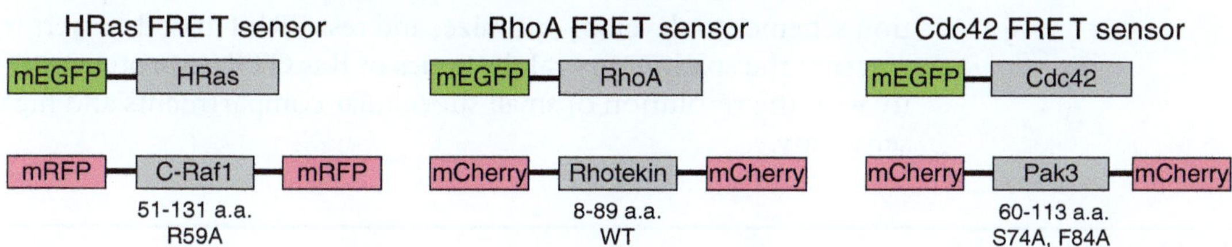

Fig. 2 Schematics of general design for small GTPase proteins. Adapted from ref. 13 with permission

Thus, it is possible to calculate the fraction of each component by curve fitting [4]. Combination of FLIM with 2-photon laser scanning microscopy (2-photon FLIM or 2pFLIM) further allows high sensitivity imaging in light scattering tissue, such as brain slices [8]. Optimized fluorescence lifetime systems provide submicron spatial resolution and temporal resolution on the order of seconds, with enough sensitivity to detect less than ~1 % change in FRET efficiency in single dendritic spines [11, 12].

Based on the FLIM-FRET technique, highly sensitive sensors for several signaling proteins including Ras GTPase proteins, such as HRas, RhoA, and Cdc42, have been developed [8, 13]. Ras GTPase sensors optimized for FLIM share a similar design: mEGFP-tagged Ras GTPase protein of interest and mCherry- or mRFP- tagged RBD of its effector, which binds selectively to the GTP-bound form of the Ras GTPase protein (Fig. 2). Two acceptors are fused to both ends of the RBD to improve the signal-to-noise ratio (SNR) (Fig. 2). When mEGFP-Ras GTPase is activated, mCherry (or mRFP)-RBD-mCherry binds to mEGFP-Ras GTPase, increasing FRET. This type of bimolecular design can be used to decrease basal FRET and, thus, increase SNR, because FRET signal measured with FLIM does not depend on the ratio between donor and acceptor concentrations [8].

Optimization of the sensors for sensitivity can be done in two steps. The first step consists of in vitro screening for RBDs with optimum binding affinity. When necessary, the binding affinity must be adjusted by mutating the RBDs [8]. The second step requires imaging with FLIM in cell lines (such as HeLa cells) transfected with sensors with various RBDs or their mutants. To further test the sensors in neurons, it is useful to image Ras GTPase activity while activating neurons with KCl or NMDA in cultured hippocampal neurons. The effect of overexpressing sensors on signaling dynamics can also be evaluated by measuring the relationship between the decay kinetics and overexpression level. Generally, Ras GTPase signaling sensors optimized and tested with this method show sufficient sensitivity for imaging in small subcellular compartments like dendritic spines [13].

This protocol describes some general principles and optimization schemes to develop, optimize, and test FRET-based sensors to measure the spatiotemporal dynamics of Ras GTPase protein activity with the resolution of small subcellular compartments and high sensitivity.

2 Materials

2.1 His-Tagged Protein Purification and In Vitro Binding Assay

1. Plasmids including cDNAs of Ras GTPases or RBD in bacterial expression vectors (such as pET).
2. *Escherichia coli* BL21(DE3)pLysS.
3. 37 °C water bath.
4. 42 °C water bath.
5. 37 °C shaking incubator.
6. Ice bucket with ice.
7. Lysogeny broth (LB).
8. Ampicillin.
9. Chloramphenicol, 35 mg/ml (in 100 % ethanol).
10. Isopropylthio-β-galactoside (IPTG) stock, 100 mM in H_2O.
11. Nickel sulfate ($NiSO_4$) hexahydrate solution, 2 M.
12. Protease inhibitors (e.g., complete mini-EDTA tablet, Roche).
13. DNAse I.
14. Triton X-100, 20 %.
15. NaCl, 5 M.
16. Tris–HCl, 1 M (pH 8.0).
17. $MgCl_2$, 1 M.

18. Standard buffer: 50 mM Tris–HCl (pH 8.0), 100 mM NaCl, 3 mM $MgCl_2$, 1 mM dithiothreitol (DTT).
19. Lysis buffer for *E. coli*: standard buffer including 1 mg/ml lysozyme, 5 U/ml DNAse I, and protease inhibitors.
20. Binding buffer: standard buffer including 400 mM NaCl.
21. Imidazole stock solution, 2 M in H_2O (pH 7.4).
22. Ni^+-NTA column for affinity chromatography (such as HisTrap, GE Life Sciences).
23. 10 ml syringes.
24. Desalting column (such as PD10, GE Life Sciences).
25. Washing: H_2O, 50 mM EDTA (pH 8.0).
26. 500 mM EDTA (pH 8.0).
27. Spectrofluorometer.
28. 2′,3′-*O*-*N*-methyl anthraniloyl–GppNHp (Gpp(NH)p), 10 mM.
29. Guanosine diphosphate (GDP), 10 mM.
30. PD-miniTrap column (GE Healthcare).
31. 35 mm tissue culture dishes.

2.2 Cell Line Culture and Transfection

1. 35 mm culture dishes.
2. Cell line culture medium: Dulbecco's Modified Eagle Medium (DMEM), 10 % fetal bovine serum (FBS), pH 7.4. Warm up the media to 37 °C before adding it to cells and store the remaining media at 4 °C.
3. HeLa cells (ATCC: CCL-2).
4. Incubator (37 °C, 5 % CO_2).
5. cDNA for donor and acceptor (2 μg total).
6. Lipofectamine 2000 (Invitrogen).
7. Opti-MEM (Invitrogen).
8. Imaging solution: HEPES-buffered artificial cerebral spine fluid (H-ACSF)—30 mM HEPES (pH 7.3), 130 mM NaCl, 2.5 mM KCl, 1 mM $CaCl_2$, 1 mM $MgCl_2$, 2 mM $NaHCO_3$, 1.25 mM NaH_2PO_4, and 25 mM glucose [11]. Store at 4 °C.

2.3 Organotypic Slice Culture Preparation

1. Dissection medium: 1 mM $CaCl_2$, 5 mM $MgCl_2$, 10 mM glucose, 4 mM KCl, 26 mM $NaHCO_3$, and 248 mM sucrose.
2. 60 mm culture dishes.
3. Tissue culture medium: 20 % (v/v) horse serum and 80 % (v/v) HEPES base MEM (8.4 g/L) supplemented with 1 mM L-glutamine, 1 mM $CaCl_2$, 2 mM $MgSO_4$, 12.9 mM D-GLUCOSE, 5.2 mM $NaHCO_3$, 30 mM HEPES, 0.075 % ascorbic acid, and 1 μg/ml insulin.

4. Dissection tools: scissors, scalpel, forceps, smoothly curved iris spatula.
5. Tissue chopper.
6. Animals: P6–7 Sprague–Dawley rats.
7. Plating material: 6-well plates, 0.4 μm/30 mm diameter cell culture insert membrane (such as PICMORG50, Fisher Scientific).
8. Imaging solution: Artificial cerebral-spine fluid (ACSF)—127 mM NaCl, 2.5 mM KCl, 25 mM $NaHCO_3$, 1.25 mM NaH_2PO_4, and 25 mM glucose. Add 4 mM $CaCl_2$, 4 mM $MgCl_2$ at the time of imaging.
9. High KCl stimulation solution: 62.5 mM KCl, 67 mM NaCl, 25 mM $NaHCO_3$, 1.25 mM NaH_2PO_4 and 25 mM glucose. Add 4 mM $CaCl_2$, 4 mM $MgCl_2$, 5 μM 2,3-dihydroxy-6-nitro-7-sulfamoyl-benzo(f)quinoxaline (NBQX), and 5 μM of R-(−)-3-(2-carboxypiperazine-4-yl)-propyl-1-phosphonic acid (D-CPP) at the time of imaging.
10. *N*-methyl-D-aspartate (NMDA) stimulation solution: 127 mM NaCl, 2.5 mM KCl, 25 mM $NaHCO_3$, 1.25 mM NaH_2PO_4, and 25 mM glucose. Add 15 μM NMDA, 4 mM $CaCl_2$, and 1 μM TTX at the time of imaging.

2.4 cDNA Bullet Preparation and Ballistic Gene Transfer

1. cDNA bullet preparation: 20 mg/ml polyvinylpyrrolidone (PVP) in 100 % ethanol, 1 mM $CaCl_2$, sensor cDNA, 100 % ethanol, 50 mM spermidine, 1.6 μm gold particles (Bio-Rad), plastic tubing, cDNA bullet prep station, tubing cutter, vial flasks, capsule dehydrators containing desiccant, parafilm.
2. Helium tank.
3. Helios-gene gun (Bio-Rad).

2.5 Imaging Apparatus and Software

1. Two-photon laser scanning microscope.
2. Ti:sapphire pulsed laser (MaiTai; Spectra-Physics, Fremont, CA; also available from other companies, such as Coherent) tuned to 900–920 nm to image mEGFP- and mRFP-tagged constructs.
3. Data output and acquisition board (such as PCI-6110, National Instruments) for scanning and red signal.
4. Data output board to produce pixel, line and frame clocks (such as PCI-6713, National Instruments).
5. Time-correlated single photon counting (TCSPC) board (such as SPC-150, Becker-Hickl).
6. Software for TCSPC data acquisition (*see* **Note 1**).

3 Methods

3.1 General Design of FRET Sensors for Small GTPase Activity

Bimolecular 2pFLIM sensors for HRas, RhoA and Cdc42 have previously been successfully developed and proven to be highly sensitive for imaging Ras GTPase signaling in small compartments of neurons, such as dendritic spines [8, 10, 13]. In view of these data, in principle a similar bimolecular design scheme can be applied to many Ras GTPase proteins. Specifically, the sensor design uses Ras GTPases tagged with mEGFP (*see* **Notes 2** and **3**) and a binding partner (RBD) tagged with two red fluorescent proteins (mCherry or mRFP) (Fig. 2) (*see* **Note 4**).

1. For most Ras GTPases, the membrane association domain is located on the C-terminus, but exceptions to this rule exist (*see* **Note 3**). To prevent disruption of membrane targeting, tag the Ras GTPase of interest with mEGFP at the opposite terminus (usually the N-terminus) (*see* **Note 3**).
2. Select the corresponding RBD for the Ras GTPase of interest and tag it with mCherry on its N-terminus and another mCherry on its C-terminus (*see* **Notes 4–7**).

3.2 In Vitro Binding Assay

The dissociation constant between purified superfolder GFP (sfGFP) tagged-Ras GTPase and mCherry-RBD can be measured in vitro with fluorescence lifetime measurements (Fig. 3). Using sfGFP instead of mEGFP and removing the membrane targeting domain of small GTPase proteins improves expression and folding in *E. coli* [14]. A poly-histidine (His_6) tagged version of the construct in a bacterial expression vector (such as pET) can be used for purification with Ni^+-NTA affinity chromatography.

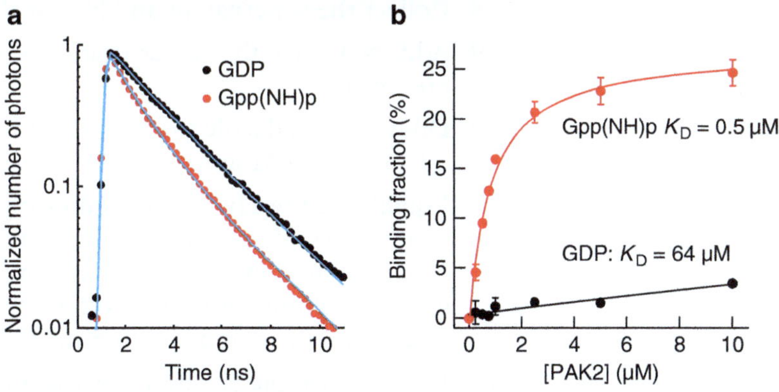

Fig. 3 Binding between purified sfGFP-Cdc42 and mCherry-PAK2 (R72C). (**a**) Fluorescence lifetime curves. *Blue lines* are fits to a double exponential. (**b**) The binding fraction measured as a function of PAK2 concentration. Lines are fits to a Michaelis–Menten function. Adapted from ref. 13 with permission

3.2.1 His-Tagged Protein Purification

1. Transform BL21 competent cells (*E. coli*) with a protein expression vector (such as pET) containing the cDNA of interest. In order to do this, gently mix 3 μl of cDNA (30–100 ng) in 25 μl of BL21 competent cells by tapping. Then, incubate the competent cells with the plasmid at 42 °C for 30 s, followed by 2 min of incubation on ice. Then, add 200 μl of Super Optimal broth with Catabolite repression (SOC) medium and plate the transformed competent cells on LB agarose plates with ampicillin for selection of the transformed bacteria. Incubate at 37 °C for 12–17 h.
2. Select a single colony of transformed competent cells on the LB agarose plate and grow this colony in 5 ml of LB containing ampicillin for approximately 12 h, at 37 °C, with agitation (225 rpm). Transfer 2.5 ml of the bacterial culture to 250 ml of LB containing ampicillin and 35 μg/ml chloramphenicol and incubate at 37 °C with agitation (225 rpm). Use the remaining LB containing the transformant to prepare a glycerol stock to be stored at −80 °C.
3. Check the optical density of the bacterial culture with the transformant in a spectrofluorometer by measuring the absorbance at 600 nm. When the OD value is 0.4, induce protein synthesis by adding 0.4 mM IPTG to the culture and agitate (250 rpm) for 16 h at 16 °C (*see* **Note 8**).
4. Spin down the culture at 6,000 × *g*, at 4 °C. Carefully discard the supernatant and keep the bacterial pellet on ice.
5. Resuspend the bacterial pellet completely in 6 ml of ice-cold lysis buffer and sonicate the bacteria at 4 °C.
6. Add Triton X-100 to a final concentration of 0.5 %.
7. Separate the insoluble constituents from the bacterial lysate by centrifuging at 15,000 × *g*, at 4 °C.
8. Collect the supernatant and filter it through a 0.22 μm filter.
9. Add NaCl to the supernatant to a final concentration of 500 mM.
10. Dilute the imidazole stock (2 M) in binding buffer to obtain 5, 50 and 500 mM imidazole.
11. Equilibrate the HisTrap column for protein purification. First, attach a 10 ml syringe to the HisTrap column, then wash the column with 5 ml of H_2O. Load 5 ml of 0.1 M $NiSO_4$ solution into the column. Wash the column with 5 ml of H_2O and then equilibrate the column with 5 ml of 5 mM imidazole.
12. Load the sample (filtered supernatant) into the column. Wash the column with 5 mM imidazole solution and then wash the column with 50 mM imidazole solution. Lastly, elute the protein with a 500 mM imidazole solution. Collect the purified

protein in a 1.5 ml tube. After this procedure, the HisTrap column can be washed and reused (*see* **Note 9**).

13. To remove the imidazole from the protein sample, use a PD10 column for salt exchange. First, equilibrate the PD10 column by washing it with 5 ml of H_2O, followed by 5 ml of equilibration buffer. Then, load the protein solution purified with HisTrap onto the PD10 column and elute the protein with standard buffer. The imidazole will be retained in the PD10 column. Do not reuse PD10 columns.
14. Measure the absorbance of the fluorophore and calculate the concentration using the known extinction coefficient of the protein (sfGFP, $A_{489} = 83{,}000\ cm^{-1}\ M^{-1}$ [15]; mCherry, $A_{587} = 72{,}000\ cm^{-1}\ M^{-1}$ [16]).

3.2.2 Measurements of the Dissociation Constant of Ras GTPase Sensors

In this step, the binding constant between GTPase and RBD is determined by measuring the fluorescence lifetime of GTP- or GDP-bound sfGFP-Ras GTPase (0.3 μM) with different concentrations (0–10 μM) of mCherry-RBD (Fig. 3b).

1. Place 10 nmol (typically ~50 μM × ~200 μl) of purified sfGFP-Ras GTPase into a 1.5 ml tube and add 250 nmol (10 mM × 25 μl) of Gpp(NH)p or GDP. Add EDTA to a final concentration of 15 mM. Incubate at 37 °C for 30 min.
2. Terminate the reaction by adding $MgCl_2$ to a final concentration 15 mM and mix gently with the pipette.
3. Remove the unbound Gpp(NH)p or GDP from the mix by using a PD-miniTrap column. First, wash the column with H_2O and then equilibrate it with standard buffer. Load the protein in the column and elute the protein with standard buffer.
4. Measure the concentration of sfGFP-Ras GTPase loaded with either Gpp(NH)p or GDP by determining the absorbance at 489 nm ($A_{489} = 83{,}000\ cm^{-1}\ M^{-1}$ for sfGFP [15]).
5. Mix sfGFP-GTPase with mCherry-RBD gently with a pipette. Keep the concentration of sfGFP-GTPase constant (0.3 μM) and perform serial dilutions of mCherry-RBD (0–10 μM). Use standard buffer as a solvent for these mixtures.
6. Incubate the mixtures at room temperature for 30 min.
7. Measure the fluorescence lifetime curve of the mixture (Fig. 3a, b) (*see* **Note 10**).
8. To quantify the binding fraction, fit the fluorescence lifetime curve with a double exponential curve:

$$F(t) = F_0\left[P_D \bullet H(t, t_0, \tau_D, \tau_G) + P_{AD} \bullet H(t, t_0, \tau_{AD}, \tau_G)\right], \quad (1)$$

where H is given by [8]:

$$H(t, t_0, \tau_D, \tau_G) = F_0 \frac{1}{2} \exp\left(\frac{\tau_G^2}{2\tau_D^2} - \frac{t - t_0}{\tau_D}\right) \operatorname{erfc}\left(\frac{\tau_G^2 - \tau_D(t - t_0)}{\sqrt{2}\tau_D \tau_G}\right) \quad (2)$$

where t is the photon arrival time for each photon collected by the PMT, t_0 is the offset time between the start of a laser pulse and the start of photon arrival, τ_D is the fluorescence lifetime of the free donor (~2.6 ns for mEGFP or sfGFP), τ_G is the standard deviation of the Gaussian pulse response function (0.1 ns), and F_0 is the peak fluorescence. P_{AD} and P_D are the fraction of GFP bound and unbound to acceptor, respectively ($P_{AD} + P_D = 1$) (Fig. 3a) [8, 10, 11]. To obtain the dissociation constant K_D, plot the binding fraction (P_{AD}) as a function of the concentration of free mCherry-RBD, and fit with a Michaelis–Menten function (Fig. 3b):

$$P_{AD} = \frac{P_{AD}^{max}}{1 + K_D / [\text{RBD}]}. \quad (3)$$

3.3 Testing the Sensitivity and Selectivity of the Sensor in HeLa Cells

After selecting the RBDs with an optimum binding affinity in vitro, the next step is to test the candidate RBDs to obtain those with the best sensitivity and specificity in cells. To do so, HeLa cells (or other cell lines) transfected with mEGFP-Ras GTPase and mCherry-RBD-mCherry are imaged to compare the signal of Ras GTPase sensors with constitutively active and dominant negative mutations in the conserved GTPase domain (Fig. 4) [17]. The specificity of the FRET sensors can be addressed by measuring sensor signal in cells co-expressing sensors together with GEFs and GAPs (Fig. 4).

1. Plate HeLa cells on 35 mm culture dishes in cell line culture medium, at 37 °C in 5 % CO_2.
2. One day after plating, when the confluence of cells is about 60 %, transfect the cells with plasmid DNA for the donor (wild type, dominant negative and constitutively active forms) and acceptor components of the sensor (*see* **Note 11**), using Lipofectamine 2000 (Invitrogen), according to the manufacturer's instructions. Mix 2 μg of cDNA in 100 μL of Opti-MEM (Invitrogen). In a separate tube, mix 3 μL of Lipofectamine 2000 in 100 μL of Opti-MEM and incubate for 5 min at room temperature. Then, mix the Opti-MEM containing the cDNA with the Opti-MEM containing the Lipofectamine 2000 and incubate this mixture for 20 min. Add dropwise into the media. Incubate the cells for 24 h, at 37 °C in 5 % CO_2.
3. Image the transfected cells in a solution containing H-ACSF [11], under 2pFLIM (or single photon FLIM).

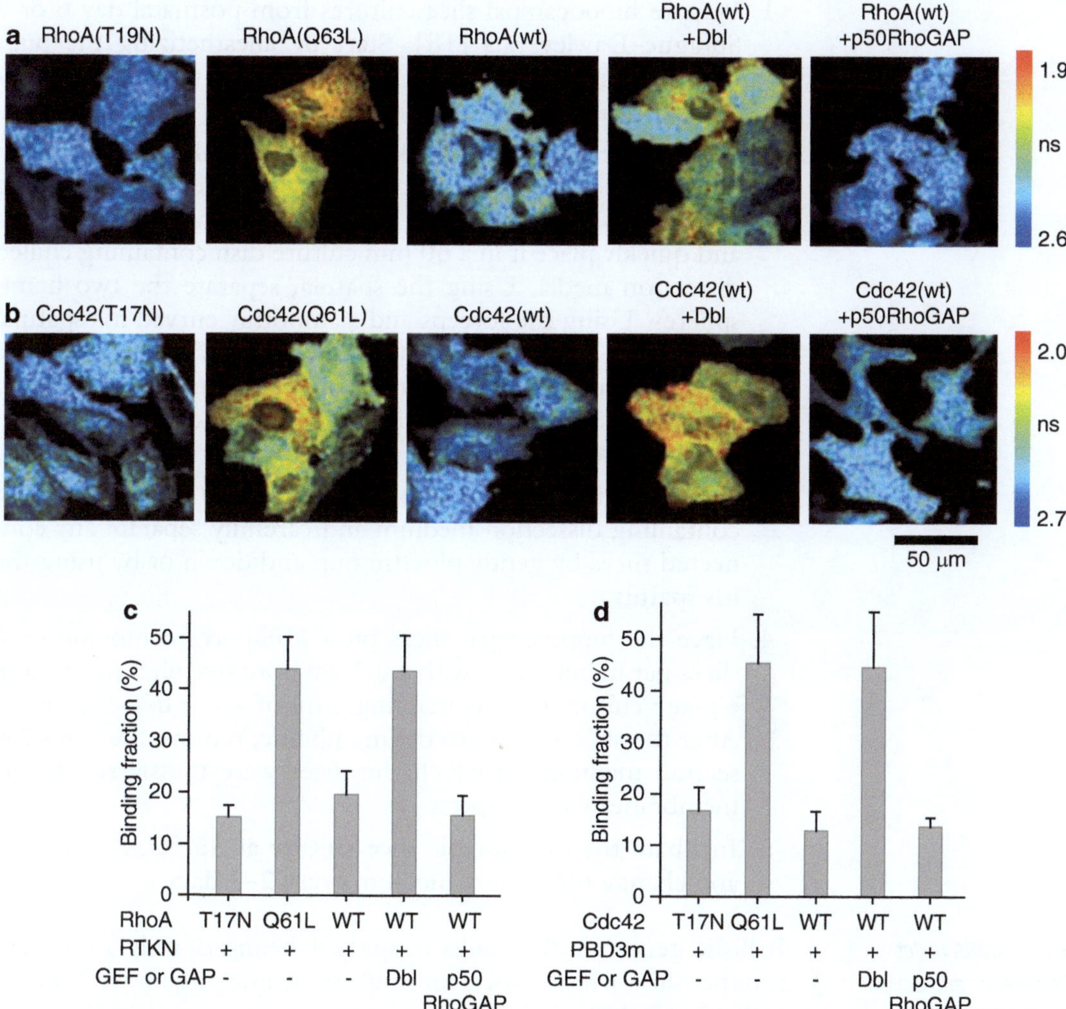

Fig. 4 Characterization of Cdc42 and RhoA sensors in HeLa cells. (**a**, **b**), Representative fluorescence lifetime images in HeLa cells transfected with mutant RhoA, Cdc42 FRET sensors or wild-type sensors co-expressed with positive (Dbl) or negative (p50 RhoGAP) RhoA and Cdc42 regulators. Warmer colors indicate shorter lifetimes and higher levels of Cdc42 or RhoA activation. Fluorescence lifetime imaging was performed 18–24 h after transfection. (**c**, **d**) Fraction of mEGFP-Rho GTPases bound to mCherry-RBD-mCherry calculated by fitting the fluorescence lifetime curve with a double exponential function. Adapted from ref. 13 with permission

4. Calculate the binding fraction by fitting fluorescence lifetime decay averaged over each cell with a double exponential curve Eqs. 1 and 2.

3.4 Imaging Ras GTPase Protein Activity in Neurons

3.4.1 Organotypic Slice Culture Preparation

Organotypic slice cultures of the hippocampus are useful preparations that allow one to transfect neurons with cDNA plasmids via ballistic gene transfer or by infection with virus bearing cDNA plasmids of interest.

1. Prepare hippocampal slice cultures from postnatal day 6 or 7 Sprague–Dawley rats [18]. Start by anesthetizing the pups using isofluorane and then quickly decapitating the anesthetized animals with surgical scissors.
2. Using a scalpel, cut the skin and the skull sagittally from between the eyes to the occipital bone while laterally folding back both sides of the skull. Then, lift the brain with a spatula and quickly place it in a 60 mm culture dish containing chilled dissection media. Using the spatula, separate the two hemispheres. Using the forceps and a smoothly curved iris spatula, remove the striatum and unfold the cortex. Then, separate the hippocampus and transfer it to a tissue chopper. Set the step interval to 350 μm on the tissue chopper and slice the hippocampus coronally.
3. Place the hippocampus slices onto a new 60 mm culture dish containing dissection medium and carefully separate any connected slices by gently pipetting up and down or by using the iris spatula.
4. Place the hippocampal slices on a Millipore membrane (3–5 slices per membrane), with a 0.2 μm pore size filter in a well of a tissue culture plate containing 1 ml of warm tissue medium. After the slice adheres to the membrane, remove the extra dissection medium (in which the slices were transferred to the membrane) with a pipette.
5. Incubate the organotypic slice culture at 35 °C, 3–5 % CO_2 and change the culture medium every 2–3 days.

3.4.2 cDNA Bullet Preparation and Ballistic Gene Transfer

Ballistic gene transfer results in sparse labeling of only a few neurons per slice, which is ideal for single spine imaging under 2pFLIM [10, 12, 13, 19].

1. After 10–15 days in culture, transfect the slices ballistically with cDNA bullets using a Helios gene gun.
2. To prepare the cDNA bullets, weigh 8–11 mg of 1.6 μm gold particles (Bio-Rad) and add 100 μl of 50 mM spermidine solution. Sonicate for at least 5 min and vortex to mix well.
3. Add a total of 40–50 μg of plasmids containing cDNA (*see* **Note 11**), briefly sonicate and vortex.
4. Add 100 μl of 1 M $CaCl_2$ solution while mixing in the vortex.
5. Incubate at room temperature for 10 min.
6. In the meantime, cut a piece of plastic tubing slightly longer than the length of the tubing station (Bio-Rad) and insert it into the tubing station. Dry the tubing with nitrogen gas at a pressure regulated to 1–2 psi in the tubing station.
7. Spin down the previous bullet mix at 16,000 × *g* with a tabletop centrifuge for 2 min.

8. Discard the supernatant and add 1 ml of 100 % ethanol. Briefly sonicate and vortex to mix well. Spin down at 16,000 × *g*, for 2 min. Repeat this step three times.
9. Add 1 ml of 100 % ethanol and 8 μl of 20 mg/ml PVP. Briefly sonicate and vortex. Then, add this mix to a tube containing 2 ml of 100 % ethanol and vortex to mix well.
10. Load the mix into the tubing using a syringe to apply suction. Do not detach the syringe and let the mix sit in the tubing for approximately 10 min.
11. Carefully use the syringe to remove the ethanol from the tubing and note that a line of gold covers the length of the tubing.
12. Roll the tubing several times to evenly distribute the bullets in the tubing. Let dry for approximately 20 min with nitrogen gas, after inserting the tubing into the tubing station.
13. Cut the tubing with the bullet tool provided by Bio-Rad into small pieces of tubing and store in a small vial flask containing a desiccant material. Seal the flask with parafilm and store at 4 °C for up to 3 months.
14. To perform transfection, put each small piece of plastic tubing containing the bullets in the bullet holder and introduce it into the appropriate compartment of the gene gun. Load a battery into the gene gun. Connect the gene gun to a helium tank and set the tank pressure to 180 psi. Attach a barrel containing a diffusion barrier to the gene gun and "shoot" the hippocampal slice cultures. Depending on the sensor used, hippocampal slices can be imaged 1–4 days after transfection.

3.4.3 Imaging Small GTPase Activity in Neurons

Depolarization of neurons with high concentration KCl solution can open voltage-gated calcium channels (VGCCs) and activate Ras GTPase proteins.

1. At the time of imaging, add $CaCl_2$ (4 mM) and $MgCl_2$ (4 mM) to ACSF and perform all hippocampal slice 2pFLIM experiments in this imaging solution, aerating with 95 % O_2 and 5 % CO_2, at a constant temperature of ~25 °C. Acquire approximately five 2pFLIM images for baseline activity of the sensor for a given Ras GTPase (Fig. 5a).
2. To activate Ras GTPases with KCl, exchange the extracellular solution to high KCl stimulation solution. Ten minutes later, exchange the solution back to imaging solution (Fig. 5a, b).
3. To determine the time course of the sensor activation, acquire 2pFLIM images every 1–2 min.

Alternatively, Ras GTPases can be activated following NMDA receptor stimulation with NMDA.

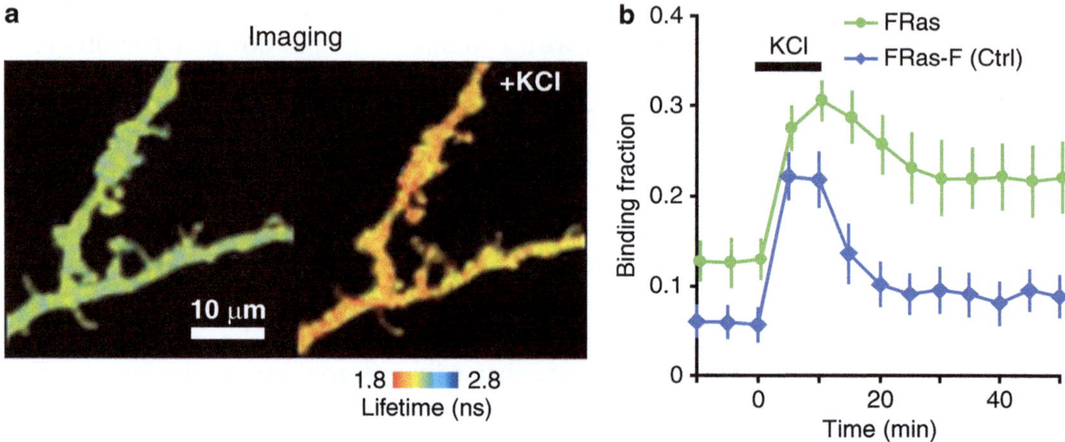

Fig. 5 Ras activation in the apical dendrites of CA1 neurons in response to depolarization. (**a**) Fluorescence lifetime images before and after the application of KCl. (**b**) Time course of Ras activation measured as the fraction of mEGFP-Ras bound to mRFP-RBD-mRFP (binding fraction). FRas includes WT RBD, whereas FRas-F includes RBD^{R59A}. *Error bars* represent S.E.M. over 7–8 cells each. Adapted from 8 with permission

1. At the time of imaging, add $CaCl_2$ (4 mM) and $MgCl_2$ (4 mM) to ACSF and perform all hippocampal slice 2pFLIM experiments in this imaging solution, aerating with 95 % O_2 and 5 % CO_2, at a constant temperature of ~25 °C. Acquire approximately five 2pFLIM images for baseline activity of the sensor for a given Ras GTPase (Fig. 6a, b).
2. To activate Ras GTPases with NMDA, exchange the extracellular solution to the NMDA stimulation solution. Two minutes later, exchange the solution back to imaging solution (Fig 6a–c).
3. To determine the time course of sensor activation, acquire 2pFLIM images every 1–2 min.

3.5 Effects of the Overexpression of Ras GTPase Protein Sensors on the Spatiotemporal Dynamics of Ras GTPases

The concentration of overexpressed sensor can be measured by comparing the green fluorescence intensity and the red fluorescence intensity of a known concentration of purified mEGFP and mCherry proteins under 2-photon or confocal microscopes (Fig. 6d, e) (*see* **Note 12**).

1. Purify mEGFP and mCherry proteins following the procedure described in Subheading 3.2.1.
2. Measure the concentration of the purified mEGFP and mRFP proteins, as explained in Subheading 3.2.1.
3. After performing FLIM experiments in live cells, measure the green and red fluorescence intensity of the primary apical dendrite. Place ~100 μl of purified mEGFP or mCherry under the objective and measure its fluorescence intensity under the microscope using the same laser power used for live cell experiments.

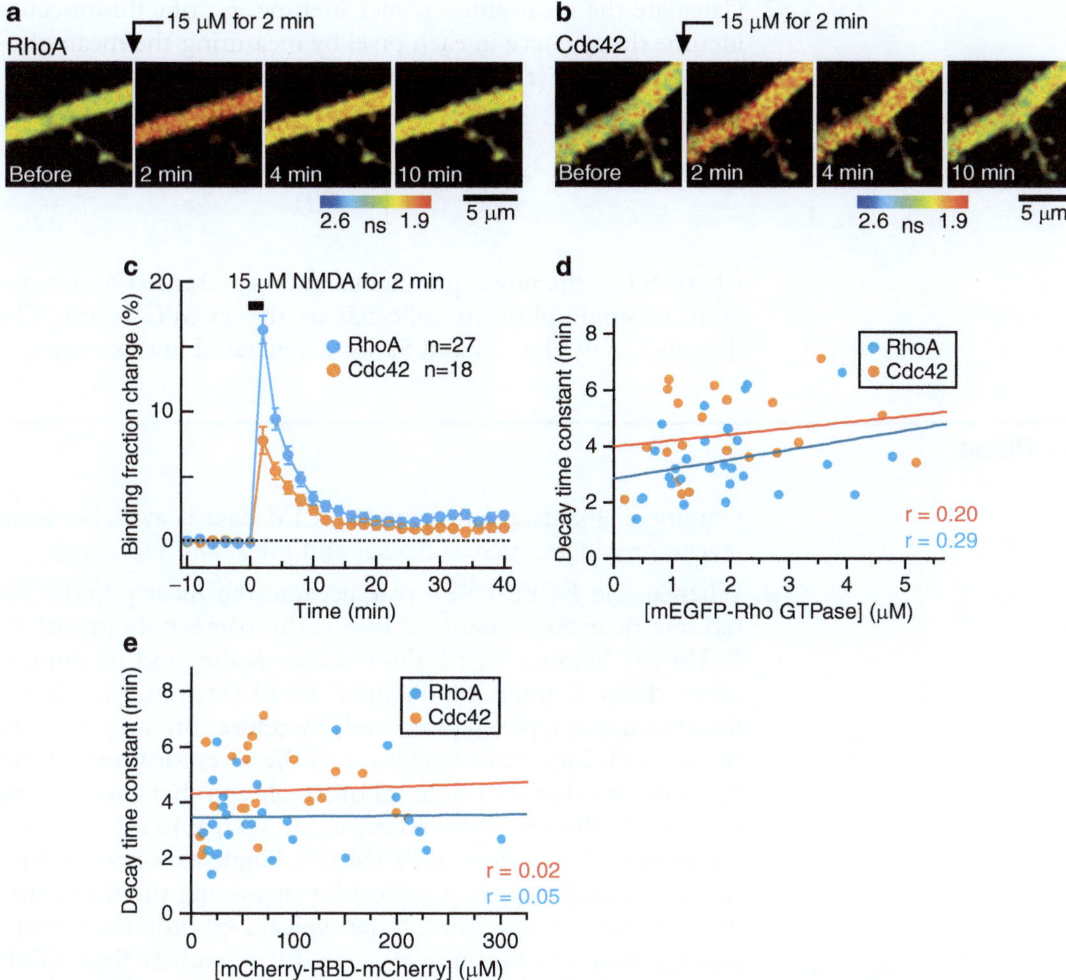

Fig. 6 Characterization of Rho GTPase FRET sensors in neurons. (**a**, **b**) Fluorescence lifetime images of RhoA (**a**) and Cdc42 (**b**) sensors in the primary dendrite of CA1 pyramidal neurons in response to bath-applied NMDA. Scale bar, 5 μm. (**c**) Time courses of the RhoA and Cdc42 activity after NMDA application. (**d**, **e**) The time constants of the signal decay plotted as a function of the concentrations of mEGFP-Rho GTPase (**d**) or mCherry-RBD-mCherry (**e**). The time constant was obtained by fitting the decay of the binding fraction (**c**) with a single-exponential function. Adapted from ref. 13 with permission

4. Calculate the concentration of both mEGFP and mCherry for each cell based on the fluorescence intensity measured.

3.6 Fluorescence Lifetime Analysis

The obtained fluorescence lifetime images are analyzed as follows:

1. Fit the fluorescence lifetime decay curve, $F(t)$, integrated over all pixels in the image to the equations described in ref. 8, Eqs. 1 and 2. To improve the stability of this fitting, all pixels in an image can be summed and fitted with Eq. 1. Furthermore, τ_D and τ_G are predetermined and these values are fixed for fitting. F_0 and t_0 are obtained by fitting.

2. Calculate the mean fluorescence lifetime, τ_m, of a fluorescence lifetime decay curve in each pixel by measuring the mean photon arrival time $\langle t \rangle$ as follows [8]:

$$\tau_{\mathrm{m}} = \langle t \rangle - t_0 = \frac{\int \mathrm{d}t \cdot tF(t)}{\int \mathrm{d}t \cdot F(t)} - t_0, \tag{4}$$

where $F(t)$ is the raw experimental data represented by a histogram of single photons collected by the TCSPC board. The data shown in Figs. 4a and 5a were generated in this way.

4 Notes

1. Commercial software to acquire FLIM data is available from several providers, such as Becker and Hickl and Picoquant.
2. Whereas the ECFP-EYFP pair became the most popular for ratiometric measurements of FRET, this pair is not optimal for FLIM [8] because ECFP fluorescence is dim and its fluorescence decay is multi-exponential. In FLIM, only the donor fluorescence is typically used and, therefore, the brightness of the donor is important while that of the acceptor is not. Thus, fluorophores that are bright, photostable and that have a single exponential fluorescence lifetime decay, such as EGFP or EYFP, are preferred as donors in FLIM [8]. Single exponential fluorescence lifetime decay is critical for measuring the fraction of donor bound to acceptor. Under typical 2-photon microscopy settings with a Ti:Sapphire laser, EGFP is brighter than EYFP, making EGFP the best choice for a donor. Also, it is best to use the monomeric version of EGFP ($\mathrm{EGFP}^{\mathrm{A206K}}$ or mEGFP) to avoid dimerization.
3. For most Ras GTPase proteins, mEGFP should be tagged to their N-termini to maintain the function of their membrane targeting motifs at the C-termini [20, 21]. However, Arf proteins have their membrane target domain on their N-termini and, therefore, should be tagged at either the C-terminus or an internal loop [21–24].
4. For FRET acceptors, monomeric red fluorescent protein (mRFP) or mCherry (a brighter variant of mRFP) [15] have been used because of their good spectral separation from mEGFP [4, 5]. The relatively low folding efficiency of mRFP and mCherry decreases the sensitivity of the sensor, since unfolded acceptor does not contribute to the FRET signal. To overcome this problem and improve the sensitivity of the sensor, a protein can be tagged with two acceptors (Fig. 2). This

approach increases the chance of having at least one mature acceptor [8]. The recently developed sREACh (non-radiative YFP REACh variant with improved solubility, environmental insensitivity and folding) has higher folding efficiency compared to mRFP and mCherry, and has been reported to be useful as a FLIM acceptor under conditions where one does not need to measure acceptor concentration [11, 25].

5. While screening RBDs for Ras GTPase sensors, one important parameter for optimal sensitivity is the affinity between a target Ras GTPase protein and its corresponding RBD. First, the affinity of RBD for the GTPase protein has to be significantly higher for the active GTP form compared to the inactive GDP form. Visualization of small compartments requires fluorophore concentrations of a few micromolar. RBD concentration needs to be higher than this to avoid saturation. Thus, for optimal sensitivity, the dissociation constant between RBD and the active Ras GTPase must be lower than a few μM, and that between RBD and the inactive GTPase protein must be higher than ~50 μM. Second, because RBD binding to active GTPase protein competes with the activity of GAPs and slows down inactivation [8], the dissociation constant between the inactive GTPase protein and RBD should not be too low. Quantitatively, the inactivation time constant, τ_{GAP}, is a function of [RBD] and the dissociation constant K_D:

$$\tau_{GAP} = \left(1 + \frac{[RBD]}{K_D}\right)\tau_{GAP}^{0} \tag{5}$$

where τ^0_{GAP} is τ_{GAP} in the absence of RBD [8]. Thus, to minimize this effect, K_D must be higher than [RBD]. However, to obtain optimal binding signal, K_D has to be lower than [RBD]. Thus, it is recommended to choose RBDs with K_D similar to or slightly lower than the expression level of RBD (micromoles). To obtain Ras GTPase-RBD combinations with optimum dissociation constants and high specificity, it may be necessary to mutate amino acids of RBD located at the interface of the interaction [4, 8–9]. In the case of the HRas sensor, which measures binding between HRas and the RBD of cRaf (cRaf[51–131]), the sensor is almost irreversibly active when wild type cRaf[51–131] (K_D ~ 0.1 μM) is used. When a low affinity mutant cRaf[51–131]^{R59A} (K_D ~ 4 μM) is used, the sensor inactivates as quickly as endogenous HRas (Fig. 5b) [8, 10]. Thus, it is recommended to screen for RBDs and mutants with dissociation constants of 1–5 μM for the Ras GTPase of interest (and lower than other Ras GTPase proteins), and that also have a high K_D (~50 μM) with the inactive GTPase protein.

Table 1
Small GTPase proteins and their RBDs for activity sensors that show high signal-to-noise ratio in HeLa cells and neurons [8, 13]

Subfamily	Small GTPase protein	RBD
Ras (mEGFP-Ras)	H-Ras	cRaf(R59A)
Rho (mEGFP-Rho)	RhoA Cdc42p, Cdc42b	Rhotekin PAK3 (S74A, F84A)

Practically, this requires in vitro measurement of the affinity between fluorescently tagged GTPase proteins and various RBDs and their mutants using fluorescence lifetime measurements. Table 1 summarizes the Ras GTPase proteins and their RBDs for previously developed sensors [8, 13 23, 26–29].

6. Sensors may be less sensitive in neurons (or other primary cell cultures) than in HeLa cells, and this may cause false negative results. To reduce the possibility of false negative results, try 2–3 different GTPase-RBD combinations with different affinities.
7. In general, for the bimolecular design, FRET signal is not very sensitive to the linker between fluorescent proteins and the target proteins [8].
8. This step of protein purification may require optimization for each particular protein. For some proteins, IPTG is not required and some proteins require lower temperatures for proper folding.
9. The HisTrap column can be reused multiple times upon washing. To wash the HisTrap column, load 1 ml of 500 mM Imidazole solution onto the column and then wash with 5 ml H_2O. Following this, wash the column with 5 ml of 50 mM EDTA solution and, finally, with 5 ml H_2O. Store the HisTrap column at 4 °C.
10. When working with 2pFLIM, excitation at 900–940 nm and emission at 490–550 nm should be used. When working with single photon FLIM, excitation at 450–470 nm should be used.
11. The optimal ratio between the donor cDNA and acceptor cDNA must be emprically determined for each sensor. Typically the donor–acceptor ratio is between 1:1 and 1:4.
12. Overexpression of mEGFP-Ras GTPase could affect the spatiotemporal dynamics of Ras GTPases. Therefore, it is necessary to assess the time constant of FRET signal decay in cells after Ras GTPase stimulation at different expression levels of mEGFP-Ras GTPase and mCherry-RBD-mCherry (Fig. 6d, e).

If there is no correlation between the decay time constant and the concentration of mEGFP-Ras GTPase or mCherry-RBD-mCherry, the overexpression of the sensor has a minimum impact on the decay kinetics of the Ras GTPase (Fig. 6d, e). When significant correlation is found, one can extrapolate the time constant (or other measured parameters) to zero expression level.

Acknowledgments

We thank E. Park, J. Nishiyama, and other members of the Yasuda lab for critical reading and discussion.

References

1. Takai Y, Sasaki T, Matozaki T (2001) Small GTP-binding proteins. Physiol Rev 81(1):153–208
2. Ye X, Carew TJ (2010) Small G protein signaling in neuronal plasticity and memory formation: the specific role of ras family proteins. Neuron 68(3):340–361
3. Miyawaki A (2003) Visualization of the spatial and temporal dynamics of intracellular signaling. Dev Cell 4(3):295–305
4. Lakowicz JR (2006) Principles of fluorescence spectroscopy, 4th edn. Springer, New York, NY
5. Mochizuki NYS, Kurokawa K, Ohba Y et al (2001) Spatio-temporal images of growth-factor-induced activation of Ras and Rap1. Nature 411(6841):1065–1068
6. Itoh RE, Kurokawa K, Ohba Y et al (2002) Activation of Rac and Cdc42 video imaged by fluorescent resonance energy transfer-based single-molecule probes in the membrane of living cells. Mol Cell Biol 22(18):6582–6591
7. Nakamura T, Aoki K, Matsuda M (2005) Monitoring spatio-temporal regulation of Ras and Rho GTPase with GFP-based FRET probes. Methods 37(2):146–153
8. Yasuda R, Harvey CD, Zhong H et al (2006) Supersensitive Ras activation in dendrites and spines revealed by two-photon fluorescence lifetime imaging. Nat Neurosci 9(2):283–291
9. Yasuda R (2006) Imaging spatiotemporal dynamics of neuronal signaling using fluorescence resonance energy transfer and fluorescence lifetime imaging microscopy. Curr Opin Neurobiol 16(5):551–561
10. Harvey CD, Yasuda R, Zhong H et al (2008) The spread of Ras activity triggered by activation of a single dendritic spine. Science 321(5885):136–140
11. Murakoshi H, Lee SJ, Yasuda R (2008) Highly sensitive and quantitative FRET-FLIM imaging in single dendritic spines using improved non-radiative YFP. Brain Cell Biol 36(1–4): 31–42
12. Lee SJ, Escobedo-Lozoya Y, Szatmari EM et al (2009) Activation of CaMKII in single dendritic spines during long-term potentiation. Nature 458(7236):299–304
13. Murakoshi H, Wang H, Yasuda R (2011) Local, persistent activation of Rho GTPases during plasticity of single dendritic spines. Nature 472(7341):100–104
14. Pedelacq JD, Cabantous S, Tran T et al (2006) Engineering and characterization of a superfolder green fluorescent protein. Nat Biotechnol 24(1):79–88
15. Cabantous S, Rogers Y, Terwilliger TC et al (2008) New molecular reporters for rapid protein folding assays. PLoS One 3(6):e2387
16. Shaner NC, Campbell RE, Steinbach PA et al (2004) Improved monomeric red, orange and yellow fluorescent proteins derived from *Discosoma* sp. red fluorescent protein. Nat Biotechnol 22(12):1567–1572
17. Wennerberg K, Der CJ (2004) Rho-family GTPases: it's not only Rac and Rho (and I like it). J Cell Sci 117(Pt 8):1301–1312
18. Stoppini L, Buchs PA, Muller D (1991) A simple method for organotypic cultures of nervous tissue. J Neurosci Methods 37(2):173–182
19. McAllister AK (2000) Biolistic transfection of neurons. Sci STKE 2000(51):pl1
20. Choy E, Chiu VK, Silletti J et al (1999) Endomembrane trafficking of ras: the CAAX motif targets proteins to the ER and Golgi. Cell 98(1):69–80
21. Heo WD, Meyer T (2003) Switch-of-function mutants based on morphology classification of Ras superfamily small GTPases. Cell 113(3): 315–328

22. Beemiller P, Hoppe AD, Sawnson JA (2006) A phosphatidylinositol-3-kinase-dependent signal transition regulates ARF1 and ARF6 during Fcgamma receptor-mediated phagocytosis. PLoS Biol 4(6):e162
23. Gillingham AK, Munro S (2007) The small G proteins of the Arf family and their regulators. Annu Rev Cell Dev Biol 23:579–611
24. Hall B, McLean MA, Davis K et al (2008) A fluorescence resonance energy transfer activation sensor for Arf6. Anal Biochem 374(2):243–249
25. Ganesan S, Ameer-Beg SM, Ng TT et al (2006) A dark yellow fluorescent protein (YFP)-based resonance energy-accepting chromoprotein (REACh) for Förster resonance energy transfer with GFP. Proc Natl Acad Sci 103(11): 4089–4094
26. Wennerberg K, Rossman KL, Der CJ (2005) The Ras superfamily at a glance. J Cell Sci 118(Pt 5):843–846
27. Kiel C, Foglierini M, Kuemmerer N et al (2007) A genome-wide Ras-effector interaction network. J Mol Biol 370(5): 1020–1032
28. Fukuda M (2003) Distinct Rab binding specificity of Rim1, Rim2, rabphilin, and Noc2. Identification of a critical determinant of Rab3A/Rab27A recognition by Rim2. J Biol Chem 278(17):15373–15380
29. Fukuda M, Kanno E, Ishibashi K et al (2008) Large scale screening for novel rab effectors reveals unexpected broad Rab binding specificity. Mol Cell Proteomics 7(6):1031–1042

Chapter 10

Imaging Kinase Activity at Protein Scaffolds

Maya T. Kunkel and Alexandra C. Newton

Abstract

Kinase signaling is under tight spatiotemporal control, with signaling hubs within the cell often coordinated by protein scaffolds. Genetically encoded kinase activity reporters afford a unique tool to interrogate the rate, amplitude, and duration of kinase signaling at specific locations throughout the cell. This protocol describes how to assay kinase activity at a protein scaffold in live cells using a fluorescence resonance energy transfer (FRET)-based kinase activity sensor for protein kinase D (PKD) as an example.

Key words FRET, Kinase activity reporter, DKAR, Protein kinase D, Scaffold protein, NHERF

1 Introduction

Phosphorylation of substrate proteins by protein kinases affords one of nature's most effective mechanisms to reversibly regulate protein function. The simple addition of phosphate alters the chemical properties of the targeted surface, thus altering protein function by many mechanisms. For example, phosphorylation can modulate the intrinsic catalytic activity of the phosphorylated substrate; this includes other kinases and even the kinase, itself, via autophosphorylation. In addition, protein phosphorylation can regulate the subcellular localization of the substrate protein by affecting its association with other proteins or with lipids, either by altering the protein conformation or by altering the electrostatic properties of the interacting interface. Control of localization is particularly critical in cell signaling, where activation of kinases occurs at precise locations to effect localized signaling. Protein phosphatases oppose protein kinases, allowing acute regulation of the time period during which a protein is modified by phosphate. Thus, phosphorylation events are usually transient. Signaling by protein kinase D (PKD) family members affords one example of tight regulation of the spatial and temporal dynamics of kinase activity.

Jin Zhang et al. (eds.), *Fluorescent Protein-Based Biosensors: Methods and Protocols*, Methods in Molecular Biology, vol. 1071, DOI 10.1007/978-1-62703-622-1_10,

The PKD family plays a role in numerous processes, including cell proliferation and survival, immune cell signaling, gene expression, vesicle trafficking, and neuronal development [1]. The role this family plays thus depends on cell type (e.g., immune versus cancer cells) and subcellular localization (e.g., regulation of vesicle transport at the Golgi). The family comprises three members, PKD1, PKD2, and PKD3, each consisting of a conserved catalytic core, an amino-terminal regulatory domain containing tandem C1 domains, and, for PKD1 and PKD2, a PDZ-binding motif at the C-terminus [2]. The C1 domains bind diacylglycerol (DAG), a lipid second messenger that recruits PKD isozymes to membranes, a first step in PKD activation. Binding of the regulatory domain to membrane-embedded DAG results in a conformational change that poises PKD for subsequent phosphorylation by novel protein kinase C (PKC) family members at two sites within its catalytic core; this event is followed by PKD autophosphorylation at a site within its C-terminal tail [3, 4]. Because phosphorylation is a hallmark of PKD activation, as it is for many other kinases, activity is traditionally demonstrated via Western blotting using phospho-specific antibodies to these activating sites. However, both the temporal and spatial resolution of this method are poor, limiting the approach for assessing kinase signaling in cells. Furthermore, while the sites probed are indicative of kinase activation, there may be other means of activating the kinase or opposing inactivating phosphorylations elsewhere on the kinase, neither of which will be taken into account when probing a specific phosphorylated site. These problems are all circumvented by use of genetically encoded, fluorescence resonance energy transfer (FRET)-based kinase activity reporters.

Genetically encoded, FRET-based kinase activity reporters enable real-time monitoring of localized kinase activity within cells. Such reporters often utilize a modular design whereby a FRET pair flanking a phospho-peptide binding domain and a substrate sequence undergoes a conformational change following phosphorylation of a consensus substrate sequence (Fig. 1). Considerations in reporter design involve selection of a suitable FRET pair, identification of a kinase-specific substrate sequence, and selection of a compatible phosphoamino-binding module that binds efficiently to the phosphorylated substrate sequence, yet not with such high affinity that the phosphorylation cannot be reversed by phosphatases (detailed in ref. 5). For some kinases, additional modules that facilitate recognition by the kinase may be necessary; for example, the reporter of ERK activity includes a docking domain for ERK on its C-terminus [6]. The prototypical kinase activity reporters were designed in 2001 to read out activity from the tyrosine kinases Src, Abl, and EGFR [7] and PKA [8]. Since then, many new reporters have been developed based on this modular design; those reporters designed for protein kinases A through D (PKA through PKD) as well as their variants (usually improvements made to increase their sensitivity) are depicted in Table 1 [8–16].

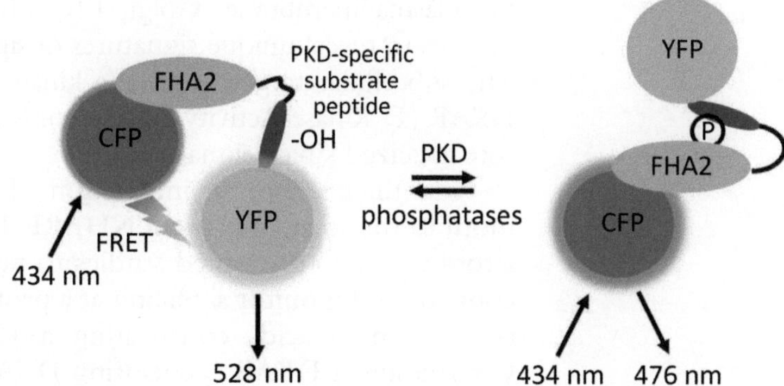

Fig. 1 Schematic diagram showing the modular structure of DKAR. Kinase activity reporters consist of a FRET donor (CFP) and acceptor (YFP) flanking a phospho-amino-acid binding domain (FHA2) and a consensus phosphorylation sequence (substrate peptide). In DKAR, the unphosphorylated reporter exists in a conformation wherein the FRET pair is undergoing FRET. When PKD is active, it phosphorylates the threonine within the consensus sequence (–OH), inducing a conformational change as the FHA2 domain binds the newly phosphorylated (*circle* with P) substrate peptide sequence. This intramolecular association alters the distance and/or relative orientation between the FRET pair, resulting in a decrease in FRET

Table 1
Evolution of kinase activity reporters for the serine/threonine protein kinases A through D

Kinase	Reporter	Notes	References
PKA	AKAR	First serine/threonine kinase reporter, irreversible	[8]
	AKAR2	Reversible response	[9]
	AKAR3	Increased response range	[10]
PKB/Akt	Aktus	First PKB/Akt reporter	[11]
	BKAR	Increased sensitivity	[12]
	AktAR	Increased response range	[13]
PKC	CKAR	First PKC reporter	[14]
	δCKAR	Isozyme-specific (PKCδ) reporter	[15]
PKD	DKAR	First PKD reporter	[16]

Genetically encoded reporters are introduced into cells by simple transfection, where they read out the rate, amplitude, and duration of endogenous (or exogenous) kinase activity in response to specific stimuli. Because they are genetically encoded, they can be targeted to subcellular regions through the addition of short targeting sequences. Such targeting allows determination of localized kinase activity occurring at the subcellular region being targeted. A variety of genetically encoded reporters have been targeted to subcellular locations such as the plasma membrane, Golgi, ER, mitochondria, and within the nucleus to reveal unique signatures of signaling at each location [5]. The subcellular targeting of the kinase activity reporter for PKD, DKAR (D Kinase Activity Reporter), is an example of how to measure localized kinase signaling.

Identification of an interaction of PKD via its PDZ-binding motif to the scaffold protein NHERF-1 (Na^+/H^+ exchanger regulatory factor-1) prompted studies to target DKAR to this protein complex and monitor signaling at a protein scaffold [17]. Addition of ten amino acids constituting a PDZ-binding motif to the C-terminus of DKAR (generating DKAR-PDZ) served to relocalize DKAR in a NHERF-1-dependent manner to the apical surface of the polarized epithelial cells, MDCK cells (Fig. 2b). Thus, in cells where DKAR-PDZ is visibly localized to the NHERF scaffold, one can monitor PKD activity at NHERF, revealing highly enriched PKD signaling at this protein complex (Fig. 3 [17]). Below, we describe how we utilized this DKAR-PDZ targeted to the NHERF protein scaffold to measure PKD signaling at this subcellular location in live cells.

2 Materials

2.1 Cell Culture

1. Madin Darby canine kidney (MDCK) cells (ATCC).
2. Dulbecco's Modification of Eagle's Medium/Ham's F12 50/50 Mix (DMEM/F12) with 10 % fetal bovine serum (FBS) (Cellgro).
3. 35 mm, sterile, glass-bottom culture dishes, No. 1.0 (MatTek Corporation).
4. Effectene transfection reagent (QIAGEN).
5. DNA encoding DKAR, DKAR-PDZ, NHERF-1.

2.2 Microscope Setup

Details listed here are specific to the experimental setup that our laboratory uses:

1. Axiovert 200 M microscope (Zeiss) with Xenon lamp (XBO 75/2 OFR, Zeiss).
2. MicroMAX 512BFT CCD camera (Roper Scientific).
3. Filters from Chroma Technology.
   - For cyan fluorescent protein (CFP): 420/20 nm excitation, 450 nm dichroic, 475/40 nm emission.
   - For yellow fluorescent protein (YFP): 495/10 nm excitation, 505 nm dichroic, 535/25 nm emission.
   - 10 % neutral density filter (22000A ND filter 1.0).
4. Metafluor software (Molecular Devices) (*see* **Note 1**).

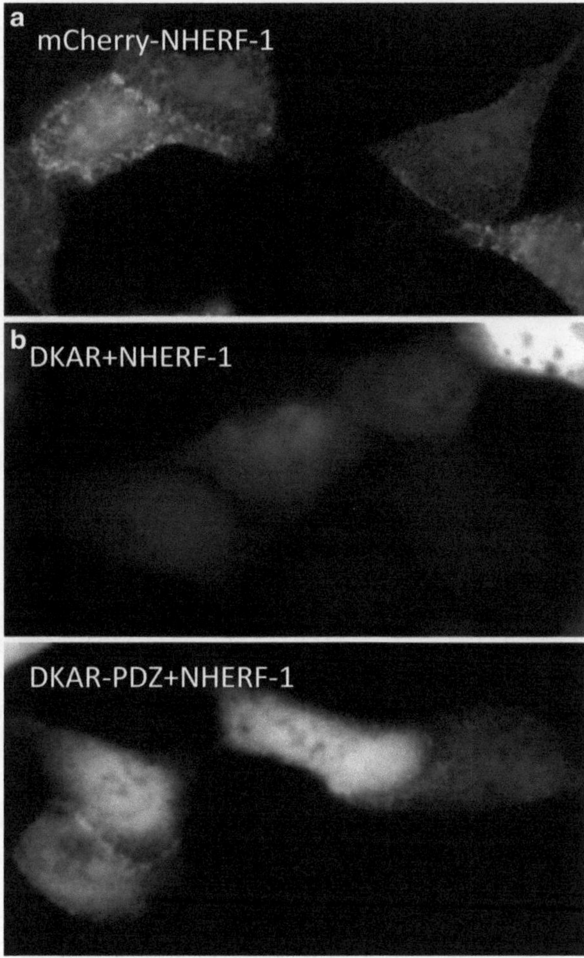

Fig. 2 Fluorescent images of MDCK cells. (**a**) mCherry-tagged NHERF-1 localizes to NHERF complexes at the apical membrane of polarized epithelial cells (e.g., MDCK cells) and this presents in a punctate pattern. (**b**) Untargeted DKAR is present throughout the cytosol and nucleus of the cell (*top*). Addition of the PDZ-targeting motif to DKAR (DKAR-PDZ) localizes it to the NHERF scaffold in NHERF-1-overexpressing MDCK cells (*bottom*); as not all cells display relocalization of DKAR-PDZ to the NHERF scaffold, care should be taken to select those that do

5. Lambda 10-2 filterwheel shutter controller (Sutter) (*see* **Note 2**).
6. 40×/1.3 NA oil-immersion objective (Zeiss).
7. Immersion oil, Type DF (Cargille Labs).

2.3 Data Acquisition and Analysis

1. Hanks' Balanced Salt Solution (HBSS, Cellgro) supplemented with 1 mM Ca^{2+} on the day of imaging.
2. Phorbol-12,13-dibutyrate (PDBu, Millipore).
3. Gö 6976 (Millipore).
4. Excel (Microsoft) or equivalent.

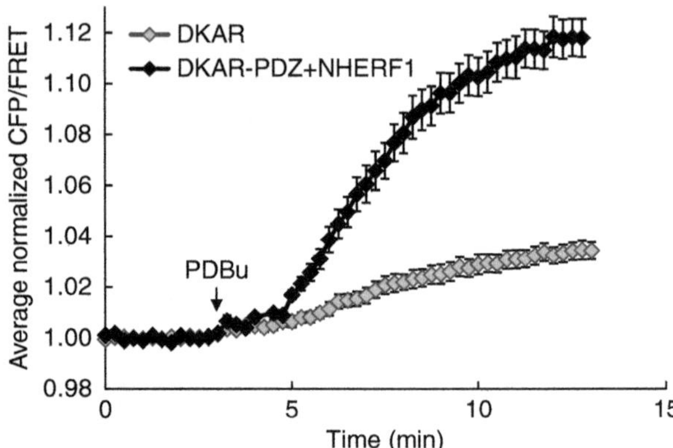

Fig. 3 Plot of the average normalized CFP/FRET ratio from DKAR or DKAR poised at NHERF-1 (DKAR-PDZ+NHERF1) following addition of PDBu. Adapted from ref. 17

3 Methods

3.1 Cell Culture

1. MDCK cells are propagated in DMEM/F12 media containing 10 % FBS; these cells adhere well to the uncoated, glass coverslip of the imaging dish. Two days prior to imaging, a confluent 10 cm dish of MDCK cells should be split 1:40 into a 35 mm imaging dish.
2. Twenty-four hour after plating cells into the 35 mm imaging dish (when they are at approximately 75 % confluence), the MDCK cells should be transfected using the Effectene transfection reagent according to the manufacturer's protocol. One microgram of DKAR DNA and 0.1 μg of NHERF-1 DNA should be transfected to attain proper relative expression levels for imaging the following day (*see* **Note 3**).

3.2 Data Acquisition

We perform all of our imaging experiments at room temperature.

1. On the day of imaging (one day post-transfection), turn on the lamp, microscope, filterwheel changer, and camera. Open the Metafluor software application.
2. Set up (or load) the imaging protocol. For our setup, this protocol consists of acquiring CFP (420/20 nm excitation, 450 nm dichroic, 475/40 nm emission, 200 ms exposure), FRET (420/20 nm excitation, 450 nm dichroic, 535/25 nm emission, 200 ms exposure) and YFP (495/10 nm excitation, 505 nm dichroic, 535/25 nm emission, 100 ms exposure) once every 15 s through the 40× objective and with a 10 % neutral density filter in place. The CFP/FRET ratio is plotted to assess the progress of the experiment in real-time and the intensity of YFP is plotted as a measure of fluorophore photobleaching (*see* **Notes 4** and **5**).

3. Clean the 40× oil-immersion objective. Apply one drop of oil onto the objective.
4. Aspirate media from the cells, rinse cells in Hanks' balanced salt solution containing 1 mM $CaCl_2$ (HBSS/1 mM $CaCl_2$), and replace with 2 ml HBSS/1 mM $CaCl_2$ for imaging.
5. Place the imaging dish on the microscope stage and secure it (we use small pieces of modeling clay) to prevent subtle movements of the dish when adding drugs during the experiment.
6. Focus on the cells and identify those reflecting proper localization as well as optimal expression levels of reporter. For NHERF-overexpressing cells, proper localization of DKAR-PDZ consists of DKAR localized to the apical membrane of the MDCK cells as shown in the bottom panel of Fig. 2b. In addition, optimal levels of DKAR are those in which the kinase activity reporter is expressed at levels within the range of cellular substrates (~1 μM) (*see* **Note 6**).
7. Acquire one series of images.
8. Select regions from the selected cells for analysis throughout the course of the experiment.
9. Subtract background levels estimated from areas with no cells (*see* **Note 7**).
10. Within Metafluor, save the log file that will contain data (intensities and ratios) from each region during the course of the experiment, and more importantly, save the images so that one can reanalyze the experiment once complete (*see* **Note 8**).
11. Begin the experiment, plotting both the FRET ratio (CFP/FRET) and YFP intensity (as a control for photobleaching) in real time. Continue acquiring until a stable baseline FRET ratio is established.
12. Addition of PDBu to stimulate PKD activity is performed by first removing 0.5–1 ml of the imaging buffer from the imaging dish during the experiment, resuspending the drug into this volume, and then adding it back drop-wise to the dish. A final concentration of 200 nM PDBu is used to activate PKD.
13. Pharmacologic inhibition of PKD activity, and thus reversal of the DKAR response, can be observed by addition of the inhibitor Gö 6976 (using the same method as described in **step 12**) at a final concentration of 500 nM (*see* **Notes 9** and **10**).

3.3 Data Analysis

1. Data within the saved log file can be opened from within Excel. The file will contain time as the first column, followed by four columns from each region analyzed: CFP intensity, FRET intensity, YFP intensity, CFP/FRET ratio calculation.
2. The baseline FRET ratio varies from cell to cell, thus, in order to compare the relative magnitude of response from the kinase

activity reporter, the traces should be normalized to the average baseline FRET ratio from each region and then referenced about the time of drug addition. The averages of these normalized CFP/FRET ratios are then plotted with respect to time (Fig. 3).

4 Notes

1. Metafluor software drives the acquisition of the CFP, FRET and YFP channels as well as a real-time analysis of the user-defined FRET ratio (CFP/FRET for DKAR) during the course of the experiment. Thus, one can evaluate the progress of the experiment and acquire images until a steady baseline is reached before addition of drugs.
2. The Lambda 10-2 filter changer is driven through the imaging software; this changer controls the excitation, dichroic and emission filters individually.
3. 0.1 μg of NHERF-1 DNA is transfected as this results in NHERF-1 expression levels that properly localize to the apical membrane in MDCK cells (*see* Fig. 2a); transfection of too much DNA results in overexpression of NHERF-1, yielding significant levels of untargeted, non-localized NHERF.
4. DKAR is basally undergoing FRET (*see* Fig. 1); therefore, we opt to plot the CFP/FRET ratio rather than the canonical FRET/CFP ratio to visualize PKD signaling.
5. Each imaging setup should be calibrated to control against fluorophore photobleaching; that is, for our setup, we determined that we can acquire a series of images (CFP, FRET, YFP) up to every 7 s through the 40× objective and a 10 % neutral density filter with excitation exposure times of 200 ms for CFP and FRET and 100 ms for YFP with no observed bleaching.
6. We calibrated the brightness of the signal from the reporter such that its expression level is within the range of endogenous substrates (~1 μM) [5]. This is important as too high a concentration of exogenous reporter could buffer the endogenous enzymatic capacity of the kinase and thus alter the activity readout.
7. One can subtract background levels of the saved images for each channel during reanalysis of the images saved during the experiment.
8. Saving the images is critical; from these one can reanalyze the experiment by defining new regions or adjust for region movement if this was observed during the experiment.
9. Reversal of responses from kinase activity reporters using kinase-specific inhibitors (e.g., Gö 6976 for PKD) further corroborates that the response from the reporter is due to signaling by the kinase of interest.

10. A simple control for any experiment is to monitor FRET changes from the reporter construct in which the phosphoacceptor site (e.g., threonine in the example of DKAR) is mutated to a non-phosphorylatable residue (e.g., alanine). This reporter should not undergo a FRET change and thus serves as a negative control in experiments where there is concern that the reporter is physically relocalizing within the cell, or where the cell is changing morphology; such instances may appear as FRET changes as the local reporter concentration change impacts basal intermolecular FRET.

References

1. Toker A (2005) The biology and biochemistry of diacylglycerol signalling. Meeting on molecular advances in diacylglycerol signalling. EMBO Rep 6(4):310–314
2. Sanchez-Ruiloba L, Cabrera-Poch N, Rodriguez-Martinez M, Lopez-Menendez C, Jean-Mairet RM, Higuero AM, Iglesias T (2006) Protein kinase D intracellular localization and activity control kinase D-interacting substrate of 220-kDa traffic through a postsynaptic density-95/discs large/zonula occludens-1-binding motif. J Biol Chem 281(27): 18888–18900
3. Rozengurt E (2011) Protein kinase D signaling: multiple biological functions in health and disease. Physiology 26(1):23–33
4. Matthews SA, Rozengurt E, Cantrell D (1999) Characterization of serine 916 as an in vivo autophosphorylation site for protein kinase D/ Protein kinase Cmu. J Biol Chem 274(37): 26543–26549
5. Kunkel MT, Newton AC (2009) Spatiotemporal dynamics of kinase signaling visualized by targeted reporters. Curr Protoc Chem Biol 1(1):17–18
6. Sato M, Kawai Y, Umezawa Y (2007) Genetically encoded fluorescent indicators to visualize protein phosphorylation by extracellular signal-regulated kinase in single living cells. Anal Chem 79(6):2570–2575
7. Ting AY, Kain KH, Klemke RL, Tsien RY (2001) Genetically encoded fluorescent reporters of protein tyrosine kinase activities in living cells. Proc Natl Acad Sci U S A 98(26): 15003–15008
8. Zhang J, Ma Y, Taylor SS, Tsien RY (2001) Genetically encoded reporters of protein kinase A activity reveal impact of substrate tethering. Proc Natl Acad Sci U S A 98(26): 14997–15002
9. Zhang J, Hupfeld CJ, Taylor SS, Olefsky JM, Tsien RY (2005) Insulin disrupts beta-adrenergic signalling to protein kinase A in adipocytes. Nature 437(7058):569–573
10. Allen MD, Zhang J (2006) Subcellular dynamics of protein kinase A activity visualized by FRET-based reporters. Biochem Biophys Res Commun 348(2):716–721
11. Sasaki K, Sato M, Umezawa Y (2003) Fluorescent indicators for Akt/protein kinase B and dynamics of Akt activity visualized in living cells. J Biol Chem 278(33):30945–30951
12. Kunkel MT, Ni Q, Tsien RY, Zhang J, Newton AC (2005) Spatio-temporal dynamics of protein kinase B/Akt signaling revealed by a genetically encoded fluorescent reporter. J Biol Chem 280(7):5581–5587
13. Gao X, Zhang J (2008) Spatiotemporal analysis of differential Akt regulation in plasma membrane microdomains. Mol Biol Cell 19(10):4366–4373
14. Violin JD, Zhang J, Tsien RY, Newton AC (2003) A genetically encoded fluorescent reporter reveals oscillatory phosphorylation by protein kinase C. J Cell Biol 161(5):899–909
15. Kajimoto T, Sawamura S, Tohyama Y, Mori Y, Newton AC (2010) Protein kinase C {delta}-specific activity reporter reveals agonist-evoked nuclear activity controlled by Src family of kinases. J Biol Chem 285(53):41896–41910
16. Kunkel MT, Toker A, Tsien RY, Newton AC (2007) Calcium-dependent regulation of protein kinase D revealed by a genetically encoded kinase activity reporter. J Biol Chem 282(9): 6733–6742
17. Kunkel MT, Garcia EL, Kajimoto T, Hall RA, Newton AC (2009) The protein scaffold NHERF-1 controls the amplitude and duration of localized protein kinase D activity. J Biol Chem 284(36):24653–24661

Chapter 11

Using a Genetically Encoded FRET-Based Reporter to Visualize Calcineurin Phosphatase Activity in Living Cells

Sohum Mehta and Jin Zhang

Abstract

Calcineurin is an evolutionarily conserved, ubiquitously expressed protein phosphatase that serves as a major effector of Ca^{2+} signals, regulating diverse biological processes such as gene expression, tissue differentiation, immune responses, and neural plasticity. The following method describes how to monitor real-time calcineurin activity in cultured mammalian cells using a fluorescence resonance energy transfer (FRET)-based activity reporter.

Key words Calcineurin, Phosphatase activity, Biosensor, Calmodulin, NFAT, FRET, Live-cell imaging

1 Introduction

1.1 Calcineurin

The Ca^{2+}- and calmodulin (CaM)-dependent, serine/threonine protein phosphatase calcineurin (aka PP2B/PP3) is an evolutionarily conserved signaling enzyme found in virtually all eukaryotes [1]. Calcineurin is comprised of a stable heterodimer between a catalytic A subunit (CNA) and a regulatory B subunit (CNB). The CNA subunit includes an N-terminal catalytic domain, which shares homology with the conserved PPP family of serine/threonine phosphatases, as well as a C-terminal regulatory arm containing a CNB-binding domain, a CaM-binding domain, and an autoinhibitory domain. Calcineurin activation occurs in response to elevated Ca^{2+} levels; Ca^{2+} binds to CNB and Ca^{2+}/CaM binds to CNA, leading to conformational rearrangements that release autoinhibition [2]. A major effector of Ca^{2+} signaling, calcineurin regulates a number of critical biological processes including tissue differentiation, cardiac development, neuronal function, and immune activation [2].

Jin Zhang et al. (eds.), *Fluorescent Protein-Based Biosensors: Methods and Protocols*, Methods in Molecular Biology, vol. 1071, DOI 10.1007/978-1-62703-622-1_11,

The importance of calcineurin is further underscored by the involvement of dysregulated calcineurin signaling in a number of human pathologies, such as heart disease, diabetes, Alzheimer's disease, and Down syndrome [3–6].

1.2 A FRET-Based Reporter for Calcineurin Activity

The introduction of genetically encoded fluorescence resonance energy transfer (FRET)-based enzyme activity reporters markedly enhanced our ability to directly visualize the native, real-time dynamics of signaling enzymes with high spatiotemporal resolution in living cells [7, 8]. Typically, these reporters utilize a molecular switch, capable of undergoing a conformational change in response to a specific enzymatic activity, sandwiched between a pair of fluorescent proteins, such that the conformational change results in a change in FRET between the fluorescent protein pair. FRET is a photophysical process in which excitation of a donor fluorophore (e.g., a cyan fluorescent protein, CFP) results in non-radiative energy transfer to an acceptor fluorophore (e.g., a yellow fluorescent protein, YFP), followed by acceptor emission. Efficient transfer depends on several factors, such as spectral overlap between donor and acceptor. Given the significant overlap between the CFP emission spectrum and YFP excitation spectrum, and the minimal overlap between CFP excitation and YFP excitation, CFP and YFP are well suited to this purpose and are a commonly used FRET pair.

Energy transfer also depends on the relative proximity (e.g., <10 nm) and orientation of the fluorophores, rendering FRET highly sensitive to the conformation of the molecular switch. In the case of protein kinase activity reporters, this switch is typically comprised of a kinase-specific consensus phosphorylation motif fused to a phosphoamino acid-binding domain (PAABD). Phosphorylation of the substrate motif promotes binding by the PAABD, inducing a conformational change in the reporter. This modular design has been widely used to generate activity reporters specific to a variety of different protein kinases [9–15]. However, whereas kinase activity reporters have gained widespread use, the development of similar reporters for protein phosphatase activity has lagged. Unlike a kinase activity reporter, which responds to the addition of a phosphate group, a phosphatase activity reporter must already be phosphorylated in its basal state, so that it will respond to the phosphate group's removal. Therefore, one reason for the comparatively slow development of phosphatase activity reporters is the challenge of designing just such a "dephosphorylation-competent" molecular switch.

In designing a FRET-based calcineurin activity reporter (CaNAR), we chose to address this problem by utilizing a well-studied substrate of calcineurin, nuclear factor of activated T-cells (NFAT) [16]. NFAT proteins are a family of transcription factors that are ubiquitously co-expressed alongside calcineurin in vertebrates, serving as the main molecular target of calcineurin signaling

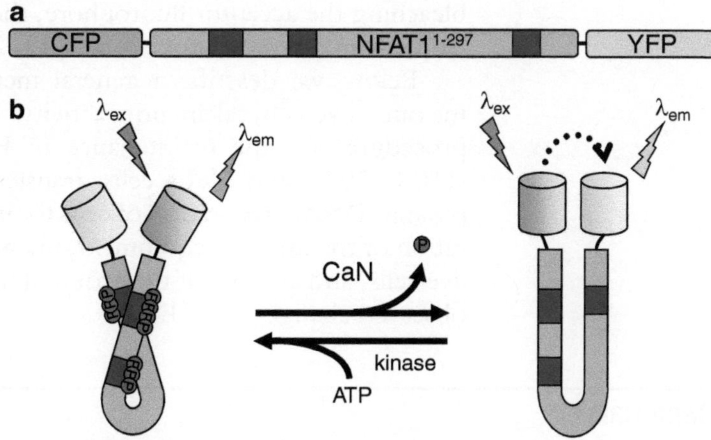

Fig. 1 The design of CaNAR. (**a**) CaNAR features the N-terminal 297 amino acids of NFAT1 sandwiched between the FRET donor CFP and the FRET acceptor YFP. Shaded boxes indicate the sites where NFAT is constitutively phosphorylated. (**b**) Dephosphorylation of the NFAT N-terminal domain by CaN results in a conformational change that exposes an NLS. In CaNAR, this serves as a molecular switch that induces a change in the proximity and relative orientation of CFP and YFP, leading to a FRET change (*dotted arrow*). This is reversed upon rephosphorylation of the reporter by cellular kinases

in a multitude of cellular processes [17–19]. In CaNAR, the N-terminal 297 amino acids from NFAT1 are sandwiched between CFP and YPF (Fig. 1a). The N-terminal portion of each NFAT isoform functions as a regulatory domain that is constitutively phosphorylated at multiple serine-rich sites by cytosolic kinases [18, 20], making CaNAR "competent" for dephosphorylation without additional manipulation. The phosphorylated residues are thought to interact with positive charges found in the nuclear localization signal (NLS) within the NFAT regulatory domain, with dephosphorylation by calcineurin inducing a conformational change that exposes the NLS and promotes transit of NFAT into the nucleus [20, 21]. This conformational change forms the basis of the calcineurin-dependent molecular switch found in CaNAR, resulting in a FRET increase in response to calcineurin activity (Fig. 1b).

CaNAR can be used to directly visualize calcineurin activity dynamics in live cells by monitoring FRET changes in response to various stimuli capable of eliciting increases in the cytosolic Ca^{2+} concentration (*see* **Note 1**). Using epifluorescence microscopy, FRET can be observed as a decrease in donor (CFP) fluorescence intensity combined with an increase in acceptor (YFP) fluorescence intensity, often expressed numerically as an acceptor-to-donor emission ratio. Since CaNAR is a unimolecular reporter (i.e., donor and acceptor are present in fixed amounts), the value of the emission ratio serves as a convenient readout of FRET [22]. It is also possible to quantitatively determine the FRET efficiency by bleaching the acceptor fluorophore, thereby abolishing FRET and dequenching donor fluorescence [22, 23].

Below, we describe a general method for using CaNAR to measure live-cell calcineurin activity, including specific, detailed procedures for the maintenance of Human Embryonic Kidney (HEK) 293T and HeLa cells, transfection of cells with CaNAR plasmid DNA, preparation of cells for imaging experiments, preparation of the imaging equipment, imaging of calcineurin activity in live cells, and analysis of the acquired imaging data to quantify any observed changes in FRET.

2 Materials

2.1 Cell Culture and Transfection

1. Cell lines: Human Embryonic Kidney—SV40 T Antigen (HEK 293T) and HeLa (American Type Culture Collection, Manassas, VA).
2. Dulbecco's Phosphate Buffered Saline—without Mg^{2+} and Ca^{2+} (DPBS, Gibco).
3. 35-mm glass-bottom imaging dishes (MatTEK, Ashland, MA).
4. Dulbecco's Modified Eagle's Medium (DMEM, Gibco/BRL, Bethesda, MD) supplemented with 10 % fetal bovine serum (FBS, Sigma) and 1 % penicillin-streptomycin (Sigma-Aldrich) (DMEM-HEK293T) for use with both HEK293T and HeLa cells (*see* **Note 2**).
5. Solution of trypsin (0.05 % for HEK 293T, 0.25 % for HeLa) and ethylenediamine tetraacetic acid (EDTA, 0.53 mM) (Invitrogen, Carlsbad, CA).
6. Lipofectamine 2000 (Invitrogen).
7. OPTI-MEM I Reduced Serum Medium (Opti-MEM, Gibco).
8. CaNAR plasmid DNA.

2.2 Preparing Epifluorescence Microscope

1. All of the experiments described below are performed on an Axiovert 200 M inverted microscope using a 40×/1.3NA oil-immersion objective lens equipped with an Aqua Stop to prevent liquid from running down the objective (Zeiss, Thornwood, NY). Images are captured using a MicroMAX BFT512 cooled charge-coupled device camera (Roper Scientific, Trenton, NJ).
2. Xenon lamp: XBO 75 W (Zeiss).
3. Neutral density filters 0.6 and 0.3 (Chroma Technology, Bellows Falls, VT).
4. Filter sets for individual channels:

FRET—420DF20 excitation filter, 450DRLP dichroic mirror, 535DF25 emission filter.

CFP—420DF20 excitation filter, 450DRLP dichroic mirror, 475DF40 emission filter.

YFP—495DF10 excitation filter, 515DRLP dichroic mirror, 535DF25 emission filter.

YFP photobleaching—525DF40 excitation filter, 560DRLP dichroic mirror (All from Chroma Technology).

A Lambda 10-2 filter changer (Sutter Instruments, Novato, CA) alternates the filters being used.

5. Immersion oil "Immersol" 518F fluorescence free (Zeiss).
6. METAFLUOR 7.7 software (Molecular Devices, Sunnvale, CA) (*see* **Note 3**).

2.3 Prepare Cells for Imaging

1. Hanks' Balanced Salt Solution for Imaging (HBSS*): 10× Hanks' Balanced Salt Solution (Gibco), 20 mM HEPES (Invitrogen), 2.0 g/L D-glucose (Sigma). Adjust pH to 7.4, then filter sterilize using a 0.22 μm filter (*see* **Note 4**). Store a 50 mL aliquot at room temperature in the microscope room and store the remainder at 4 °C (*see* **Note 5**).

2.4 Acquiring Images and Data

1. Ionomycin (iono, Calbiochem) and thapsigargin (TG, Sigma) are both dissolved to 1 mM in dimethyl sulfoxide (DMSO) and stored at −20 °C.

2.5 Analyzing Images and Data

1. Spreadsheet application (e.g., Microsoft Office Excel).

3 Methods

3.1 Cell Culture and Transfection

1. The cells are maintained in T-25 cm^2 flasks in a humidified 37 °C incubator with a 5 % CO_2 atmosphere. Cells should be passaged whenever they reach 70–80 % confluency (every 2–3 days) into flasks for maintenance or 35-mm dishes for imaging.
2. To passage cells, aspirate culture media from the flask and wash cells gently with 2 mL DPBS. Add 300 μL of trypsin/EDTA, gently rocking the flask from side to side to disperse the solution, and let sit 2–5 min (*see* **Note 6**). Add 4.7 mL of fresh media into the flask and mix (*see* **Note 7**). For imaging, perform a 1:10 split of cells into the 35 mm glass-bottom dishes. Cell should reach 60–70 % confluence after approximately 24 h (*see* **Note 8**). Transfect the cells at this confluence.
3. For each 35-mm dish being transfected, prepare 2 separate microcentrifuge tubes. Tube 1 contains 1 μg CaNAR plasmid

DNA and 50 μL Opti-MEM. Tube 2 contains 2 μL Lipofectamine 2000 and 50 μL Opti-MEM. Allow tubes to incubate at room temperature for 5 min, then add contents from Tube 1 drop-wise into Tube 2 and mix well with pipet (*see* **Note 9**). Incubate this solution at room temperature for 20 min.

4. Gently add the CaNAR transfection solution to the cells drop-wise, and carefully rock the dish back and forth to evenly disperse the solution. Incubate at 37 °C with 5 % CO_2 for 48 h (*see* **Note 10**).

3.2 Preparing Epifluorescence Microscope

1. Turn on lamp, microscope, filter changer, camera, and computer. Load the METAFLUOR 7.7 application plus an appropriate protocol for acquiring a time series of sets of images for the FRET, CFP, and YFP channels (*see* **Note 11**). Confirm that all of the appropriate filters are in place.
2. Set the excitation exposure times for the FRET, CFP, and YFP channels to 500, 500, and 50 ms, respectively (*see* **Note 12**). In addition, select the time interval between each set of acquisitions. This is typically 30 s, but may be any number between 10 and 120 s.
3. Apply a small drop of immersion oil directly onto the objective. Make sure not to use an excess amount (i.e., 1 drop from the attached applicator). Avoid touching the objective with the applicator tip.

3.3 Preparing Cells for Imaging

1. Aspirate off media from transfected cells in imaging dish and wash twice with 1 mL HBSS*.
2. Gently add 1–2 mL HBSS* to the imaging dish, while holding the dish on a slight angle. Slowly return the dish to a level position and place securely on microscope stage (*see* **Note 13**).
3. Raise the objective until the immersion oil comes fully into contact with the glass cover-slip and then bring the cells into focus while viewing through the eyepiece.
4. In the dark, use either the FRET or CFP channel to select cells with good morphology and good CaNAR expression, meaning intermediate emission intensity and uniformly distributed fluorescence (*see* **Note 14**).

3.4 Acquiring Images and Data

1. Select several regions of interest to follow during the course of the experiment (*see* **Note 15**). A background region consisting of an untransfected cell must also be selected to correct for cell autofluorescence and other background fluorescence.
2. Record a baseline by acquiring 3–5 min of data (all three channels) from unstimulated cells. Remove ~ 300 μL of HBSS* from the imaging dish and mix with a 1–2 μL aliquot

of 1 mM iono or TG in a 1.5 mL tube, then gently pipet this solution back into the imaging dish. The final concentration of drug should be 1 μM. Be sure to record the time of drug addition. The yellow-to-cyan emission ratio (FRET channel intensity/CFP channel intensity) should increase, indicating a change in calcineurin activity (*see* **Note 16**).

3. At the end of the experiment, remove all neutral density filters, use the YFP photobleaching excitation filter, and then excite for 5 min. This should sufficiently photobleach YFP, although it is important to verify this by acquiring the YFP channel. The acquired data can be used to calculate absolute FRET efficiency with the following formula:

$$\text{FRET Efficiency} = 1 - \frac{\text{CFP Emission}_{(\text{before YFP photobleaching})}}{\text{CFP Emission}_{(\text{after acceptor photobleaching})}}.$$

3.5 Analyzing Images and Data

1. Use METAFLUOR 7.7 to generate pseudo-colored images for each acquisition, where pseudo-coloring is used to indicate the yellow-to-cyan emission ratio (FRET channel intensity/CFP channel intensity) (*see* **Note 17**). These images can be strung together as a movie, or a subset can be used to illustrate the observed real-time FRET changes. An example is shown in Fig. 2.
2. In a spreadsheet application, calculate emission ratio at each time point using the logged emission intensity data and the following formula:

$$\text{Yellow-to-Cyan Emission Ratio} = \frac{\text{FRET channel Emission Intensity} - \text{FRET channel Emission Intensity of Background}}{\text{CFP channel Emission Intensity} - \text{CFP channel Emission Intensity of Background}}.$$

3. Plot the ratio time course (yellow-to-cyan emission ratio versus time).

4 Notes

1. It is important to verify effective drug concentrations and other conditions affecting drug function. For example, when using receptor agonists to elevate cytosolic Ca^{2+}, be sure to confirm receptor expression in the cell line being used. Western blots using antibodies against the receptor of interest are helpful in this regard. Additionally, effective drug concentrations can be verified by monitoring the phosphorylation of CaNAR via gel mobility shift on western blots employing an anti-GFP antibody (eBioscience, San Diego, CA) [16].

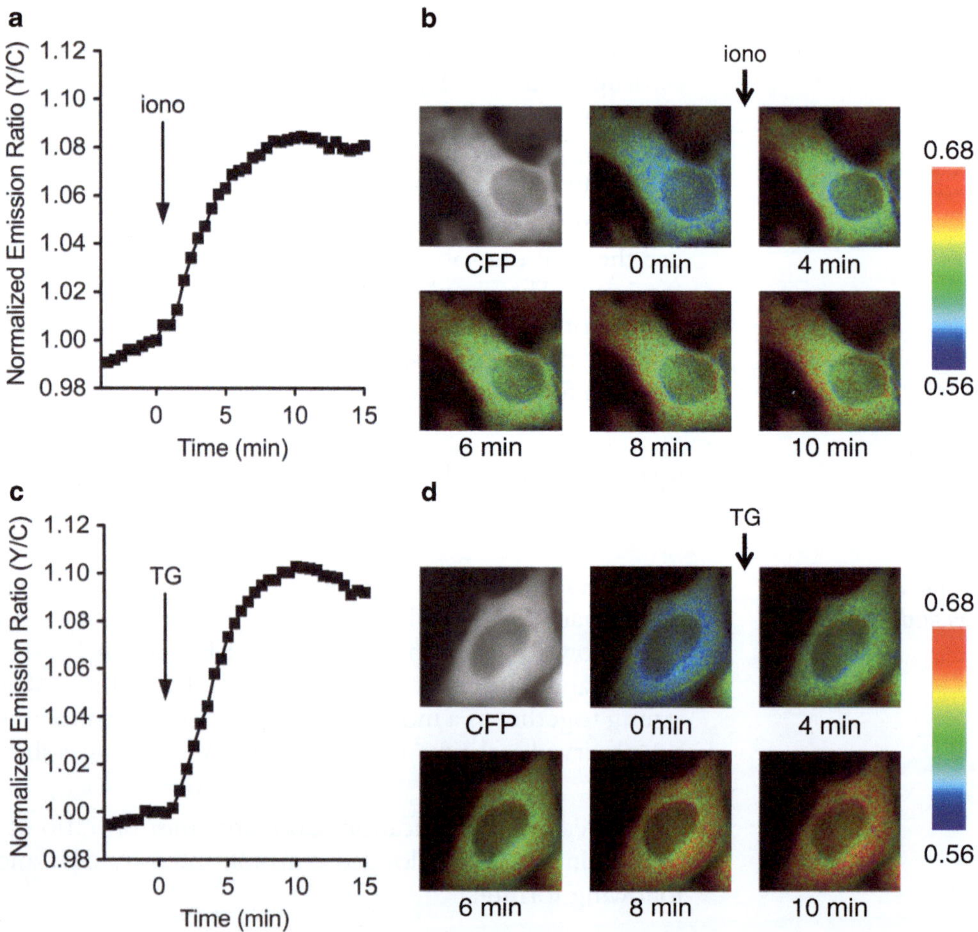

Fig. 2 CaNAR response in live cells. HeLa cells were transiently transfected with CaNAR plasmid DNA and imaged after 48 h. The cells were treated with (**a**, **b**) 1 μM ionomycin, a Ca^{2+} ionophore, or (**c**, **d**) 1 μM thapsigargin, a SERCA pump inhibitor. (**a**, **c**) Plotting the yellow-to-cyan emission ratio versus time shows the change in CaN activity upon stimulation. (**b**, **d**) Pseudo-colored images showing the CaNAR response

2. All solutions should be made under sterile conditions in a tissue culture hood and cell culture media should be warmed to 37 °C before using with cells.

3. Metafluor requires a PC running Microsoft Windows XP, Windows Vista, or Windows 7. Other imaging software with similar or equivalent functionality is also suitable for these experiments.

4. All solutions should be prepared with water that has an 18.2 MΩ-cm resistivity unless otherwise noted.

5. Imaging is typically performed at room temperature. However, FRET responses in certain cells lines are enhanced by imaging at 37 °C using an optional Heatable Insert P for Scanning Stage and Mechanical Stage (Zeiss). When using this setup, an aliquot of HBSS* should be preheated to 37 °C prior to imaging.

6. Be sure the cells are fully detached before continuing. Gently swaying the flask side to side should help.
7. Gently pipet cells up and down to break up cell clumps, while being careful not to over-pipet and damage the cells.
8. This protocol can be adapted for other cell lines by following appropriate cell culture and transfection guidelines for the cell line of choice. Additionally, splitting times may vary, so it is important to verify doubling times for each cell line used.
9. Be sure to mix gently. Do not vortex the solution.
10. CaNAR generally requires longer expression times compared to other FRET-based reporters [24], possibly owing to its size or need to be pre-phosphorylated by cells. Optimal CaNAR expression time should be determined empirically for each cell type used.
11. The FRET channel logs sensitized YFP emission intensity upon CFP excitation, the CFP channel logs direct CFP emission intensity upon CFP excitation, and the YFP channel logs direct YFP emission intensity upon YFP excitation. The YFP channel serves to control for YFP photobleaching and does not factor into emission ratio calculations.
12. These exposure times are a good starting point for most experiments. Strict adherence to these numbers is not required, and exposure times can be adjusted as needed, based on the observed brightness of the reporter.
13. Securing the dish to the stage is essential to minimize slight movement of the dish that may occur while imaging.
14. This step focuses on the key criteria for proper cell selection. First, cell morphology should be verified before starting an experiment, as healthy cells are required for successful imaging experiments. For instance, when imaging HEK 293 cells, select cells that are spread out and lying flat rather than balled-up and rounded, as the latter could indicate unhealthy cells. Second, the fluorescence intensity of CaNAR should be closely monitored, though a recommended range cannot be given as the intensity values will vary between microscope setups. However, cells with a moderate intensity level are typically used. Cells with very dim fluorescence intensities will have a low signal-to-noise ratio, rendering changes in FRET difficult to visualize, whereas cells with very high fluorescence intensities may have perturbed endogenous signaling pathways because of excessive expression of CaNAR.
15. The selected regions of interest will need to remain in the same cellular region throughout the time series, and may be manually readjusted should the cells move. Alternatively, cell tracking software (e.g., Imaris Track) can be used to overcome this problem.

16. It is important to confirm that changes in FRET from CaNAR are specifically due to calcineurin activity. However, as CaNAR is highly phosphorylated, generating a CaNAR variant that cannot be dephosphorylated is impractical. Rather, specificity can be determined by pretreating cells with a specific calcineurin inhibitor, such as cyclosporin A, prior to stimulation [16].
17. If the software being used lacks this feature, an image processing application (e.g., ImageJ) can be used to create pseudo-colored ratiometric images from the raw emission intensity images from the individual channels.

References

1. Hilioti Z, Cunningham KW (2004) Calcineurin: roles of the Ca^{2+}/calmodulin-dependent protein phosphatase in diverse eukaryotes. Top Curr Genet 5:73–90
2. Rusnak F, Mertz P (2000) Calcineurin: form and function. Physiol Rev 80(4):1483–1521
3. Harris CD, Ermak G, Davies KJ (2005) Multiple roles of the DSCR1 (Adapt78 or RCAN1) gene and its protein product calcipressin 1 (or RCAN1) in disease. Cell Mol Life Sci 62(21):2477–2486. doi:10.1007/s00018-005-5085-4
4. Heineke J, Molkentin JD (2006) Regulation of cardiac hypertrophy by intracellular signalling pathways. Nat Rev Mol Cell Biol 7(8): 589–600. doi:10.1038/nrm1983
5. Heit JJ (2007) Calcineurin/NFAT signaling in the beta-cell: from diabetes to new therapeutics. Bioessays 29(10):1011–1021. doi:10.1002/bies.20644
6. Xie CW (2004) Calcium-regulated signaling pathways: role in amyloid beta-induced synaptic dysfunction. Neuromolecular Med 6(1):53–64. doi:10.1385/NMM:6:1:053
7. Mehta S, Zhang J (2011) Reporting from the field: genetically encoded fluorescent reporters uncover signaling dynamics in living biological systems. Annu Rev Biochem 80: 375–401. doi:10.1146/annurev-biochem-060409-093259
8. Newman RH, Fosbrink MD, Zhang J (2011) Genetically encodable fluorescent biosensors for tracking signaling dynamics in living cells. Chem Rev 111(5):3614–3666. doi:10.1021/cr100002u
9. Fosbrink M, Aye-Han NN, Cheong R, Levchenko A, Zhang J (2010) Visualization of JNK activity dynamics with a genetically encoded fluorescent biosensor. Proc Natl Acad Sci U S A 107(12):5459–5464. doi:10.1073/pnas.0909671107
10. Fuller BG, Lampson MA, Foley EA, Rosasco-Nitcher S, Le KV, Tobelmann P, Brautigan DL, Stukenberg PT, Kapoor TM (2008) Midzone activation of aurora B in anaphase produces an intracellular phosphorylation gradient. Nature 453(7198):1132–1136. doi:10.1038/nature06923
11. Gao X, Zhang J (2008) Spatiotemporal analysis of differential Akt regulation in plasma membrane microdomains. Mol Biol Cell 19(10):4366–4373. doi:10.1091/mbc.E08-05-0449
12. Kunkel MT, Toker A, Tsien RY, Newton AC (2007) Calcium-dependent regulation of protein kinase D revealed by a genetically encoded kinase activity reporter. J Biol Chem 282(9):6733–6742. doi:10.1074/jbc.M608086200
13. Violin JD, Zhang J, Tsien RY, Newton AC (2003) A genetically encoded fluorescent reporter reveals oscillatory phosphorylation by protein kinase C. J Cell Biol 161(5):899–909. doi:10.1083/jcb.200302125
14. Wang Y, Botvinick EL, Zhao Y, Berns MW, Usami S, Tsien RY, Chien S (2005) Visualizing the mechanical activation of Src. Nature 434(7036):1040–1045. doi:10.1038/nature03469
15. Zhang J, Hupfeld CJ, Taylor SS, Olefsky JM, Tsien RY (2005) Insulin disrupts beta-adrenergic signalling to protein kinase A in adipocytes. Nature 437(7058):569–573. doi:10.1038/nature04140
16. Newman RH, Zhang J (2008) Visualization of phosphatase activity in living cells with a FRET-based calcineurin activity sensor. Mol Biosyst 4(6):496–501. doi:10.1039/b720034j
17. Crabtree GR, Olson EN (2002) NFAT signaling: choreographing the social lives of cells. Cell 109(Suppl):S67–S79

18. Hogan PG, Chen L, Nardone J, Rao A (2003) Transcriptional regulation by calcium, calcineurin, and NFAT. Genes Dev 17(18): 2205–2232
19. Horsley V, Pavlath GK (2002) NFAT: ubiquitous regulator of cell differentiation and adaptation. J Cell Biol 156(5):771–774
20. Okamura H, Aramburu J, Garcia-Rodriguez C, Viola JP, Raghavan A, Tahiliani M, Zhang X, Qin J, Hogan PG, Rao A (2000) Concerted dephosphorylation of the transcription factor NFAT1 induces a conformational switch that regulates transcriptional activity. Mol Cell 6(3):539–550
21. Porter CM, Havens MA, Clipstone NA (2000) Identification of amino acid residues and protein kinases involved in the regulation of NFATc subcellular localization. J Biol Chem 275(5):3543–3551
22. Ananthanarayanan B, Ni Q, Zhang J (2008) Chapter 2: molecular sensors based on fluorescence resonance energy transfer to visualize cellular dynamics. Methods Cell Biol 89:37–57. doi:10.1016/S0091-679X(08)00602-X
23. Miyawaki A, Tsien RY (2000) Monitoring protein conformations and interactions by fluorescence resonance energy transfer between mutants of green fluorescent protein. Methods Enzymol 327:472–500
24. Depry C, Zhang J (2011) Using FRET-based reporters to visualize subcellular dynamics of protein kinase A activity. Methods Mol Biol 756:285–294. doi:10.1007/978-1-61779-160-4_16

Chapter 12

Genetically Encoded FRET Indicators for Live-Cell Imaging of Histone Acetylation

Kazuki Sasaki and Minoru Yoshida

Abstract

Histone acetylation is dynamically and reversibly controlled by histone acetyltransferases and deacetylases during cellular events such as cell division and differentiation. However, the dynamics of histone modifications in living cells are poorly understood because of the lack of experimental tools to monitor them in a real-time fashion. Herein, we introduce Förster/fluorescence resonance energy transfer (FRET)-based indicators to visualize acetylation of histone H4, and describe a protocol for live-cell imaging with high spatiotemporal resolution.

Key words Histone acetylation, Bromodomain, Förster/fluorescence resonance energy transfer (FRET), Live-cell imaging, Fluorescent protein

1 Introduction

Acetylation of lysine residues (K) on the N-terminal tails of histones is one of the most well-known post-translational histone modifications. The acetylation of histones contributes to interaction with chromatin-associated proteins containing bromodomains. The bromodomain is an evolutionarily conserved acetylated lysine-binding domain that is present in many nuclear proteins, including BET family members such as BRDT, BRD2, and BRD4; histone acetyltransferases such as TAFII250, PCAF, and GCN5; and ATP-dependent chromatin-remodeling factors [1]. The diversity of bromodomains confers unique acetylation-site specificity on the proteins that contain them, which is useful as an acetylated histone-recognizing tool having acetylation-site specificity.

FRET is a process in which energy shifts from an initially excited donor fluorophore to an acceptor fluorophore only when the donor's emission spectrum and the acceptor's excitation spectrum overlap, and the molecules are close together in an appropriate orientation. FRET is applicable as a general imaging technique for studying the dynamics of protein modification in living cells [2].

Jin Zhang et al. (eds.), *Fluorescent Protein-Based Biosensors: Methods and Protocols*, Methods in Molecular Biology, vol. 1071, DOI 10.1007/978-1-62703-622-1_12, © Springer Science+Business Media, LLC 2014

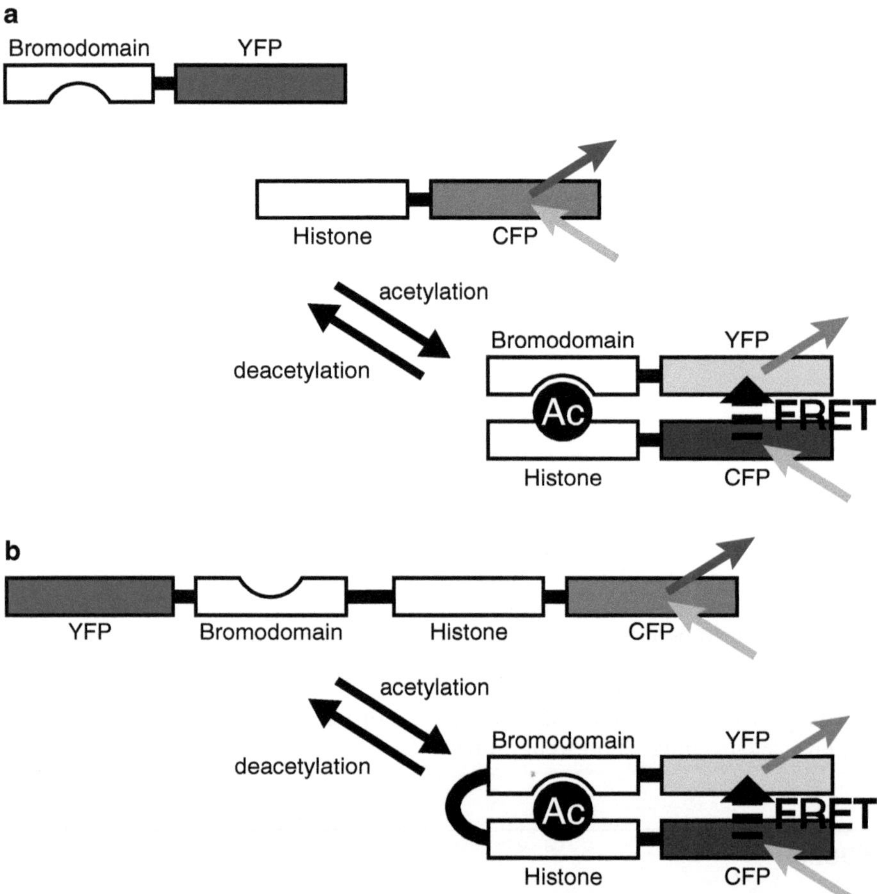

Fig. 1 Schematic representation of a bimolecular (intermolecular) FRET indicator (**a**) and a unimolecular (intramolecular) FRET indicator (**b**) for study of histone acetylation. Bimolecular FRET has an uncertain stoichiometry of CFP and Venus, since it is difficult to express two proteins at equal levels inside cells. On the other hand, in the case of the unimolecular FRET indicator, it is difficult to predict the direction of the acetylation-dependent conformational change. Histac, which is a FRET-based indicator for acetylation of histone H4 at K5 and K8 [3], behaves in the opposite manner shown in (**b**), i.e., FRET decreases in an acetylation-dependent manner

Discovery and cloning of green fluorescent protein (GFP) from *Aequorea victoria* and the development of derivative color variants enables us to exploit genetically encoded FRET indicators. Two different approaches have been taken in the development of FRET-based indicators: bimolecular (intermolecular) (Fig. 1a) and unimolecular (intramolecular) FRET (Fig. 1b). Bimolecular FRET and unimolecular FRET rely on protein–protein interaction and conformational change, respectively. Interpretation of bimolecular FRET is more complicated, because it is difficult to express two proteins at equal levels; therefore, there is often an uncertain stoichiometry of the donor and acceptor fluorophores. In the case of unimolecular FRET, the two fluorophores are linked in a single peptide, and are therefore always present in an equimolar ratio. Furthermore, the intramolecular interaction enhances selectivity and affinity. Thus, this protocol will focus on unimolecular FRET. Many fluorescent protein pairs for FRET-based indicators use color variants of *Aequorea* GFP or GFP-like fluorescent proteins derived from other organisms; however, the most commonly used fluorescent protein pair employs a cyan fluorescent protein (CFP) variant as a donor and a yellow fluorescent protein (YFP) variant as an acceptor. Coupled evolution of the FRET dynamic range of the CFP–YFP pair has been achieved by random mutagenesis and screening of both fluorescent proteins. The evolved pair, CyPet and YPet, has a sixfold larger dynamic range in FRET, relative to the parent CFP–YFP pair [4]. Circular permutation of YFP (cpYFP), in which the original termini are connected by a short linker and new termini are introduced at different positions in the protein, allowed optimization of the relative orientation between the two fluorophores. Yellow Cameleon 3.60 (YC3.60), which uses a circularly permuted version of the YFP variant, Venus, that has a new N-terminus at Asp173 (cp173Venus), exhibits a 5.6-fold FRET dynamic range relative to the parent YC [5]. To eliminate the relative orientation-dependent FRET response, and to create a situation in which FRET depends primarily on the distance between the two fluorophores, a long and flexible linker consisting of $(SAGG)_{29}$, where 29 indicates the number of repeats, was employed; this construct is named the "EV linker" [6]. Using the EV linker, FRET indicators for PKA, ERK, JNK, EGFR/Abl, Ras, and Rac1, which were previously reported by the same group, were improved.

It remains unclear, however, whether the distance or relative orientation of two fluorophores is primarily responsible for improvement of the FRET dynamic range. Therefore, development of FRET indicators requires optimization by trial and error. Here we described the design and application of the histone acetylation reporter, Histac, to highlight some of the important considerations when using genetically encodable FRET indicators to study biological processes in living cells.

2 Materials

2.1 Transfection Materials

1. Dulbecco's Modified Eagle's Medium (DMEM).
2. Fetal Bovine Serum (FBS), Qualified, USDA Approved Regions (Gibco, Life technologies, Carlsbad, CA) (*see* **Note 1**).
3. Penicillin–Streptomycin, liquid.
4. 0.05 % Trypsin–EDTA (1×), phenol red.
5. Phosphate-Buffered Saline (PBS): 137 mM NaCl, 2.7 mM KCl, 10 mM Na_2HPO_4, 1.8 mM KH_2PO_4, pH 7.4.

6. FuGENE HD (Roche Applied Science, Indianapolis, IN).
7. Opti-MEM I Reduced-Serum Medium, liquid (Gibco, Life technologies, Carlsbad, CA).

2.2 Western Blot Analysis Materials

1. Trichostatin A.
2. 2× sample buffer: 125 mM Tris–HCl (pH 6.8), 4 % (w/v) SDS, 0.004 % (w/v) bromophenol blue, 10 % (w/v) sucrose, and 10 % (v/v) 2-mercaptoethanol (*see* **Note 2**).
3. 30 % (w/v) acrylamide solution.
4. Ammonium persulfate (APS).
5. *N*,*N*,*N*,*N′*-tetramethyl-ethylenediamine (TEMED).
6. Full-Range Rainbow Marker.
7. Immobilon-P PVDF transfer membrane (Millipore, Temecula, CA).
8. Tris-buffered saline with Tween (1× TBS-T): 10 mM Tris–HCl, pH 7.5, 100 mM NaCl, and 0.1 % (v/v) Tween-20.
9. Albumin, from bovine serum (BSA).
10. Histone H4 antibody (Cell Signaling Technology, Danvers, MA).
11. Living Colors Full-Length A. v. polyclonal antibody (anti-GFP antibody) (Clontech, Takara, Otsu, Shiga, Japan).
12. Anti-acetyl-Histone H4 (Lys5) (Millipore, Temecula, CA).
13. Anti-acetyl-Histone H4 (Lys8) (Millipore, Temecula, CA).
14. ECL-anti rabbit IgG, horseradish peroxidase-linked whole antibody (from donkey) (GE Healthcare, Little Chalfont, Buckinghamshire, UK).
15. ECL-anti mouse IgG, horseradish peroxidase-linked whole antibody (from sheep) (GE Healthcare, Little Chalfont, Buckinghamshire, UK).
16. ECL plus Western blotting detection reagents (GE Healthcare, Little Chalfont, Buckinghamshire, UK).

2.3 Imaging in Living Cells Materials

1. 35-mm Glass Base Dish (Glass 12 ϕ) (Asahi Glass, Tokyo, Japan).
2. DMEM, High Glucose, HEPES, no Phenol Red.

2.4 Instrumentation

1. Luminescent image analyzer: LAS-3000 (Fujifilm, Tokyo, Japan).
2. Inverted microscope: Olympus IX81-ZDC (*see* **Note 3**) equipped with xenon lamp (*see* **Note 4**), charge coupled device (CCD) camera (*see* **Note 5**), excitation and emission filter wheel, and a temperature-controlled chamber.
3. Illumination system: AH2-RX-T (150 w/75 w xenon burner and power supply) (Olympus, Tokyo, Japan).
4. Xenon short arc lamp: UXL-75XB (Ushio, Tokyo, Japan).

5. CCD camera: UIC-QE cool charged-coupled device (Molecular Device, Downingtown, PA).
6. Optical filter changer: Lambda 10-2 (Shutter Instrument, Novato, CA).
7. Excitation filter: 440AF21 (Omega Optical, Brattleboro, VT).
8. Emission filter: 480AF30 for CFP, 535AF26 for Venus (Omega Optical, Brattleboro, VT).
9. Dichroic mirror: 455DRLP (Omega Optical, Brattleboro, VT).
10. Neutral density (ND) filter (Olympus, Tokyo, Japan).
11. 100× oil immersion objective lens: UPLSAPO100xO Numerical Aperture (N. A.) 1.40 (Olympus, Tokyo, Japan).
12. 40× oil immersion objective lens: UPLSAPO40xO N. A. 0.95 (Olympus, Tokyo, Japan).
13. Stage chamber: MIU-IBC-IF (*see* **Note 6**) (Olympus, Tokyo, Japan).
14. Peristaltic pump: MP-1000 (Tokyo Rikakikai, Tokyo, Japan).
15. Fluorescence ratio imaging and analyzing software: Meta Imaging Series Version 7.7 (Molecular Devices, Downingtown, PA).

3 Methods

3.1 Design and DNA Preparation

FRET-based acetylation indicators are typically tandem fusion proteins consisting of a histone, a flexible linker, an acetylated histone-recognition domain, and the two different-colored fluorescent proteins, CFP and Venus [3]. Acetylation-site specificity of the indicator depends on the selective recognition potency of the bromodomain [1, 7, 8]. For example, when two bromodomains of BRDT [9], which bind to acetylated histone H4 at K5 and K8 [10], are utilized as the acetylation recognition domain of Histac, it allows imaging of the acetylation of histone H4 at K5 and K8 [3]. Using bromodomains of Brd2, which binds to acetylated histone H4 at K12 [8, 11], Histac-K12 can visualize the acetylation of histone H4 at K12 [12].

Unfortunately, it is almost impossible to predict in advance the appropriate position of two fluorophores for FRET. Thus, in practice, FRET-based indicators are improved through a process of trial and error. The fluorescent protein pair, the length of linker, the recognition domain, and the order of the domains need to be selected for optimization of FRET efficiency [13]. Constructed cDNAs expressing acetylation indicators are introduced into the appropriate expression vectors using enzyme sites, and the plasmids are then purified.

3.2 Transfection

To introduce the expression vector containing Histac into the cells, lipofection is usually used. Commercially available transfection reagents such as FuGENE and Lipofectamine should be used according to the manufacturers' instruction. Below is a typical transfection protocol using FuGENE HD for transfection into Cos7 cells in 35-mm polystyrene dish (*see* **Note** 7).

1. Two days before transfection, plate cells in 35-mm dish. Cos7 cells should be 80 % confluent at the time of transfection.
2. Change to 2 mL of medium without antibiotics.
3. Dilute 2 μg of DNA containing Histac in 400 μL of Opti-MEM and mix gently.
4. Add 5 μL of FuGENE HD to the diluted DNA mixture.
5. Mix gently and incubate for 15 min at room temperature.
6. Add the DNA–FuGENE HD complexes to the Cos7 cells (prepared in Subheading 2).
7. Rock the dish gently and incubate for 8 h or overnight at 37 °C in a 5 % CO_2 incubator.
8. Replace with 2 mL of growth medium and incubate for an additional 2–24 h at 37 °C in a 5 % CO_2 incubator.
9. Check for low toxicity to cells and nuclear localization of Histac (*see* **Note 8**).

3.3 Western Blot Analysis

To confirm that histone H4 in Histac is actually acetylated, we use Western blot using anti-acetylated histone H4 antibody.

1. One or two days before transfection, plate cells in a 35-mm dish.
2. Transfect cells as described above (*see* Subheading 3.2).
3. Treat Cos7 cells expressing Histac with a final concentration of 1 μM TSA (or 0.1 % ethanol as a control) for 3 h at 37 °C in a 5 % CO_2 incubator.
4. Remove the medium from the culture dish and wash with 2 mL of cold PBS.
5. Remove PBS, add 50 μL of 2× sample buffer, scrape the cells from the dishes and harvest into microtubes.
6. Boil the lysates for 10 min.
7. Perform electrophoresis (*see* **Note 9**) and transfer the proteins to a PVDF membrane (*see* **Note 10**) according to standard techniques.
8. Block the membrane with TBST supplemented with 2 % BSA for 1 h on an orbital shaker.
9. Add anti-acetylated histone H4 antibodies (1:1,000 dilution) or anti-GFP antibody (1:1,000 dilution) and incubate for 2 h at room temperature or overnight at 4 °C on an orbital shaker.
10. Rinse the membrane three times with TBST for 10 min each.

11. Incubate the membrane in 1:5,000 dilution of HRP-conjugated anti-rabbit antibody in TBST supplemented with 2 % BSA for 1 h at room temperature on an orbital shaker.
12. Rinse the membrane three times with TBST for 10 min each.
13. Mix ECL plus detection solutions A and B in a ratio of 40:1.
14. Cover the membrane with the mixed reagent.
15. Detect chemiluminescence from HRP conjugated to secondary antibodies using a luminescent image analyzer, LAS-3000.

3.4 Imaging in Living Cells

1. Two days before transfection, plate cells in the glass bottom region of the dish. Cos7 cells should be 80 % confluent at the time of transfection (*see* **Note 11**).
2. Change to medium without antibiotics; 200 μL of culture medium is used to cover only the glass bottom region of the dish.
3. Transfect 0.2 μg of DNA containing Histac using 0.5 μL of FuGENE HD and 40 μL of Opti-MEM (*see* Subheading 3.2).
4. Rock the dish gently and incubate for 8 h or overnight at 37 °C in a 5 % CO_2 incubator.
5. Replace with 2 mL of growth medium and incubate for an additional 2–24 h at 37 °C in a 5 % CO_2 incubator.
6. Replace with 2 mL of phenol red-free DMEM 10 % FBS for >2 h before imaging.
7. Set the glass-bottomed dish onto the inverted microscope equipped with a 5 % CO_2, 37 °C, and high-moisture incubator system.
8. Find a cell with optimal Histac expression levels (*see* **Note 12**).
9. Focus on the cell manually and register the Z position. The ZDC autofocus system corrects for Z-direction drift during long imaging periods (*see* **Note 3**).
10. Start acquisition of fluorescent images (*see* **Note 13**). The standard time interval is about 2–5 min.
11. Wait for 10–30 min before treating the cells in order to confirm that the fluorescence intensity has stabilized.
12. Treat with a final concentration of 1 μM TSA and observe for about 3 h until the response of the cells saturates (Fig. 2).
13. Remove medium containing TSA and replace with the growth medium (without phenol red) using a peristaltic pump.
14. Observe the cells for an additional 180 min to check the reversibility of Histac.

3.5 Acceptor Photobleaching

To examine the FRET efficiency and whether the indicator truly undergoes FRET, acceptor photobleaching can be used. Since photobleaching of the acceptor fluorophore abolishes its ability to

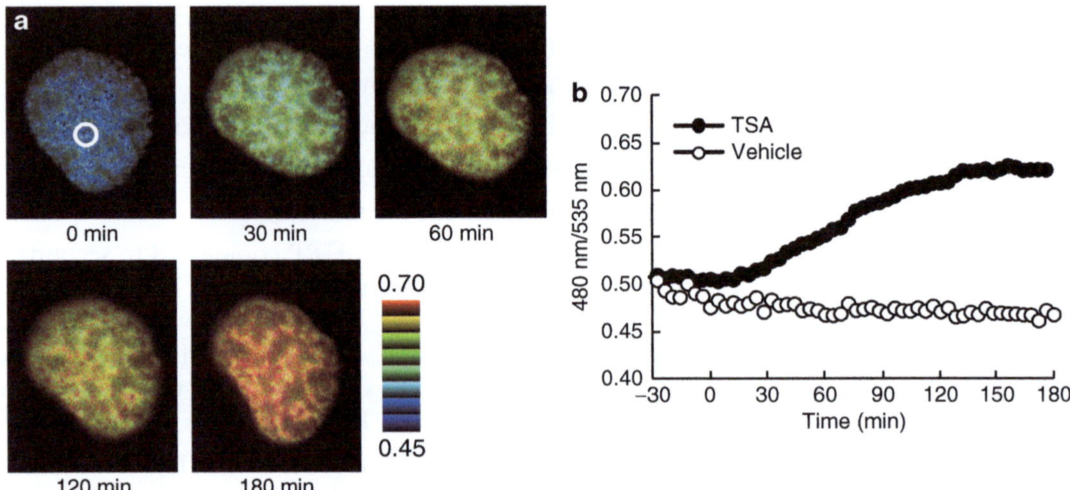

Fig. 2 FRET imaging of acetylation of histone H4. (**a**) Pseudocolor images in the nucleus of Cos7 cells expressing Histac are shown in the intensity modulated display (IMD, *see* Subheading 3.6.2). TSA at a final concentration of 1 μM was added to the medium at 0 min. The *white circle* indicates the region of interest (ROI). (**b**) A time course of FRET (480 nm/535 nm) in the ROI shown in (**a**) after treatment with 1 μM TSA and vehicle (0.1 % EtOH)

absorb the excited state energy transferred from the donor fluorophore, the fluorescent intensity of the donor should increase after acceptor photobleaching, if FRET actually occurs.

1. Replace medium with 2 mL of phenol red-free DMEM, 10 % FBS set the glass-bottomed dish on the inverted microscope.
2. Find a cell with optimal Histac expression level and focus on the cell manually.
3. Remove neutral density (ND) filters.
4. Illuminate Venus in Histac at wavelength 500–540 nm until it completely bleaches, without affecting CFP (*see* **Note 14**).
5. Calculate the FRET efficiency using the corrected intensities obtained for CFP and Venus (*see* Subheading 3.6.1 for details about background subtration).

$$\text{FRET efficiency (\%)} = \frac{\text{Donor}_{\text{after}} - \text{Donor}_{\text{before}}}{\text{Donor}_{\text{after}}} \times 100. \quad (1)$$

$\text{Donor}_{\text{after}}$ = The emission intensity of donor fluorophore after photobleaching.

$\text{Donor}_{\text{before}}$ = The emission intensity of donor fluorophore before photobleaching.

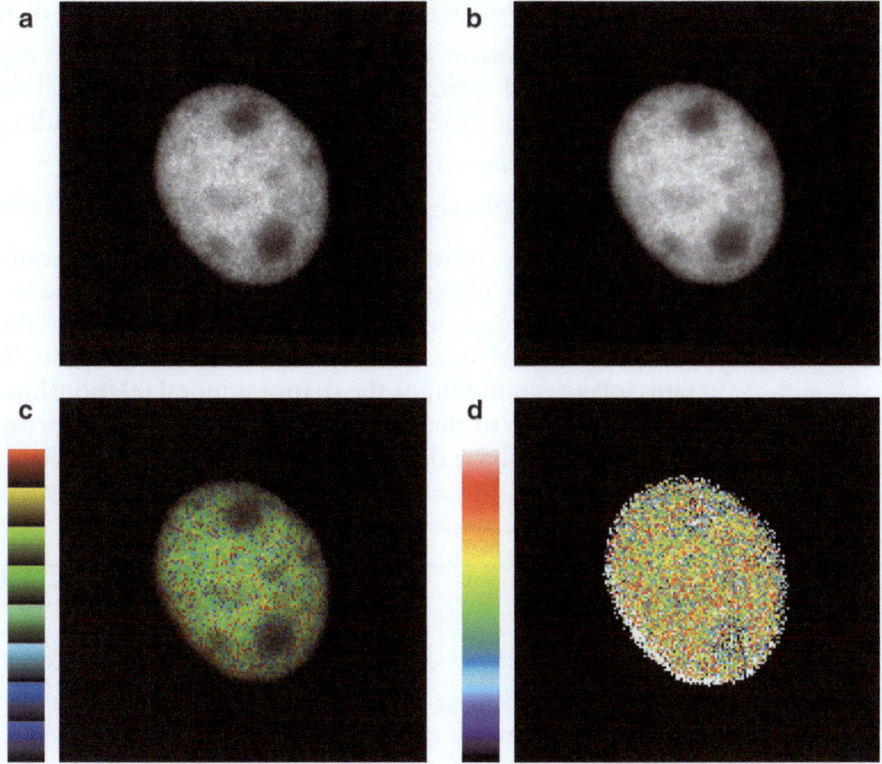

Fig. 3 Intensity-modulated display (IMD) for FRET presentation. CFP image (**a**) and YFP image (**b**) excited at 440 nm acquired as raw data. (**c**) Pseudocolor image generated by IMD. The eight ratio colors are displayed at 32 different intensity levels (IMD ratio bar). (**d**) The simple ratio image is shown

3.6 Data Analysis Using MetaMorph

Two analysis software packages, MetaFluor and MetaMorph, are usually used when FRET images are analyzed. Herein, we describe the analysis using MetaMorph ver. 7.7.

3.6.1 Background Subtraction

1. Open CFP and YFP stack files to be analyzed.
2. Create a region of interest (ROI) devoid of cells to be used as the background region from the CFP stack files.
3. Transfer the ROI from the CFP files to the YFP files using the *Transfer Region* dialog (*Regions* menu).
4. From the *Process* menu, select the *Background and Shading Correction* dialog.
5. Subtract the average value of background ROI from the original images using *Statistical correction* to create corrected images.

3.6.2 IMD Ratio Image

To incorporate information about protein concentration into the display of FRET ratio, an image called the intensity-modulated display (IMD) usually can be used. The IMD is a very useful technique invented by Dr. Roger Tsien [14]. The concentration of the protein is represented by the brightness of each color (Fig. 3). The IMD images can be generated using MetaFluor and MetaMorph.

1. From the *Process* menu, select the *Ratio Imaging* dialog.
2. Define *Numerator* and *Denominator* images (we defined the background subtracted CFP and YFP images as Numerator and Denominator images, respectively (*see* Subheading 3.6.1)).
3. Select *8 ratios with 32 intensities* in *IMD Display.*
4. Find and input appropriate *Min Ratio* and *Max Ratio.*

One of the most critical issues is how the acquired ratio images are to be effectively displayed. Herein we have introduced an IMD ratio image technique, which has represented the nuclear localization of Histac and its acetylation at the region. The IMD ratio image can highlight the spatiotemporal relationship between the enrichment of the indicator in subcellular compartments and the modification by displaying the fluorescent intensity and ratio simultaneously.

4 Notes

1. Check some lots of FBS in advance for cell growth.
2. Add the reducing agent 2-mercaptoethanol immediately prior to use because it oxidizes over time.
3. ZDC is a zero drift system. The compensation of thermal drift is performed in reference to the bottom surface of the glass bottom dish, where a 785-nm laser is focused.
4. An LED illumination system may be better than a Xenon arc lamp.
5. An electron multiplying CCD (EM-CCD) or complementary metal-oxide semiconductor (CMOS) camera can be substituted for a CCD camera.
6. IBC is mounted on the stage to maintain 5 % CO_2 conditions for long-term observation. If the stage top incubator is not equipped with a built-in gas mixer, use a 5 % CO_2/95 % air bomb.
7. Find the optimal ratio of transfection reagent to DNA because optimal expression depends on the cell type and DNA.
8. The expression level and localization of Histac can be easily checked using a fluorescent microscope. Before imaging, the researcher should find a cell with the appropriate expression condition.
9. Separate by electrophoresis through a 12 % acrylamide separating gel preceded by a stacking gel. The molecular weights of Histac and endogenous histone H4 are ~120 kDa and ~10 kDa, respectively.
10. Pre-wash the PVDF membrane in 100 % methanol for 2–3 min and equilibrate in transfer buffer for 15 min before blotting.

11. For imaging of Cos7 cells expressing Histac, glass-bottomed dishes, consisting of a 12-mm glass coverslip attached to the base of 35-mm polystyrene dish, are used. The surface area of the 12-mm glass coverslip is approximately one-tenth that of a 35-mm dish.
12. Binning of the CCD and exposure time needs to be adjusted for the fluorescent intensity of the sample.
13. FRET measurement: CFP and YFP are excited at wavelength of 440 nm; emission wavelengths are 480 nm and 535 nm for CFP and YFP, respectively. Herein, YFP images were acquired at wavelengths of 440 nm (excitation) and 535 nm (emission).
14. The bleaching time depends on the power of illumination and the photostability of the fluorophore.

Acknowledgments

This work was supported by JST, PRESTO.

References

1. Muller S, Filippakopoulos P, Knapp S (2011) Bromodomains as therapeutic targets. Expert Rev Mol Med 13:e29
2. Miyawaki A (2011) Development of probes for cellular functions using fluorescent proteins and fluorescence resonance energy transfer. Annu Rev Biochem 80:357–373
3. Sasaki K, Ito T, Nishino N, Khochbin S, Yoshida M (2009) Real-time imaging of histone H4 hyperacetylation in living cells. Proc Natl Acad Sci U S A 106:16257–16262
4. Nguyen AW, Daugherty PS (2005) Evolutionary optimization of fluorescent proteins for intracellular FRET. Nat Biotechnol 23:355–360
5. Nagai T, Yamada S, Tominaga T, Ichikawa M, Miyawaki A (2004) Expanded dynamic range of fluorescent indicators for Ca(2+) by circularly permuted yellow fluorescent proteins. Proc Natl Acad Sci U S A 101:10554–10559
6. Komatsu N, Aoki K, Yamada M, Yukinaga H, Fujita Y, Kamioka Y et al (2011) Development of an optimized backbone of FRET biosensors for kinases and GTPases. Mol Biol Cell 22:4647–4656
7. Hassan AH, Awad S, Al-Natour Z, Othman S, Mustafa F, Rizvi TA (2007) Selective recognition of acetylated histones by bromodomains in transcriptional co-activators. Biochem J 402: 125–133
8. Kanno T, Kanno Y, Siegel RM, Jang MK, Lenardo MJ, Ozato K (2004) Selective recognition of acetylated histones by bromodomain proteins visualized in living cells. Mol Cell 13:33–43
9. Pivot-Pajot C, Caron C, Govin J, Vion A, Rousseaux S, Khochbin S (2003) Acetylation-dependent chromatin reorganization by BRDT, a testis-specific bromodomain-containing protein. Mol Cell Biol 23: 5354–5365
10. Moriniere J, Rousseaux S, Steuerwald U, Soler-Lopez M, Curtet S, Vitte AL et al (2009) Cooperative binding of two acetylation marks on a histone tail by a single bromodomain. Nature 461:664–668
11. Umehara T, Nakamura Y, Jang MK, Nakano K, Tanaka A, Ozato K et al (2010) Structural basis for acetylated histone H4 recognition by the human BRD2 bromodomain. J Biol Chem 285:7610–7618
12. Ito T, Umehara T, Sasaki K, Nakamura Y, Nishino N, Terada T et al (2011) Real-time imaging of histone H4K12-specific acetylation determines the modes of action of histone deacetylase and bromodomain inhibitors. Chem Biol 18:495–507
13. Shimozono S, Miyawaki A (2008) Engineering FRET constructs using CFP and YFP. Methods Cell Biol 85:381–393
14. Tsien RY, Harootunian AT (1990) Practical design criteria for a dynamic ratio imaging system. Cell Calcium 11:93–109

Chapter 13

Genetically Encoded Fluorescent Biosensors for Live-Cell Imaging of MT1-MMP Protease Activity

Mingxing Ouyang, Shaoying Lu, and Yingxiao Wang

Abstract

The proteolytic activity of Membrane-type 1 Matrix Metalloproteinase (MT1-MMP) is crucial for cancer cell invasion and metastasis. To visualize the protease activity of MT1-MMP with high spatiotemporal resolution at the extracellular plasma membrane surface of live cancer cells, a genetically encoded fluorescent biosensor of MT1-MMP has been developed. Here we describe the design principles of the MT1-MMP biosensor, the characterization of the MT1-MMP biosensor in vitro, and the live-cell imaging protocol used to visualize MT1-MMP activity in mammalian cells. We also provide brief guidelines for observing MT1-MMP subcellular activity by fluorescence resonance energy transfer (FRET) in a cell migration assay.

Key words Fluorescent biosensor, FRET, Live-cell imaging, MT1-MMP, Matrix Metalloproteinase, Protease, Cancer cell

1 Introduction

FRET occurs when two fluorophores are in proximity, with the emission spectrum of the donor overlapping the excitation spectrum of the acceptor. Any change of the distance and/or relative orientation between these two fluorophores can affect the efficiency of FRET and therefore the ratio of acceptor to donor emission [1]. Previous studies have shown that fusion proteins with interacting peptide partners sandwiched between two fluorescent protein color variants are capable of monitoring various cellular events in live cells with high spatial and temporal resolution [2–10]. A high-efficiency FRET pair, which uses an enhanced cyan fluorescence protein (ECFP) as the donor and a variant of the yellow fluorescence protein (YPet) as the acceptor, enabled the detection of possibly moderate but physiologically important molecular activities, such as vascular endothelial growth factor (VEGF)-induced

Mingxing Ouyang and Shaoying Lu contributed equally to this work.

Jin Zhang et al. (eds.), *Fluorescent Protein-Based Biosensors: Methods and Protocols*, Methods in Molecular Biology, vol. 1071, DOI 10.1007/978-1-62703-622-1_13, © Springer Science+Business Media, LLC 2014

Src signal in endothelial cells [11]. This, however, still only allowed the visualization of one type of active molecular event in a single live cell. Biosensors with new FRET pairs need to be developed such that multiple active molecular events can be visualized simultaneously within the same live cells. Fluorescence proteins (FPs) with different excitation and emission wavelengths have been recently developed through the directed evolutionary strategy [12]. Among these FPs, mOrange2 and mCherry, with the emission spectrum of mOrange2 overlapping the excitation spectrum of mCherry, appear to have the potential to constitute another new FRET pair with spectra that are distinguishable from those of the CFP and YPet pair. Therefore, mOrange2 and mCherry could be used as a new FRET pair to enable the visualization of a second molecular event in the same cells.

The extracellular matrix (ECM) in surrounding tissue provides a barrier function against cancer growth, invasion and metastasis. One of the matrix metalloproteinase (MMP) family proteins, membrane type 1 MMP (MT1-MMP), has been shown to be crucial for tumor cells to negotiate with and invade human dermis and cross-linked type I collagen gels, which mimic the in vivo situation [13]. In clinical samples, MT1-MMP can also be detected in tumors and their surrounding tissues [14]. Although the exact mechanism is not clear, MT1-MMP may degrade type I collagen through MMP-2-dependent and -independent pathways [13, 15]. MT1-MMP can also directly digest a variety of ECM proteins, including fibronectin, vitronectin, collagen type I, II, III, and other plasma membrane receptors such as CD44 and integrins [16, 17]. Thus, MT1-MMP activity can be representative of the ability of the cancer cells to invade basement membrane or to induce metastasis of the tumor.

Here we describe the design principles of the MT1-MMP protease biosensor with the ECFP/YPet or the mOrange2/mCherry FRET pair. We provide the protocols used to characterize the ECFP/YPet-based MT1-MMP biosensor in vitro and how to visualize the MT1-MMP activity in mammalian cells by live-cell FRET imaging.

2 Materials

2.1 *In Vitro Assay*

Prepare all solutions using ultrapure water and analytical grade reagents.

1. LB (Luria Broth) medium: dissolve 25 g of Luria Broth Base (Invitrogen) containing 10 g of peptone, 5 g of yeast extract, and 10 g of sodium chloride into 1 L of water, and autoclave at 120 °C for 30 min.
2. Ampicillin (Amp) stock solution (100 mg/ml): dissolve 1 g of Amp powder into 10 ml of water, and filter sterilize with a 0.45 μm syringe filter. Keep stock solution in 1 ml aliquots at −20 °C.

3. LB-Amp agar plates: add 15 g of agar powder per 1 L of LB medium. After autoclaving, cool down the solution to 60 °C and supplement with 100 μg/ml of Amp. Pour 15–20 ml of the solution to each 10 cm-diameter Petri dish and, after solidification at room temperature, seal the plates and store at 4 °C.
4. BL21(DE3) competent *E. Coli* (Invitrogen).
5. Isopropyl β-D-1-thiogalactopyranoside (IPTG) stock solution (1 M): dissolve 2.38 g of IPTG into 10 ml of water and filter sterilize with a 0.45 μm syringe filter. Keep stock solution in 1 ml aliquots at −20 °C.
6. Phenylmethylsulphonylfluoride (PMSF) stock solution (100 mM): dissolve 0.174 g of PMSF in 10 ml of isoproponal, and store in 1 ml aliquots at −20 °C.
7. pRSETb and pDisplay vectors (Invitrogen).
8. Bacterial lysis solution: dissolve half of a protease inhibitor cocktail tablet (Merck) in 10 ml of B-PER protein extraction reagents (Thermo Scientific) and supplement with 100 μM of PMSF.
9. HisPur Ni-NTA agarose resin (Thermo Scientific).
10. Affinity purification column (Sigma) for His-tagged protein purification from bacterial lysate.
11. TBS solution: 50 mM Tris–HCl, pH 7.4, 300 mM NaCl.
12. Washing buffer: 50 mM Tris–HCl, pH 7.4, 300 mM NaCl, 10 mM imidazole.
13. Elution buffer: 50 mM Tris–HCl, pH 7.4, 300 mM NaCl, 100 mM imidazole.
14. Matrix Metalloprotease proteolysis assay buffer: 50 mM HEPES titrated with 1 M NaOH solution to pH 6.8, 10 mM $CaCl_2$, 0.5 mM $MgCl_2$, 50 μM $ZnCl_2$, and 0.01 % Brij-35.
15. Bio-Rad protein assay kit (Bio-Rad).
16. Recombinant catalytic domain of human MT1-MMP, MT2-MMP or MT3-MMP, or active human MMP-2, MMP-9 (Calbiochem).
17. SDS-PAGE gel fixing solution: 50 % methanol and 10 % glacial acetic acid in water.
18. Coomassie Blue staining solution: 0.1 % Coomassie Brilliant Blue R-250, 50 % methanol (v/v), and 10 % (v/v) acetic acid.
19. Destaining solution: 50 % (v/v) methanol in water with 10 % (v/v) acetic acid.

2.2 Mammalian Cell Culture and Imaging Process

1. MiniPrep or MaxiPrep kits (Qiagen).
2. Mammalian expression vector: PCR3.1 Uni (Invitrogen).

3. HeLa cells, human breast cancer MDA-MB-231 cells, human HT1080 fibrosarcoma cells, human fibroblasts (ATCC).
4. Mammalian cell culture medium: Dulbecco's modified Eagle's medium (DMEM) supplemented with 10 % fetal bovine serum (FBS), 2 mM L-glutamine, 100 U/ml penicillin, 100 μg/ml streptomycin, and 1 mM sodium pyruvate (Invitrogen).
5. Phosphate buffered saline (PBS) tablets (Sigma).
6. 35 mm glass-bottom dishes (Cell E&G).
7. Lipofectamine 2000 (Invitrogen).
8. Opti-MEM I reduced serum medium (Invitrogen).
9. Serum-starvation medium: DMEM culture medium containing only 0.5 % FBS.
10. CO_2-independent medium (Invitrogen).
11. Epithelial growth factor (EGF, Sigma).
12. MetaFluor 6.2 software (Universal Imaging).
13. Fibronectin (Sigma).

3 Methods

3.1 Design Principles of MT1-MMP FRET Biosensors

1. To monitor the protease activity of MT1-MMP, a FRET biosensor contains an MT1-MMP specific substrate peptide sandwiched between a FRET donor and an acceptor fluorescent protein (Fig. 1a). The substrate peptide, CPKESCNLFVLKD, is derived from the MT1-MMP cleavage site identified in its substrate molecule, proMMP-2 [18].
2. A highly efficient FRET pair, ECFP and YPet, are chosen as the donor and acceptor, respectively, in the FRET biosensor [11]. When a biosensor is intact, ECFP and YPet form a dimer which displays a high FRET signal. Upon the cleavage of the substrate peptide by active MT1-MMP, YPet separates from ECFP and diffuses away, which leads to a significant decrease of the FRET signal corresponding to an increase in the ECFP/YPet emission ratio (Fig. 1b).
3. Since MT1-MMP is known to be active at the extracellular surface [19], the MT1-MMP biosensor is subcloned into a pDisplay vector (Invitrogen) for the correct mammalian cell expression and localization (Fig. 1c). The pDisplay vector contains an N-terminal murine Ig κ-chain leader sequence, which directs the biosensor protein to the secretory pathway, and a C-terminal trans-membrane domain of the platelet derived growth factor receptor beta (PDGFR-β) at the C-terminus, which targets the biosensor protein to the plasma membrane. Since both MT1-MMP and PDGFR-β have been

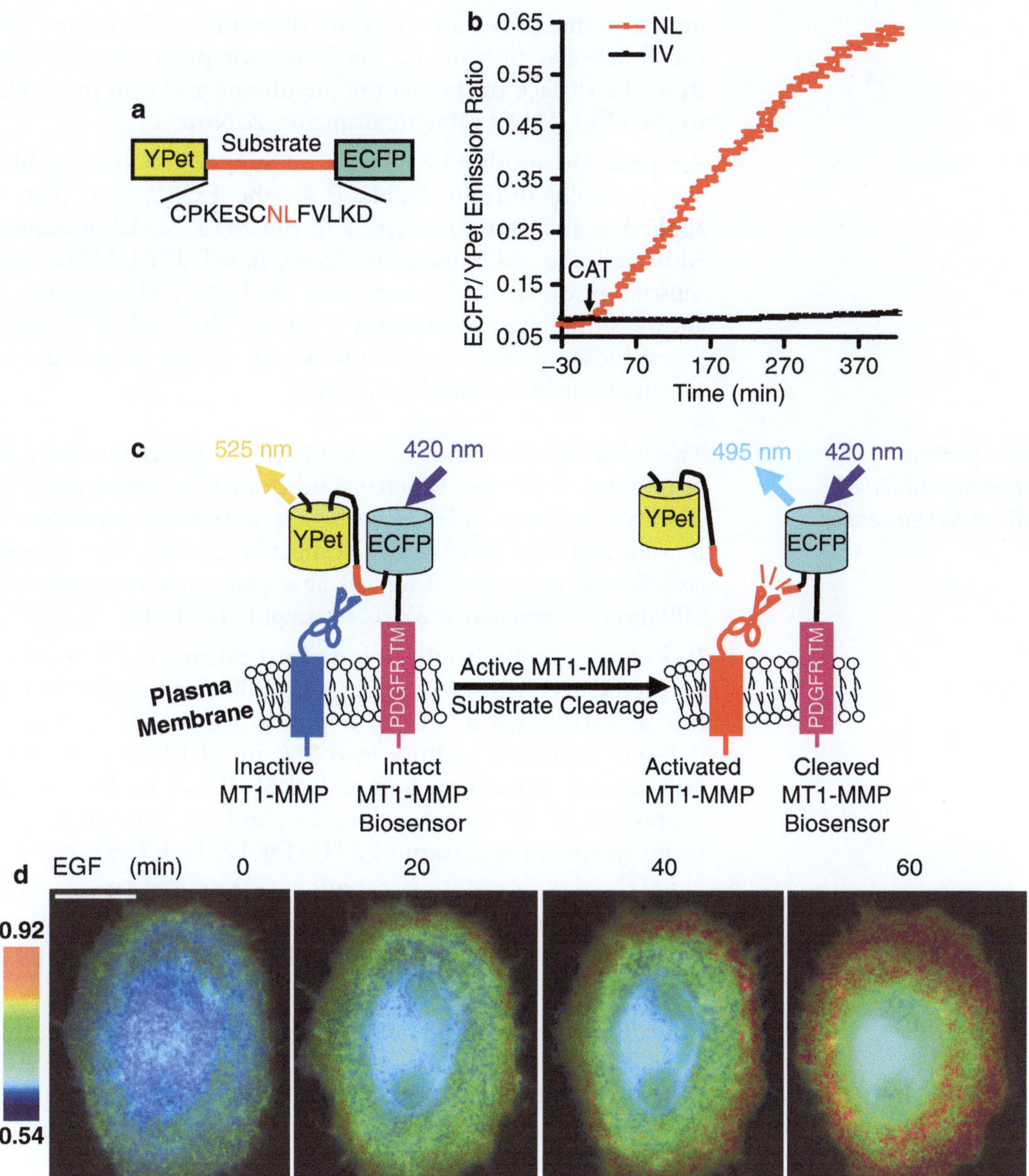

Fig. 1 Design strategy of the MT1-MMP biosensor, and its FRET response in vitro and in mammalian cells. (**a**) Domain structure of MT1-MMP biosensor with YPet and ECFP at its N- and C-termini connected by a substrate peptide of MT1-MMP. (**b**) The time courses of ECFP/YPet emission ratio (mean ± SD) of wild-type (*red line*) and mutant (NL to IV mutation; *black line*) biosensors before and after incubation with the active catalytic domain of MT1-MMP (CAT). (**c**) The activation mechanism of the cell plasma membrane-tethered MT1-MMP biosensor. The biosensor is fused to the trans-membrane domain of PDGFR to position its sensing element outside of the plasma membrane, making it accessible to MT1-MMP. Active MT1-MMP can cleave the substrate peptide to separate ECFP and YPet, which leads to a decrease in FRET. (**d**) ECFP/YPet emission ratio images of a HeLa cell co-transfected with the MT1-MMP biosensor and MT1-MMP before and after EGF stimulation. Scale bar: 20 μm. This research was originally published in Journal of Biological Chemistry. Ouyang et al. Visualization of polarized membrane type 1 matrix metalloproteinase activity in live cells by fluorescence resonance energy transfer imaging. J Biol Chem. 2008; 283(25):17740–8. © The American Society for Biochemistry and Molecular Biology

demonstrated previously to co-localize at the cell surface [20], the functional domain of the biosensor protrudes outward from the surface of the plasma membrane and is in proximity to the MT1-MMP catalytic domain (*see* **Note 1**).

4. Alternatively, another FRET pair, mOrange2/mCherry, which are spectrally distinguishable from the ECFP/YFP pair, is applied as the donor/acceptor in the MT1-MMP biosensor. Although the mOrange2/mCherry-based MT1-MMP biosensor has less dynamic range than the ECFP/YPet version, it allows simultaneous imaging with an ECFP/YFP variant-based biosensor to visualize two kinds of molecular activities simultaneously in a single cell [21].

3.2 *In Vitro* Characterization of MT1-MMP Biosensor

1. The plasmid encoding the MT1-MMP biosensor with an N-terminal *6x*-His tag, which was subcloned into vector pRSETb (Invitrogen) using *Bgl*II/*Hind*III sites for bacterial expression, is transformed into BL21(DE3) competent *E. Coli.* The bacteria are allowed to grow on an LB agar plate supplemented with 100 μg/ml ampicillin at 37 °C overnight (12–16 h).
2. Inoculate a single bright colony into 50 ml of LB medium supplemented with 100 μg/ml ampicillin, and shake the culture at 250 rpm at 37 °C for 6–8 h. When the OD_{600} reaches 0.2–0.4, dilute the culture into 200 ml of LB medium with 100 μg/ml ampicillin, add 0.4 mM IPTG to induce the expression of the biosensor protein, and continue shaking at room temperature (around 25 °C) for 12–16 h (*see* **Note 2**).
3. Spin down the bacteria at 5,000 × *g* at 4 °C for 10 min. Add 10 ml of bacterial lysis solution and gently and completely resuspend the bacterial pellet. Rock the suspension gently at room temperature for 10 min.
4. Centrifuge the lysate at 20,000 × *g* at 4 °C for 15 min, and filter the supernatant through a 0.45 μm syringe filter. Add 0.5 ml of a 50 % slurry of Ni-NTA agarose resin and gently rock the mixture in a 15 ml tube at room temperature for 1 h to allow binding (*see* **Note 3**).
5. Assemble the affinity purification column and transfer the biosensor–Ni-NTA resin mixture solution from the binding reaction to the column. Wait until all of the liquid passes through the column. The resin beads bound with biosensor proteins will remain in the column.
6. Rinse the beads in the column three times with 10 ml of TBS solution (50 mM Tris–HCl, pH 7.4, 300 mM NaCl) before washing the beads five times with 10 ml of washing buffer (*see* **Note 4**).
7. Elute the biosensor proteins from the beads with elution buffer (*see* **Note 5**).

8. Dialyze the biosensor protein solution in matrix metalloprotease proteolysis assay buffer at 4 °C overnight (*see* **Note 6**).
9. Determine the molar concentration of the biosensor protein. One standard method is as follows: (1) calculate the molecular weight of the biosensor protein based on its amino acid sequence; (2) measure the protein mass concentration using the protein assay kit (Bio Rad); (3) the molar concentration of the biosensor is equal to its mass concentration divided by its molecular weight.
10. Prepare biosensor protein solution (1 μM) in the matrix metalloprotease proteolysis assay buffer in 96-well plates at 100 μl per well. Measure the biosensor emission profile from 440 to 580 nm with 2 nm intervals following 415 nm excitation using a fluorescence plate reader (TECAN, Sapphire II). Record the emission profile at 10 min intervals at 37 °C for 0.5–1 h and continue for another 6–8 h after adding the recombinant catalytic domain of human MT1-MMP, MT2-MMP or MT3-MMP (2 μg/ml, Mol. Wt. 20 kDa, Calbiochem) or active human MMP-2, MMP-9 (6 μg/ml, Mol. Wt. 66 kDa, Calbiochem). The time course of emission ratios of ECFP/YPet (at 476 nm for ECFP and at 526 nm for YPet) is calculated from the recorded data using Excel (Microsoft) (Fig. 1b) (*see* **Note 7**).
11. The samples with or without the protease treatment are separated by 10 % SDS-PAGE gels followed by gel fixation and Coomassie Blue staining. After de-staining, the biosensor protein is visualized and the image is record using a digital camera.

3.3 Visualizing MT1-MMP Activity in Mammalian Cells by Live-Cell FRET Imaging

1. For mammalian cell expression, subclone the MT1-MMP biosensor into pDisplay (Invitrogen) using the *Bgl*II/*Pst*I restriction sites.
2. The DNA constructs for transfection into mammalian cells should be prepared by commercial MiniPrep or MaxiPrep kits. These constructs include the empty vector (PCR3.1 Uni) and expression vectors encoding wild-type and mutant MT1-MMP biosensors, and human MT1-MMP (*see* **Note 8**).
3. HeLa cells, which express only low levels of endogenous MT1-MMP, are used to characterize the MT1-MMP biosensor in mammalian cells.
4. The day before transfection, pass HeLa cells onto 35 mm glass-bottom dishes and maintain the cells in complete culture medium without antibiotics. The cell density should be around 50–80 % at the time of transfection.
5. DNA transfection can be carried out with a transfection reagent, such as Lipofectamine 2000, according to the manufacture's protocol. For co-transfection in the glass-bottom dish, gently and thoroughly mix 1.5 μg of DNA encoding the

Table 1
The settings of filters for fluorescence imaging

	Excitation filter (nm)	Dichroic mirror (long pass; nm)	Emission filter (nm)
CFP YFP (FRET)	420/20	450	475/40 535/25
GFP	495/10	515	535/25
mCherry	560/40	595	653/95

MT1-MMP biosensor with 1.5 μg of DNA encoding MT1-MMP (or another construct encoding an MT1-MMP mutant) in 200 μl of Opti-MEM I reduced serum medium before adding another 200 μl of Opti-MEM I containing 6–7.5 μl of Lipofectamine 2000. Add the DNA–Lipofectamine 2000 mixture solution to the cells and incubate at 37 °C. After 6–8 h of transfection, replace the culture medium with fresh serum-starvation medium containing only 0.5 % FBS and no antibiotics. Before conducting FRET imaging experiments, transfected cells should be incubated for 36–48 h in the starvation media (*see* **Note 9**).

6. During imaging, maintain the cells in serum-free, CO_2-independent medium (Gibco BRL) at 37 °C. Images are collected by a Zeiss axiovert inverted microscope equipped with a cooled charge-coupled device camera (Cascade 512B; Photometrics) using MetaFluor 6.2 software (Universal Imaging). Important imaging parameters, such as dichroic mirrors and excitation and emission filters used for FRET and different fluorescent proteins, are shown in Table 1.
7. When selecting cells for imaging, choose fluorescent cells that exhibit an intermediate fluorescence intensity so that the FRET change following EGF (50 μg/ml) stimulation can be readily visualized. Collect images at 2 min intervals for half an hour to obtain the basal FRET signal and for another 2–3 h after EGF stimulation to record EGF-induced FRET changes (*see* **Note 10**).
8. To correct for the autofluorescence of the cells under study, the fluorescence intensity of non-transfected cells should be quantified and subtracted from the ECFP and YPet (FRET) signals of transfected cells. The pixel-by-pixel ratio images of ECFP/YPet, representing the FRET efficiency and activation levels of the biosensor, are then calculated directly based on the background-subtracted fluorescence intensity images of ECFP and YPet by the MetaFluor software (Fig. 1d). Finally, emission ratios of ECFP/YPet are averaged within selected

regions-of-interest to allow the quantification and statistical analysis by Excel (Microsoft) or MATLAB (The MathWorks).

9. The specificity of the MT1-MMP biosensor is further characterized in HeLa cells co-transfected with different MT1-MMP mutants [22]. Imaging of FRET change can also be achieved in cell lines without co-transfection of MT1-MMP, such as MDA-MB-231 cells, which express substantial levels of endogenous MT1-MMP [22]. FRET levels of the MT1-MMP biosensors in different cell lines can be used to compare their MT1-MMP activity levels, such as among fibroblasts, MT1-MMP-knockout fibroblast, and HT1080 cells [22].

3.4 Visualizing MT1-MMP Subcellular Activity During Migration Assay by FRET Imaging

1. Coat glass-bottom dishes with 10 ng/ml Fibronectin (Fn) at 4 °C overnight (or room temperature for 2 h). By applying micro-fabrication technology, the glass surface on the glass-bottom dishes can be micro-patterned with Fn-coated strips (10–20 μm width) [22]. In this way, cells seeded on the Fn-coated strips can achieve directed migration along the strips (*see* **Note 12**). After applying Fn, rinse the coated dishes once with PBS before seeding cells (*see* **Note 11**).
2. Before seeding, HeLa cells co-transfected with the MT1-MMP biosensor and MT1-MMP should be starved in serum-starvation medium containing 0.5 % FBS for 36–48 h in tissue culture dishes. Cells are then passaged onto Fn-coated dishes for 2–6 h before beginning the imaging process (*see* **Note 12**).
3. Follow the same procedures described in Subheading 3.3 (**steps 6–8**) for imaging processing and FRET quantification.

4 Notes

1. The transmembrane domain of PDGFR-β, which targets the biosensor protein to the plasma membrane, is located at the C-terminus of the biosensor. In order to keep the FRET donor, ECFP, on the membrane after MT1-MMP cleavage, ECFP must be incorporated at the C-terminal side of the substrate peptide, while YPet is incorporated at the N-terminal side.
2. Due to leaky expression from the T7 promotor, BL21(DE3) bacteria usually express a certain level of biosensor protein after growing overnight on the LB plate. Before inoculating a colony into liquid culture, the fluorescence intensity of the colonies can be easily checked under the 488 nm excitation light with a microscope or under UV light. Try to pick a colony with bright fluorescence for efficient biosensor expression in liquid culture.

3. To protect the biosensor protein from photobleaching by light, wrap the 15 ml tube containing the mixture solution with aluminum foil during the rocking incubation step. Likewise, avoid any strong light illumination of the biosensor protein during the entirety of purification procedures.
4. The purpose of the washing steps is to remove nonspecific binding proteins from the beads and biosensor proteins. Most of the time, the washing solution can pass through the column smoothly by gravity flow. If the flow is too slow on some occasions, a small amount of air pressure can be applied to the column to facilitate the flow speed.
5. During the elution step, the concentration of the biosensor protein is highest in the first 0.5–1 ml of elution solution coming out from the column and gradually becomes lower thereafter. To obtain concentrated biosensor, try to collect the elution into separate tubes at different elution times. However, if the biosensor concentration is too high (for example, above 100 μM), it can become partially precipitated during storage, so try to maintain the biosensor protein at an appropriate concentration, such as 10–50 μM, during storage at 4 °C.
6. The purpose of the dialysis is to replace the elution buffer with the matrix metalloprotease proteolysis assay buffer. The dialysis can be done in a 2-L beaker with 1.5–1.8 L of matrix metalloprotease proteolysis assay solution outside the dialysis sacks in a cold room (4 °C) or refrigerator. Keep stirring the dialysis solution slowly with a magnetic stirrer to facilitate the dialysis procęss. To avoid the direct collision of the stir bar and the dialysis sacks, a small amount of air can be left in the dialysis sacks to keep them afloat during the dialysis process.
7. Control wells with biosensor protein but without any MMP enzyme are required to measure the biosensor stability during the experimental process. At least three parallel samples for each condition are required during the measurement. In order to avoid significant photobleaching of the biosensor, time intervals during the measurement should not be too short. To fairly compare the cleavage speed of different MMP enzymes, add all enzymes to the biosensor solution at almost the same time. Because it takes several hours to complete the measurement on the plate reader at 37 °C, water evaporation from the solution should be controlled by adding water into the empty wells in the plate and sealing the plate boundary with parafilm.
8. In order to achieve efficient and controllable expression of the exogenous proteins in mammalian cells after transfection, high DNA quality from the plasmid extraction step is required. Run

a DNA agarose gel to check whether the plasmid is intact or contaminated with other DNA.

9. To achieve high efficiency in co-transfection of two kinds of DNAs in the same cells, it is necessary to mix them thoroughly before further mixing with transfection reagent, such as Lipofectamine 2000. The expression level of the MT1-MMP biosensor in mammalian cells usually peaks 36–48 h after transfection and decreases quickly after 72 h. Therefore, the imaging process should be conducted within 36–72 h after transfection. To increase the dynamic range of MT1-MMP activation and hence FRET changes of the biosensor upon EGF stimulation, it is helpful to maintain the cells in serum starvation medium with 0.5 % FBS to reduce the basal MT1-MMP activity.
10. It is important to ensure that the imaged cells have appropriate expression levels of each biosensor. Weak expression leads to low signal–noise ratio, while overly high expression can cause abnormal cellular functions and/or intermolecular FRET. Usually weak expression of the biosensors can be detected if the fluorescence image of the cell was not clearly distinguishable from the background. On the other hand, cells with overly high expression of biosensors have bright fluorescent signals and can sometimes shrink quickly during imaging.
11. Successfully coating the glass surface with Fn is critical for migration assay of transfected cells. An easy way to confirm successful Fn coating is to put a drop of PBS solution on the surface. If the drop spreads out evenly along the surface, Fn coating is likely successful. If the drop maintains a round shape without spreading, it suggests that the glass surface is hydrophobic and that Fn coating has failed.
12. The time period after the cells are seeded on the Fn-coated glass needs to be well controlled. Long seeding times will reduce the mobility of cells during the migration assay.

Acknowledgment

This work is supported by grants from NIH HL098472, CA139272, NS063405, NSF CBET0846429 (Y.W., S. L.), and the Wallace H. Coulter Foundation and Beckman Laser Institute, Inc. (Y.W.). The funding agencies had no role in study design, data collection and analysis, decision to publish, or preparation of the manuscript.

References

1. Tsien RY (1998) The green fluorescent protein. Annu Rev Biochem 67:509–544
2. Kunkel MT et al (2005) Spatio-temporal dynamics of protein kinase B/Akt signaling revealed by a genetically encoded fluorescent reporter. J Biol Chem 280(7):5581–5587
3. Miyawaki A et al (1997) Fluorescent indicators for Ca2+ based on green fluorescent proteins and calmodulin. Nature 388(6645):882–887
4. Mochizuki N et al (2001) Spatio-temporal images of growth-factor-induced activation of Ras and Rap1. Nature 411(6841):1065–1068
5. Pertz O et al (2006) Spatiotemporal dynamics of RhoA activity in migrating cells. Nature 440(7087):1069–1072
6. Ting AY et al (2001) Genetically encoded fluorescent reporters of protein tyrosine kinase activities in living cells. Proc Natl Acad Sci U S A 98(26):15003–15008
7. Violin JD et al (2003) A genetically encoded fluorescent reporter reveals oscillatory phosphorylation by protein kinase C. J Cell Biol 161(5):899–909
8. Wang Y et al (2005) Visualizing the mechanical activation of Src. Nature 434(7036):1040–1045
9. Zhang J et al (2005) Insulin disrupts beta-adrenergic signalling to protein kinase A in adipocytes. Nature 437(7058):569–573
10. Zhang J et al (2001) Genetically encoded reporters of protein kinase A activity reveal impact of substrate tethering. Proc Natl Acad Sci U S A 98(26):14997–15002
11. Ouyang M et al (2008) Determination of hierarchical relationship of Src and Rac at subcellular locations with FRET biosensors. Proc Natl Acad Sci U S A 105(38):14353–14358
12. Shaner NC et al (2004) Improved monomeric red, orange and yellow fluorescent proteins derived from *Discosoma* sp. red fluorescent protein. Nat Biotechnol 22(12):1567–1572
13. Sabeh F et al (2004) Tumor cell traffic through the extracellular matrix is controlled by the membrane-anchored collagenase MT1-MMP. J Cell Biol 167(4):769–781
14. Itoh Y, Seiki M (2006) MT1-MMP: a potent modifier of pericellular microenvironment. J Cell Physiol 206(1):1–8
15. Deryugina EI et al (2001) MT1-MMP initiates activation of pro-MMP-2 and integrin alphavbeta3 promotes maturation of MMP-2 in breast carcinoma cells. Exp Cell Res 263(2): 209–223
16. Seiki M (2003) Membrane-type 1 matrix metalloproteinase: a key enzyme for tumor invasion. Cancer Lett 194(1):1–11
17. Seiki M, Yana I (2003) Roles of pericellular proteolysis by membrane type-1 matrix metalloproteinase in cancer invasion and angiogenesis. Cancer Sci 94(7):569–574
18. Kinoshita T et al (1996) Processing of a precursor of 72-kilodalton type IV collagenase/gelatinase A by a recombinant membrane-type 1 matrix metalloproteinase. Cancer Res 56(11):2535–2538
19. Sato H et al (1994) A matrix metalloproteinase expressed on the surface of invasive tumour cells. Nature 370(6484):61–65
20. Lehti K et al (2005) An MT1-MMP-PDGF receptor-beta axis regulates mural cell investment of the microvasculature. Genes Dev 19(8):979–991
21. Ouyang M et al (2010) Simultaneous visualization of protumorigenic Src and MT1-MMP activities with fluorescence resonance energy transfer. Cancer Res 70(6):2204–2212
22. Ouyang M et al (2008) Visualization of polarized membrane type 1 matrix metalloproteinase activity in live cells by fluorescence resonance energy transfer imaging. J Biol Chem 283(25):17740–17748

Chapter 14

Biosensor Imaging in Brain Slice Preparations

Marina Polito, Pierre Vincent, and Elvire Guiot

Abstract

Cyclic-AMP dependent protein kinase (PKA) is present in most branches of the animal kingdom, and is an example in the nervous system where a kinase effector integrates the cellular effects of various neuromodulators. The recent development of FRET-based biosensors, such as AKAR, now allows the direct measurement of PKA activation in living cells by simply measuring the ratio between the fluorescence emission at the CFP and YFP wavelengths upon CFP excitation. This novel approach provides data with a temporal resolution of a few seconds at the cellular and even subcellular level, opening a new avenue of understanding the integration processes in space and time.

Our protocol has been optimized to study morphologically intact mature neurons and we describe how simple and cheap wide-field imaging, as well as more elaborate two-photon imaging, allows real-time monitoring of PKA activation in pyramidal cortical neurons in neonate rodent brain slices. In addition, many practical details presented here also pertain to image analysis in other cellular preparations, such as cultured cells. Finally, this protocol can also be applied to the various other CFP-YFP-based FRET biosensors that are available for other kinases or other intracellular signals. It is likely that this kind of approach will be generally applicable to a broad range of assays in the near future.

Key words Protein kinase A, Biosensor, FRET, Time-lapse imaging, Fluorescence, Brain slice preparation

1 Introduction

The cyclic adenosine monophosphate (cAMP) and protein kinase A (PKA) signaling pathway is involved in virtually all physiological processes. More specifically in the nervous system, the cAMP/PKA cascade plays an important role in several integrated brain functions, such as neuronal survival, axonal regeneration, memory and cognitive functions. At the cellular level, the cAMP/PKA cascade is responsible for the modulation of various processes, including some specific forms of synaptic plasticity, control of excitability, regulation of nuclear factors and imprinting of long-term changes. One major target of cAMP is PKA, hence the importance of monitoring its activation. Besides PKA, cAMP also directly modulates other downstream effectors, such as the exchange proteins directly

Jin Zhang et al. (eds.), *Fluorescent Protein-Based Biosensors: Methods and Protocols*, Methods in Molecular Biology, vol. 1071, DOI 10.1007/978-1-62703-622-1_14,

activated by cAMP (Epac1 and Epac2) and several channels, like cyclic-nucleotide gated channels and I_h.

The cAMP/PKA signal, like many other cellular signals, takes place in the complex multidimensional space of the living cell. The temporal dimension is obviously a critical one, as the PKA signal rises following a stimulus and then declines over time as the stimulus is removed and/or feedback mechanisms cause the signal to revert towards baseline. The spatial dimension also plays a role since the concentration of signaling molecules decreases with distance. Protein-protein interactions, such as those mediated by AKAPs, maintain all partners of the signaling cascade in close proximity in so-called signaling microdomains [1]. In addition to these factors, the geometry of the cell can affect signal integration, with a high surface to volume ratio favoring the accumulation of cAMP and the activation of PKA in thin dendrites versus the cell body [2]. Therefore, since temporal and spatial measurements are of critical importance to the understanding of intracellular signaling, direct imaging of the cAMP/PKA signaling cascade is a highly desirable approach.

The first attempts to image changes in cAMP concentration in neurons made use of purified PKA chemically labeled with fluorescein and rhodamine: the dissociation of the regulatory and catalytic subunit upon cAMP binding was monitored as a decrease in fluorescence resonance energy transfer (FRET) between the two chromophores [3]. This method allowed the recording of cAMP diffusion in giant Aplysia neurons [4], then in the intact stomatogastric ganglion of the spiny lobster [5]. However, this first biochemical sensor remained impractical to use in vertebrate neurons because the recombinant and labeled holoenzyme had to be introduced into the cells of interest, a difficult task for vertebrate neurons [6]. A major breakthrough in this field resulted from the creation of FRET-based genetically encoded optical sensors, the first sensors being designed to report calcium [7, 8], eventually leading to sensors for various other biological signals.

Single wavelength sensors have been made which use the intrinsic sensitivity of GFP to pH or chloride, a property amplified by mutating the GFP sequence. They can also have an external sensor grafted to it, such as for the GCamp series of calcium biosensors. However, changes in fluorescence emission measured at a single wavelength can result from various artifacts, technical (fluctuations in illumination intensity, focus drift) or related to the cell (changes in cell volume, hence sensor concentration in the imaged volume) and this is where ratiometric quantification comes in handy. The principle of ratiometric quantification was initially demonstrated for calcium imaging using the small molecule indicator, fura2 [9]. In these studies, the measurement is done simultaneously at two wavelengths where the intensity change of the indicator goes in the opposite direction of the biological signal of interest (i.e., intracellular calcium). Importantly, since artifactual changes affect both wavelengths by the same factor, they are cancelled in the ratio. This quantification method also applies to most FRET-based biosensors.

FRET-based biosensors are commonly constituted of a domain sensitive to the biological signal of interest sandwiched between a pair of fluorophores, usually CFP and YFP or their close derivatives [10]. The signal triggers a conformational change of the biosensor, modifying the extent of FRET from CFP to YFP upon CFP excitation, leading to reciprocal changes in the CFP and YFP fluorescence intensities. The last decade has seen the creation of a number of biosensors based on this design for biological signals ranging from the detection of the neurotransmitter glutamate [11, 12] to the monitoring of apoptosis [13]. After the original CFP-YFP pair, new fluorophores are progressively being used, resulting in improved ratio changes [14–16]. In this chapter, we will focus on one series of biosensors specifically developed to monitor PKA-mediated phosphorylation events.

A-kinase activity reporter (AKAR) is a recombinant protein composed of a phosphoamino acid binding domain and a PKA-specific substrate fused between CFP and YFP. When phosphorylated by PKA, intramolecular binding of the substrate by the phosphoamino acid binding domain drives a conformational reorganization, leading to an increase in FRET between CFP and YFP. Variants with increasingly better signals were produced over the years and are reviewed in [17]. Briefly, AKAR2.2, which employs a monomeric version of enhanced CFP (ECFP) as the donor and a monomeric version of the YFP variant, Citrine, as an acceptor, exhibits a ~25 % ratio change upon PKA phosphorylation [18]. Meanwhile, AKAR3, in which Citrine is replaced by a circularly permuted form of Venus (cpV E172), exhibits a ~40 % ratio change [19]. Finally, replacing ECFP with Cerulean led to AKAR4, with ratio changes up to 68 % [20].

Although commonly dubbed "PKA sensors", one must keep in mind that the phosphorylation level of the biosensor results from the equilibrium between PKA-dependent phosphorylation and dephosphorylation performed by various phosphatases. An increase in AKAR signal can therefore result either from an activation of PKA, or from an inhibition of phosphatases in the context of a tonic PKA activity. This has been observed experimentally in neurons in brain slices where Gs-coupled receptors, as well as phosphatase inhibition, led to an increase in the extent of AKAR phosphorylation [21]. Since the ratio recovered slowly after agonist removal [21], one can assume that the kinetics of phosphatase-mediated dephosphorylation was much slower than that of PKA-mediated phosphorylation. Therefore, as long as this condition is verified, one can assume that AKAR primarily reports PKA activity.

AKAR thus opened the possibility of quantifying the relative amplitude of the PKA signal obtained in different situations. For example, we measured the response to different neuropeptides, reporting a relative order of potency of the receptors CRF1 ≅ PAC1 > VPAC1 in their ability to activate the cAMP/PKA signaling cascade in pyramidal cortical neurons [22]. AKAR imaging also provided key experimental data in the temporal and spatial dimensions: by comparing AKAR response kinetics (indicating PKA activation in the bulk cytosol) with the downstream effects of PKA using electrophysiological approaches (which report PKA activation beneath the membrane), we demonstrated that neurons have a sub-membrane domain where full PKA activation occurs in ~30 s, whereas it takes 2.5 min (5× longer) to reach the maximum phosphorylation level in the bulk cytosol [21]. The potential of optical biosensors for spatial resolution was later used in combination with two-photon microscopy, revealing that cAMP accumulates in dendrites of small diameter while its concentration fades away in the bulk cytosol, an effect which most likely results from a higher surface to volume ratio in small dendrites, favoring cAMP concentration buildup [23]. Although this had been predicted theoretically [2], imaging provided a clear experimental confirmation of the importance of cell geometry on signal integration.

The spatial dimension can also be accessed using targeting domains to target the probe towards a specific subcellular compartment. AKAR2.2 fused with a nuclear localization signal at its C-terminal was thus used to monitor the kinetics of PKA activation inside the nucleus of neurons in brain slices, and reported a much slower response than in the cytosol [21].

As described above, the spatial organization of the cell of interest directly determines signal integration processes. The biological preparation is therefore an important point, particularly in the case of neurons. Our choice has been to study mature neurons in their most physiological context, i.e., in brain slice preparations made from rodent neonates. This preparation, derived from the work of electrophysiologists, has the advantage of providing morphologically intact differentiated neurons, which maintain functional network connections and bear endogenous receptors. This chapter describes the details of AKAR imaging in brain slices, emphasizing some technical points like image acquisition and image analysis. Most of the information presented here also pertains to imaging in any other preparation, including cell cultures, and potentially to many other dual-emission biosensors.

2 Materials

2.1 Solutions

1. Cutting solution contains: 125 mM NaCl, 0.4 mM $CaCl_2$, 1 mM $MgCl_2$, 1.25 mM NaH_2PO_4, 26 mM $NaHCO_3$, and 25 mM glucose, saturated with 5 % CO_2 and 95 % O_2 (*see* **Note 1**).
2. Culture medium is made of 50 % Minimum Essential Medium, 50 % Hanks' Balanced Salt Solution, 6.5 g/L glucose, penicillin–streptomycin.
3. Recording solution contains: 125 mM NaCl, 0.4 mM $CaCl_2$, 1 mM $MgCl_2$, 1.25 mM NaH_2PO_4, 26 mM $NaHCO_3$, and 25 mM glucose, saturated with 5 % CO_2 and 95 % O_2.
4. A 10× stock solution of 1.25 M NaCl, 12.5 mM NaH_2PO_4, 260 mM $NaHCO_3$ is made weekly and kept at 4 °C. On the day of the experiment, a 1× solution is prepared by diluting the stock and adding $CaCl_2$, $MgCl_2$, and glucose to the final concentrations indicated above.

2.2 Brain Slice Preparation

Brain slices are prepared from young rodents, rats at postnatal age P10 to P14, or mice at P8-P12 (*see* **Note 2**). Brain slices are cut using a vibrating blade microtome VT1200S (Leica). Brain slices are kept on a Millicell-CM membrane PICM0RG50 (Millipore).

2.3 Biosensor and Viral Vectors

1. PKA-sensitive AKAR biosensors, as well as a control version of the AKAR2.2 sensor with a T391A point mutation in the phosphorylation site, were provided by Jin Zhang, Johns Hopkins School of Medicine, Baltimore, USA.
2. The viral vector used in our experiments is derived from the Sindbis virus [24] and provided by Invitrogen, San Diego, CA. The coding sequence of the biosensor was sub-cloned into the viral vector, pSinRep5. Viral particles are produced following the guidelines published by Invitrogen.

2.4 Wide-Field Imaging

The imaging setup uses an upright microscope (Olympus BX51WI) equipped with a 20× 0.5 NA, a 40× 0.8 NA, or a 60× 0.9 NA water-immersion objective with a piezoelectric device (P-721 PIFOC, Physik Instrumente GmbH) to remotely control image focus. Epifluorescence is excited using a halogen light source, a shutter (Uniblitz, Vincent and Associates), and a D436/20 filter for excitation (*see* **Note 3**). A 455DCXT dichroic mirror separates the emission from the excitation. The emitted light is filtered by alternating the emission filters, HQ480/40 for CFP and D535/40 for YFP, with a filter wheel (Sutter Instruments, Novato, CA, USA). Filters were obtained from Chroma Technology (Brattleboro, VT, USA) or Semrock (Rochester, NY, USA). Images are recorded with a low noise CCD camera (ORCA-AG, Hamamatsu). Images for shading correction are acquired using a 100 μM solution of Coumarin 343 in a 10 mM sodium phosphate buffer at pH 7, stored in small aliquots and kept at −20 °C (*see* **Note 4**).

2.5 Spectral Analysis

Emission spectra of the biosensor were recorded over a single neuron in the brain slice using a fibered spectrometer USB2000 (Ocean Optics) with a 600 μm fiber placed at the focal image plane of the microscope. The objective was 20× 0.5 N.A.

2.6 Two-Photon Imaging

Two-photon images were obtained with a custom-built two-photon laser scanning microscope based on an Olympus BX51WI upright microscope with a 60× 0.9 NA water-immersion objective and a Ti:sapphire laser (MaiTai HP; Spectra Physics, Mountain View, CA) tuned to 850 nm wavelength and 12 mW power for CFP excitation. Galvanometric scanners (model 6210, Cambridge Technology, Cambridge, MA) were used for raster scanning, and a piezo-driven objective scanner (P-721 PIFOC, Physik Instrumente GmbH) was used to control the depth of the focal plane in the preparation. A two-photon emission filter was used to reject residual excitation light (E700 SP, Chroma Technology, Brattleboro, VT, USA). A fluorescence cube containing 479/40 and 542/50 emission filters and a 506 nm dichroic beamsplitter (FF01-479/40, FF01-542/50, and FF506-Di02-25×36 Brightline Filters, Semrock, Rochester, NY, USA) was used for the orthogonal separation of the two fluorescence signals. Two imaging channels (H9305 photomultipliers, Hamamatsu) were used for simultaneous detection of the two types of fluorescence emission.

2.7 Image Acquisition

The wide-field imaging setup is controlled using iVision-Mac software (BioVision technologies, Exton, USA) and custom scripts. The two-photon imaging system is controlled by MPscope software [25].

2.8 Image Analysis

Image analysis is performed using Igor software (Wavemetrics, USA) and custom-written functions.

3 Methods

3.1 Brain Slice Preparation

The animal is killed by decapitation, following the guidelines of our institution. The brain is quickly removed and placed in ice-cold cutting solution. After cutting, brain slices are left for 1 h to recover in cutting solution. The Millicell membrane is placed into a 35 mm petri dish containing 1 ml of culture medium and incubated for 1 h at 35 °C in 5 % CO_2 atmosphere (*see* **Note 5**). Two to three brain slices are carefully placed onto the Millicell membrane and all cutting solution is delicately removed. The slices are then left for 1 h in the incubator before infection with a viral vector encoding the appropriate AKAR biosensor. Viral infection is performed by gently spreading 1–3 μl of virus suspension over the surface of the brain slice with a pipette. The dish is then kept overnight in the incubator.

The next day, 1 ml of recording solution is placed on the Millicell membrane and the brain slices are gently detached and transferred into a 100 ml beaker containing recording solution gassed with 95 % O_2/5 % CO_2. Slices are left for 1 h in this solution, a time needed to recover a pH/metabolic equilibrium (*see* **Note 6**).

The imaging chamber on the microscope is continuously perfused with the recording solution saturated with 95 % O_2/5 % CO_2 at a rate of 2 ml/min. Smoothly switching between different reservoirs allows the bathing solution to be changed to a solution containing drugs without mechanically disturbing the preparation.

3.2 Wide-Field Imaging

Images (one for each CFP and YFP channel) are acquired with an exposure duration of 0.2–1 s, depending on the preparation expression level, illumination power and camera sensitivity. Image pairs are recorded at time intervals from 5 s to several minutes, depending on the duration of the biological event to be monitored (*see* **Note 3**). Besides lateral drift of the brain slice, which can be easily corrected off-line (*see* below), focus drift must be corrected in real time.

3.3 Two-Photon Imaging

CFP and YFP signals are recorded simultaneously during the scanning of the preparation by the laser beam. 512×512 pixels images are acquired at a rate of 1.6 frames/s. Vertical stacks of ~50 images are recorded with a 0.5 μm increment step.

For both wide-field and two-photon imaging, a total number of 100 data points with negligible photobleaching over a 1–2 h time period is common practice.

3.4 Image Analysis (Fig. 1)

Images saved by the acquisition software are immediately processed by the analysis program and the resulting ratio measurements and pseudo-color images are displayed in real time. Image analysis comprises a series of steps, based on the description made for ratiometric calcium imaging [26]:

1. Subtract the camera offset from each image.
2. Divide the intensity values by the exposure time.
3. Divide each image by the shading image of the corresponding imaging channel (*see* **Note 4**).
4. Subtract an optional background value from the image. Since the autofluorescence of the neurons is usually low compared to the fluorescence of the biosensor, and since this value cannot be determined objectively, this parameter is usually left null.
5. Correct the images for preparation movements, typically horizontal translations of several microns (*see* **Note 7**).

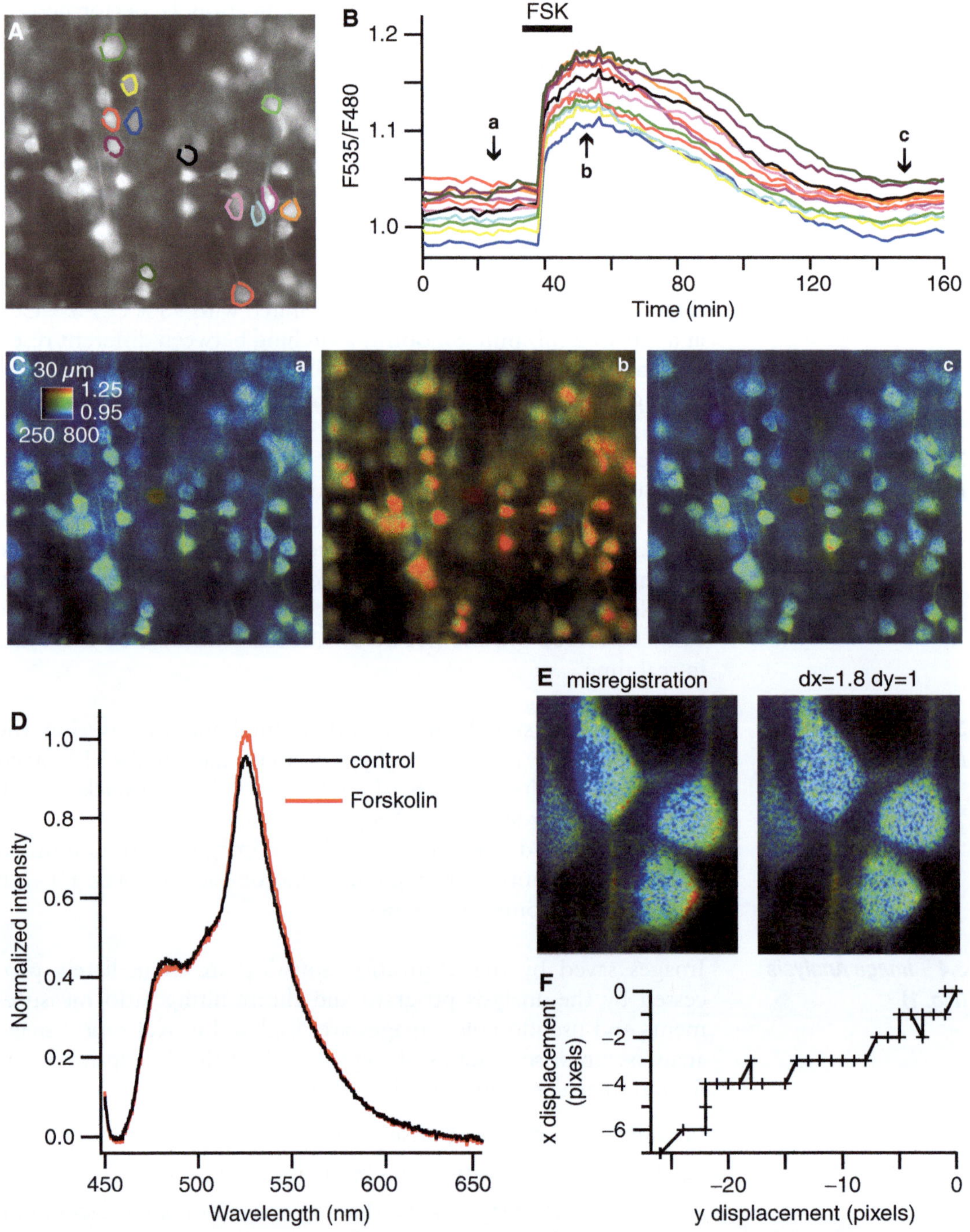

Fig. 1 Monitoring PKA activity in cortical neurons expressing AKAR2.2 by wide-field microscopy. (**a**) Fluorescence image of the brain slice expressing AKAR2.2 at 535 nm emission wavelength. (**b**) F535/F480 ratio values during the course of the recording, measured over the regions of interest of the same color and drawn on (**a**). 13 μM forskolin (FSK) was added to the bath for the time period indicated by the bar. *Arrows* indicate the time corresponding to the pseudocolor images in (**c**). (**c**) Pseudocolor images representing the F535/F480 emission ratio in control condition (*a*), in the presence of 13 μM FSK (*b*), after recovery (*c*). The calibration square indicates fluorescence intensity horizontally and the F535/F480 ratio vertically. (**d**) AKAR2.2 emission spectrum

6. Correct the images for misregistration, usually by less than 2 pixels (*see* **Note 8**).
7. Calculate the ratio pixel by pixel as Ratio = (YFP)/(CFP), using the YFP and CFP images at the two wavelengths, after the corrections described above.
8. Display the images in pseudocolor, coding the ratio value in hue and the fluorescence in intensity (*see* **Note 9**).
9. Add a calibration scale to the image. Since pseudocolor images represent two dimensions (ratio and intensity), the calibration becomes a square, showing the intensity of the pixel (in counts per pixel per second) on the horizontal axis and the ratio on the vertical axis. The size of this calibration square also defines the size scale of the image.
10. Use the regions of interest drawn on the image to calculate the average ratio as an average weighted by the intensity (*see* **Note 10**).
11. There are sometimes nonlinear deformations within the brain slice, resulting in movements which cannot be corrected by a global translation. Fine local adjustments of the regions of interest compensate for these small distortions.

3.5 Controls

CFP and YFP are far from ideal chromophores and both exhibit pH and environment sensitivities. It is therefore quite useful to check that the ratio changes recorded in an experiment are not artifactual. A control biosensor mutated in the phosphorylation site must report no ratio change in the same situation as the experiment and thus demonstrates that the ratio signal indeed reflects a phosphorylation event. The T391A mutant of AKAR2.2 is used systematically to check for a lack of ratio response in the experimental conditions of interest. The mutant biosensor also shows that cell to cell differences in ratio are not the consequence of different PKA activation level but are intrinsic to the biosensor (*see* **Note 11**) (Fig. 2).

3.6 Targeting the Biosensor to a Specific Subcellular Compartment

The fusion of a targeting domain to the biosensor allows for precise subcellular targeting into a specific cell compartment. For instance, a nuclear localization signal (NLS) fused to AKAR2 allows for the measurement of PKA activation in the nucleus [21] (Fig. 3).

Targeting the biosensor to the membrane has also been accomplished in a variety of cultured cells [20].

Fig. 1 (continued) recorded on an individual neuron in control condition and 15 min after bath application of 13 μM forskolin. (**e**) Magnification of cell bodies showing the effect of a misregistration between the F535 and F480 channels (*left*) and the corrected image (*right*). Misregistered ratio images display the characteristic "rainbow" appearance, with one edge with low ratio while the opposite edge has high ratio. The correction is calculated at the 1/5th pixel resolution. (**f**) Movement of the brain slice that occurred during the time course of the experiment

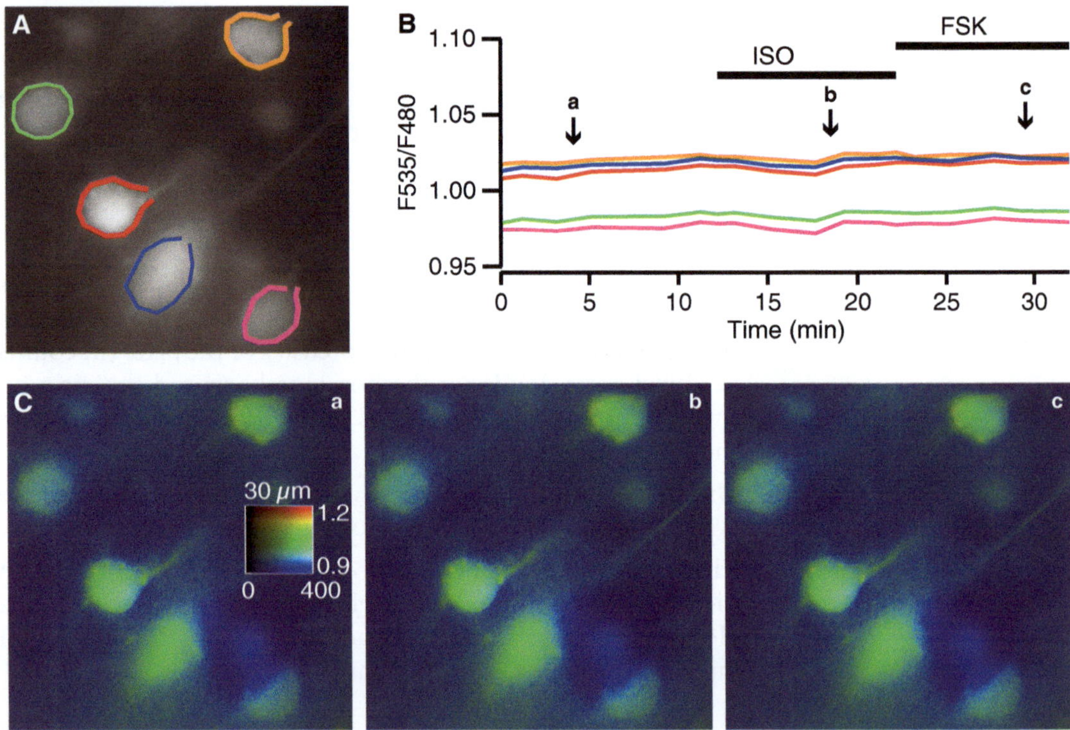

Fig. 2 Lack of response when the phosphorylation site of AKAR2.2 is mutated. (**a**) Fluorescence image of the brain slice expressing AKAR2.2mut (T391A) viewed by wide-field microscopy at the 535 nm emission wavelength. (**b**) Ratio values during the course of the recording, measured over the regions of interest of the same color and drawn on (**a**). 1 μM isoproterenol (ISO) and 13 μM forskolin (FSK) were applied in the bath for the durations indicated by the bars on the graph. *Arrows* indicate the time corresponding to the images in (**c**). (**c**) Pseudocolor images in control condition (*a*), in the presence of 1 μM isoproterenol (*b*), and in the presence of 13 μM forskolin (*c*). The calibration square indicates fluorescence intensity horizontally and F535/F480 ratio vertically

3.7 Monitoring cAMP with Epac-Based Biosensors

There are several biosensors for cAMP with slightly differing properties and a rather similar affinity for cAMP in the micromolar range [27–29]. More recently, a cAMP biosensor containing a cAMP binding pocket with increased affinity (~0.3 μM) was derived [30]. Sequential use of such biosensors allows for a rough estimate of the concentration range of cAMP reached during a given cellular response. In pyramidal cortical neurons, Epac1-camps, with micromolar affinity, shows no response to a β-adrenergic stimulation, while the Epac2-camps300, with 0.3 μM affinity shows a small but clear response (Fig. 4, from ref. 23). The sensitivity difference between these two sensors is also visible during the forskolin response, which almost saturates with Epac2-camps300 while it is only half of the maximum with Epac1-camps (Fig. 4).

3.8 Two-Photon Imaging

Two-photon image acquisition and image analysis is performed in a manner similar to wide-field imaging with the following differences: (1) laser scanning provides images that are intrinsically registered,

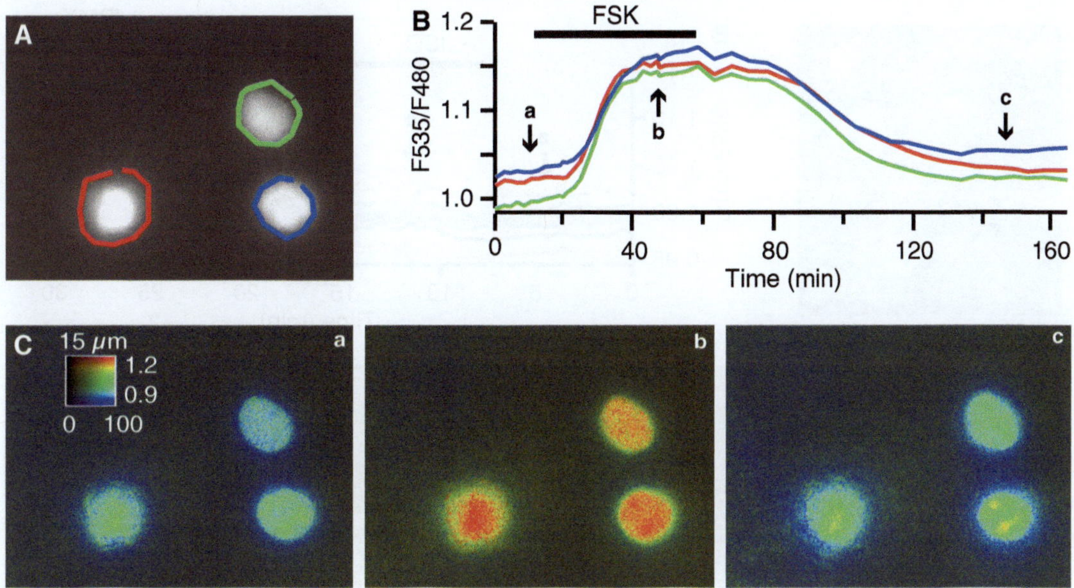

Fig. 3 Wide-field imaging of the PKA activity inside the nucleus of cortical neurons expressing AKAR2-NLS. (**a**) Fluorescence image of the brain slice at 535 nm emission wavelength. (**b**) F535/F480 ratio values during the course of the recording, measured over the regions of interest of the same color and drawn on (**a**). 13 μM forskolin (FSK) was applied in the bath for the duration indicated by the bar on the graph. (**c**) Pseudocolor images in control condition (*a*), in the presence of 13 μM forskolin (*b*) and after recovery (*c*). The calibration square indicates fluorescence intensity horizontally and F535/F480 ratio vertically

so there is no need for misregistration correction; (2) for each data point, a stack of images is recorded on a thickness of up to 50 μm at 0.5 μm intervals; and (3) corrections for preparation movements are applied to the *X*, *Y and Z* dimensions, thereby correcting for possible focus drift.

While wide-field images are intrinsically fuzzy (Fig. 5a), two-photon microscopy provides much sharper images (Fig. 5a) allowing for the measurement in specific subcellular domains such as dendritic branches and even spines (Fig. 5c). With this improved spatial resolution, the response to receptor activation can be monitored specifically over the somatic cytosol (Fig. 6) as well as within thin dendritic branches (Fig. 7).

3.9 Response Quantification

The basal ratio is often quite variable from cell to cell, and this variability is generally observed with the control biosensor which is insensitive to the biological signal of interest (Fig. 2). This suggests that some yet to be determined effect influences the biosensor optical properties, including the maturation of the chromophore [16] and absolute calibration of ratio values is therefore impossible at the present time. The ratio response must therefore be calibrated between the Rmin and Rmax values, as originally described for calcium dyes [9] and later adapted for cAMP biosensors [31].

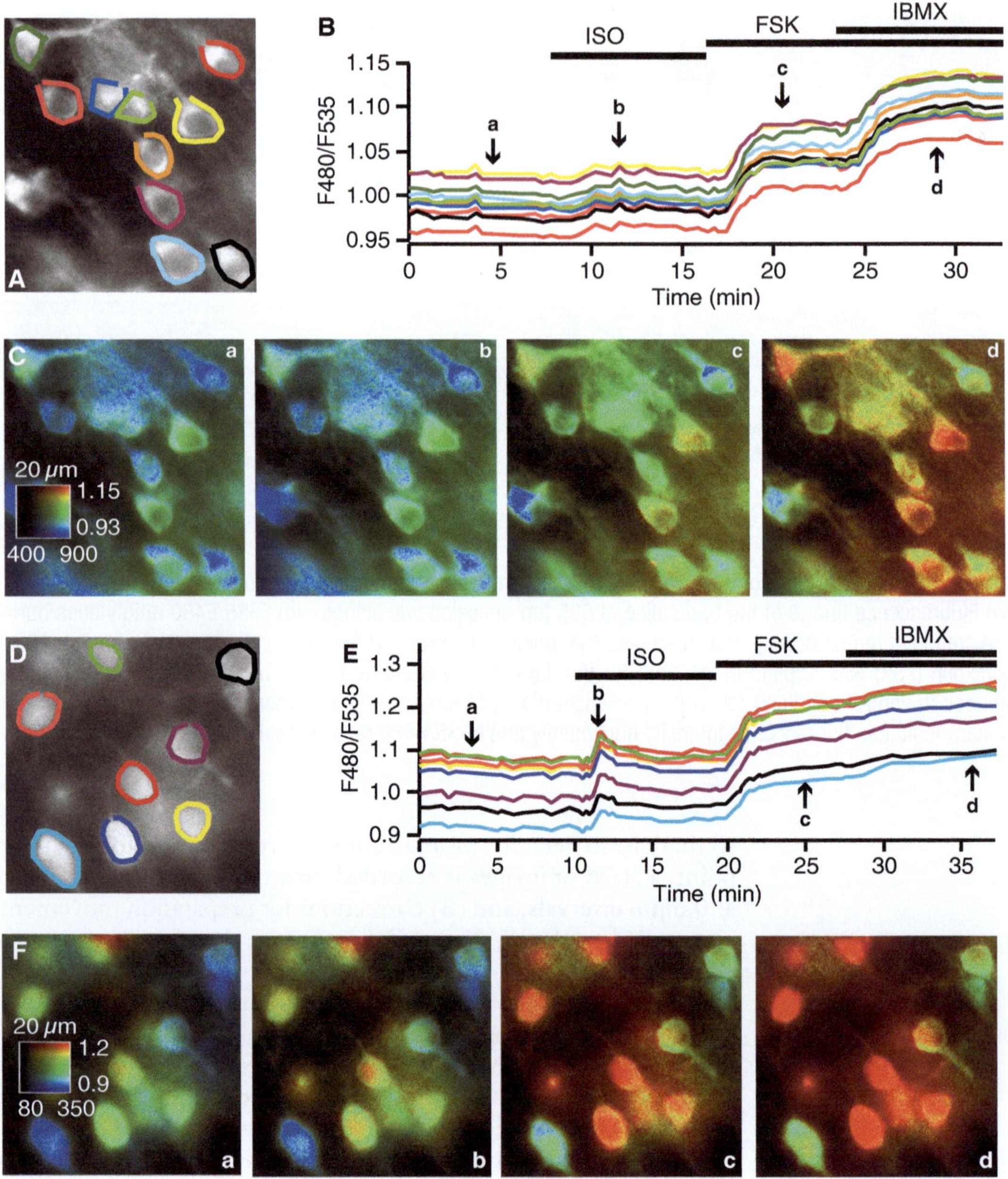

Fig. 4 Monitoring changes in cAMP concentration in response to β-adrenergic stimulation with Epac1-camps and the more sensitive Epac2-camps300 using wide-field microscopy. (**a**, **d**), Wide-field fluorescence images of brain slices expressing Epac1-camps (**a**) and Epac2-camps300 (**d**) at the 535 nm emission wavelength. (**b**, **e**), Each trace indicates the F480/F535 emission ratio measured on the regions indicated by the color contour drawn on the pseudocolor images. 100 nM isoproterenol (ISO), 13 μM forskolin (FSK) and 200 μM IBMX were applied in the bath for the period of time indicated by the bars. (**c**, **f**), Pseudocolor images presenting fluorescence emission ratios indicative of changes in cAMP concentration, as measured with Epac1-camps (**c**) and Epac2-camps300 (**f**) in control condition (*a*), in the presence of isoproterenol (*b*), forskolin (*c*), and forskolin + IBMX (*d*)

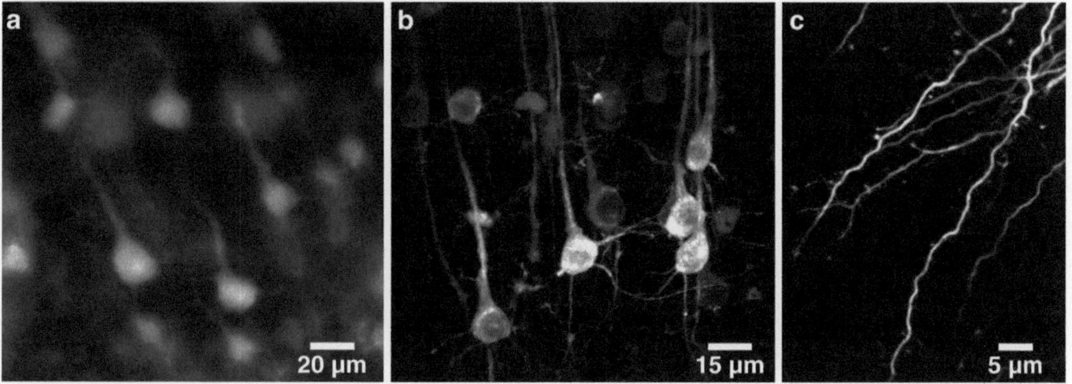

Fig. 5 Fluorescence images of cortical brain slices expressing AKAR2.2 at the 535 nm emission wavelength recorded (**a**) by wide-field microscopy, (**b**) by two-photon microscopy. (**c**) High resolution imaging of the thin dendrites and dendritic spines by two-photon microscopy

3.10 Improved Biosensors for the cAMP/PKA Signaling Cascade

AKAR2.2 was the first biosensor to be routinely used in brain slice preparations, as it provided sizable ratio changes that were reversible on the time-scale of an experiment. This sensor was later improved by replacing the citrine acceptor with a cpVenus, producing AKAR3 with a larger maximal ratio change (Fig. 8) [19]. Further improvement of this biosensor was obtained by changing the CFP donor into a cerulean, having less bleaching and even larger ratio change [20]. This AKAR4 biosensor, as well as other new similar constructs [32], has not yet been tested in brain slices.

An improved version of the cAMP sensor, ^{T}Epac^{VV}, with an amazing 80 % ratio change has also been released recently but to our knowledge has not yet been tested in brain slices [15].

Further development in the field of biosensor development will undoubtedly change the landscape of biosensor imaging with improved signal resolution in space and time, thereby opening new windows on the amazing complexity of intracellular signal integration in neurons.

4 Notes

1. There are various recipes for the cutting buffer, which would maintain the neurons in better shape, containing kynurenate, lower sodium, etc. The precise recipe used for an experiment may depend on the brain region and should be tested and optimized for each brain region.
2. Although brain slices from young animals are more transparent and have a higher infection rate, both conditions that are favorable for imaging, it is still possible to do imaging on slices from adult rodents.

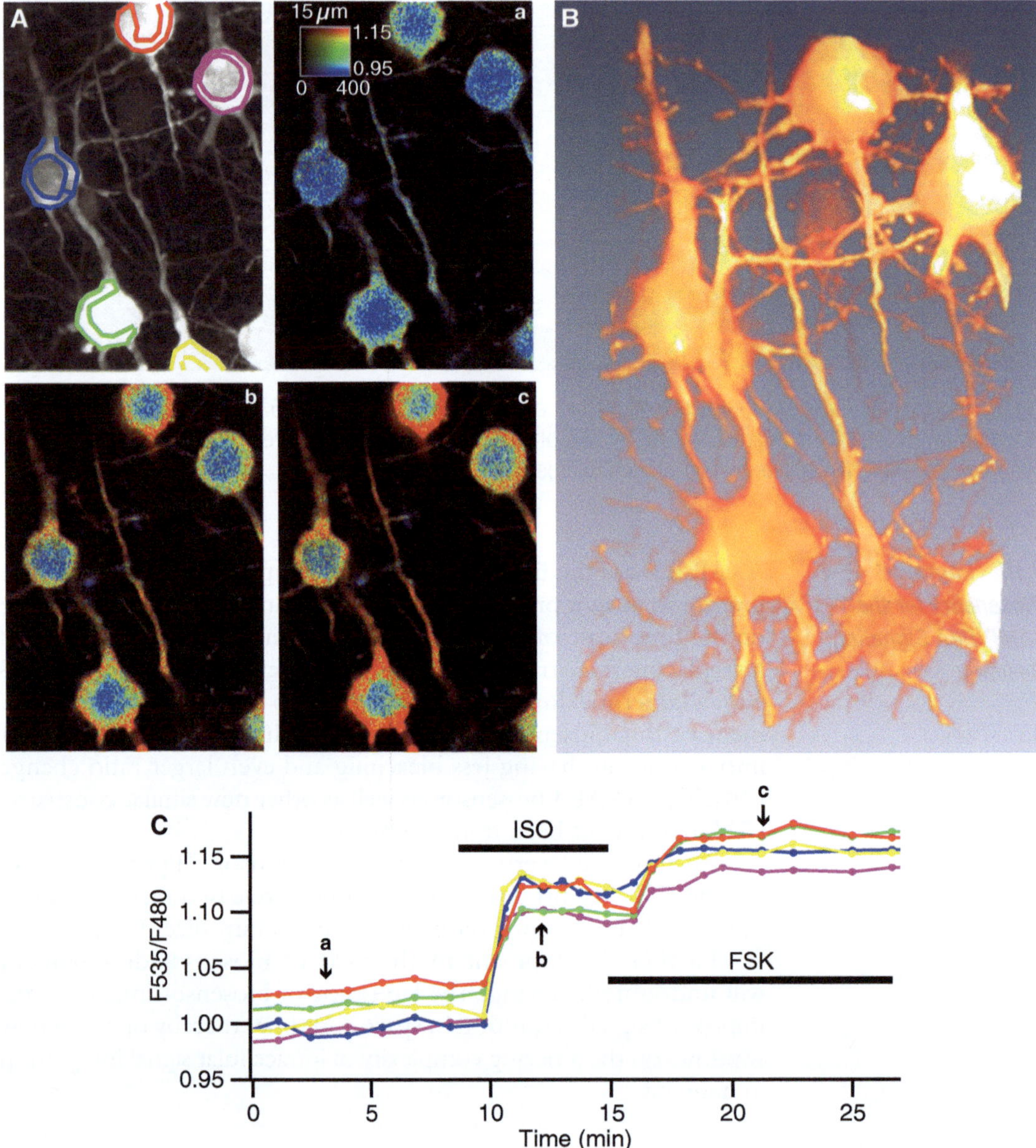

Fig. 6 Monitoring PKA signaling in the soma of cortical neurons expressing AKAR2.2 by two-photon microscopy. (**a**) *Top left*: vertical projection of the brain slice image stack (20 μm thick, 0.5 μm interval) recorded at the 535 emission wavelength. Pseudocolor images in control condition (*a*), in the presence of 100 nM isoproterenol (*b*) and in the presence of 13 μM forskolin (*c*). (**b**) 3D visualization of the image stack (**c**) Ratio values during the course of the recording, measured over the somatic cytosol of the cells. The nucleus, which has different response kinetics, was excluded from the measurement. Each trace corresponds to the cell selected by the region of interest of the same color and drawn on (**a**). *Arrows* indicate the time corresponding to the images in (**a**). 100 nM isoproterenol (ISO) and 13 μM forskolin (FSK) were applied in the bath for the durations indicated by the bars

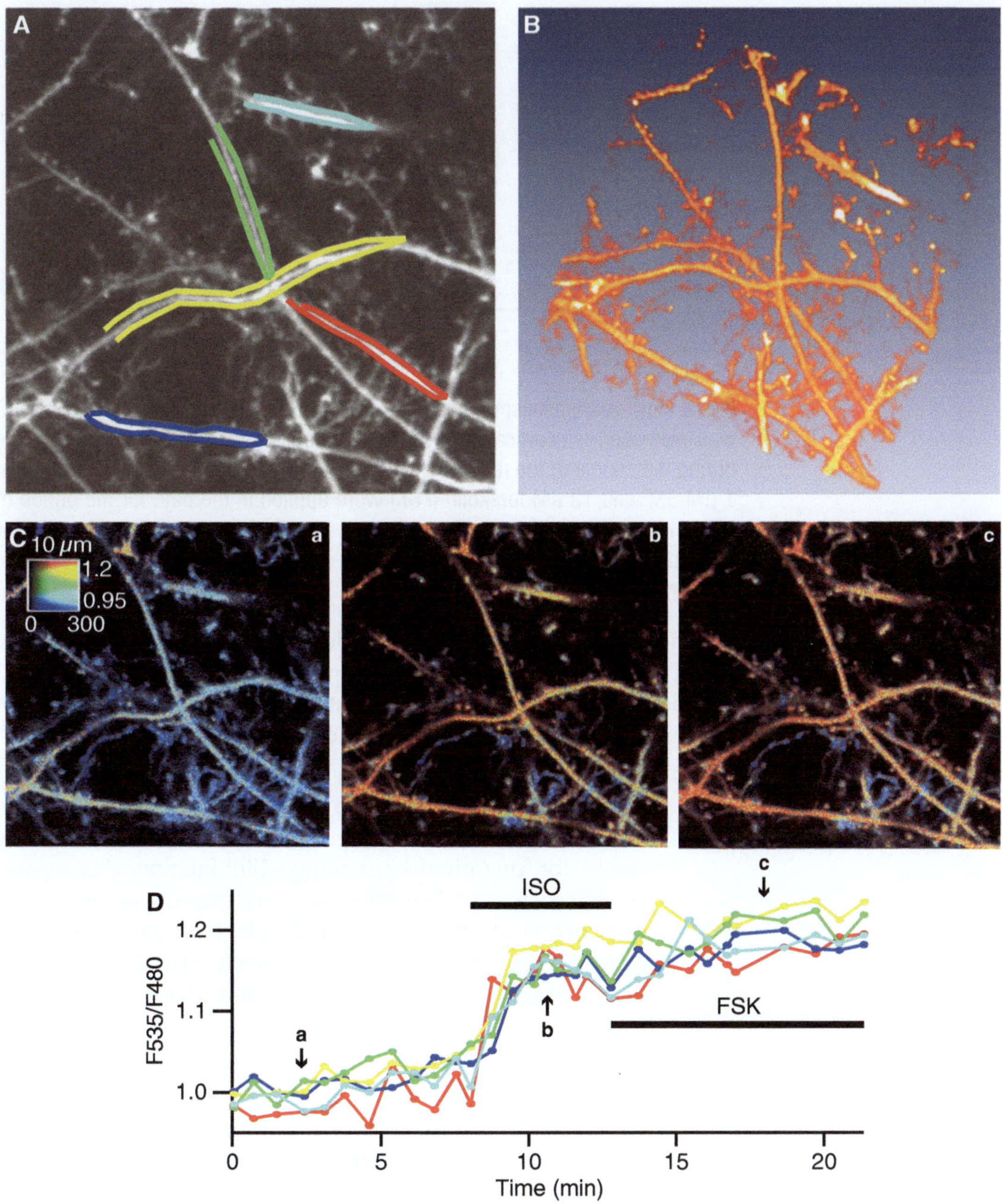

Fig. 7 AKAR2.2 response measured in thin dendrites of cortical neurons by two-photon microscopy. (**a**) Vertical projection of the brain slice image stack (8 μm thick, 0.5 μm interval) recorded at 535 nm emission wavelength. (**b**) 3D visualization of the image stack (**c**) Pseudocolor images in control condition (*a*), in the presence of 100 nM isoproterenol (*b*) and in the presence of 13 μM forskolin (*c*). (**d**) Ratio values during the course of the recording, measured on the regions indicated by the color contour on the grayscale image in (**a**). *Arrows* indicate the time corresponding to the images in (**a**). 100 nM isoproterenol (ISO) and 13 μM forskolin (FSK) were applied in the bath for the durations indicated by the bars

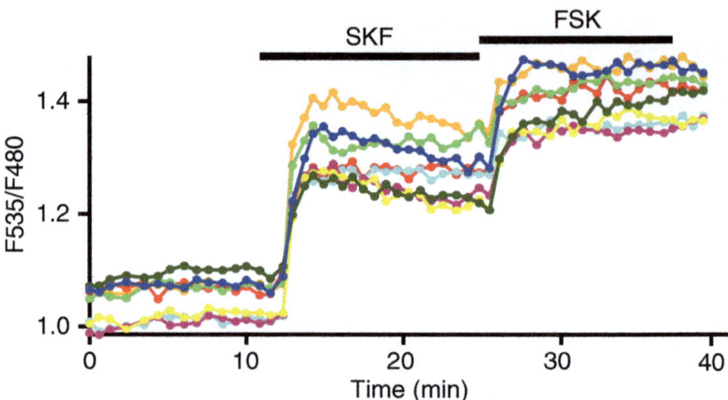

Fig. 8 Improved biosensors result in much larger signal over noise ratio. AKAR3 was expressed in a cortical brain slice. Each trace corresponds to the ratio values during the course of the recording, measured on the soma of individual neurons. 1 μM SKF and 13 μM forskolin (FSK) were applied in the bath for the period of time indicated by the bars. The response amplitude is larger than the similar experiment shown in Fig. 6

3. Bleaching of the biosensor may dramatically limit the experiment. Therefore, unnecessary illumination of the biosensor should be avoided. Below we have listed some common sources of useless damage to the fluorophores:
   (a) Excessively high illumination power: epifluorescence imaging is performed quite well with a 100 W halogen light source, which is sufficient to distinguish the cells by eye and is very stable for quantitative imaging. Illumination that is too strong, such as arc lamps that are used improperly (i.e., without correct attenuation), can bleach the fluorophores and suppress the ratio change. An indication of this phenomenon is a rapid drift of the ratio baseline as a function of the cumulated illumination duration, associated with very small ratio responses.
   (b) Poor illumination control: the light path used for epifluorescence must be electronically controlled by a fast shutter, only allowing illumination of the preparation when acquiring images.
   (c) Too strong transillumination while looking at the biological sample: simply put a yellow tainted glass in the transillumination light path in order to prevent bleaching by the blue component of the transillumination light.
4. Shading images are images of a uniformly fluorescent object, such as a fluorescent solution placed in a hemocytometer. Coumarin 343 (100 μM) is convenient because it emits in both CFP and YFP channels. Since the nonuniformity of the field may be slightly different for the two channels, a shading

image is acquired for each detection channels. Data images are divided by the shading image of the corresponding wavelength, which serves two purposes. First, dividing the acquired image by the shading image corrects for the nonuniformity of the imaging system, usually the center of the image being brighter than the edges. This correction is sometimes called flat-field correction. Shading correction serves a second purpose, as well: since the images from each channel are divided by their respective shading, by definition, the ratio of the shading solution equals 1. Shading correction thus defines the ratio value of 1 as equal to the ratio measured for the specific shading chromophore (Coumarin 343, in our case) which can thus be used to roughly compare ratio values across different imaging setups.

5. It has now been widely reported (although not explained) that brain slices survive better at 35 °C than at 37 °C. A specific incubator should be dedicated to this temperature. Infections also succeeded at room temperature when performed directly inside a 100 ml beaker. In this case, the solution level should be adjusted so that the slices are held on a mesh close to the surface, at the interface with air just for the duration of the infection (30 min), then later submerged again and kept overnight, with continuous gassing with 95 % O_2/5 % CO_2.
6. Drifts of the baseline ratio are sometimes observed during the course of the experiment. This effect can also be observed with AKARmut, showing an effect independent of phosphorylation events, possibly reflecting changes in chromophore properties. Since this effect was reduced when the slices were pre-incubated in the recording medium for ~1 h, it is suspected to result from changes in pH or chloride concentration, which eventually stabilize in the recording solution.
7. During the course of the experiment, the biological object often moves as a result of mechanical instability of the microscope or intrinsic movements of the biological object. A brain slice typically moves by tens of microns, both horizontally and vertically. While the vertical movement must be corrected by adjusting the focus of the microscope, horizontal translations are easy to correct using software, translating the acquired image until it is registered with a reference image. Both image registration and translation are calculated automatically using built-in Igor Pro routines [33].
8. Since all microscopes exhibit some level of chromatic aberration (objective, filter thickness and wedge effect), it is often necessary to register the images from the two channels at the sub-pixel level. This correction is usually determined once for an imaging system and applied to all images acquired on this system.

9. The color of each pixel is coded in tridimensional space and we use the "hue-saturation-value" coding, representing the ratio with a hue from blue to red, and the fluorescence level with the intensity [26]. Saturation is set to 100 %. This results in more "natural looking" images, with the most intense regions of the image appearing bright, and the image gradually fading into darker hues and eventually black as the fluorescence intensity in the original image declines towards background. This display mode thus avoids the artificial thresholding of images which show arbitrary sharp edges.
10. The ratio is evaluated as a weighted average: each pixel in the region of interest is weighted by the intensity of the fluorescence [26]. As a consequence, bright pixels, corresponding to regions of the image containing a large amount of biosensor will contribute more to the average than dim pixels where the biosensor is scarce. This has the advantage that background pixels accidentally contained within a region of interest will be cancelled-out in the average. When fluorescence intensity is close to background, the ratio converges towards "0/0" and values can randomly reach arbitrary values from $-\infty$ to $+\infty$. To remove these infinite values, all ratio values are cropped from $0.1 \times R_{min}$ to $10 \times R_{max}$.
11. The cell to cell differences in the ratio value observed with this control biosensor indicate that different cells "process" the biosensor differently. Sometimes, aggregates where the ratio value is maximal can be seen in the cytosol and demonstrates that rough absolute ratio values cannot be used to evaluate basal cAMP concentrations or PKA phosphorylation levels. The ratio response must be evaluated compared to the R_{min} and R_{max} determined experimentally for each cell.

References

1. Tasken K, Aandahl EM (2004) Localized effects of cAMP mediated by distinct routes of protein kinase A. Physiol Rev 84:137–167
2. Neves SR, Tsokas P, Sarkar A, Grace EA, Rangamani P, Taubenfeld SM, Alberini CM, Schaff JC, Blitzer RD, Moraru II, Iyengar R (2008) Cell shape and negative links in regulatory motifs together control spatial information flow in signaling networks. Cell 133:666–680
3. Adams SR, Harootunian AT, Buechler YJ, Taylor SS, Tsien RY (1991) Fluorescence ratio imaging of cyclic AMP in single cells. Nature 349:694–697
4. Bacskai BJ, Hochner B, Mahaut-Smith M, Adams SR, Kaang BK, Kandel ER, Tsien RY (1993) Spatially resolved dynamics of cAMP and protein kinase A subunits in Aplysia sensory neurons. Science 260:222–226
5. Hempel CM, Vincent P, Adams SR, Tsien RY, Selverston AI (1996) Spatio-temporal dynamics of cyclic AMP signals in an intact neural circuit. Nature 384:166–169
6. Vincent P, Brusciano D (2001) Cyclic AMP imaging in neurones in brain slice preparations. J Neurosci Methods 108:189–198
7. Miyawaki A, Llopis J, Heim R, McCaffery JM, Adams JA, Ikura M, Tsien RY (1997) Fluorescent indicators for Ca2+ based on green fluorescent proteins and calmodulin. Nature 388:882–887
8. Romoser VA, Hinkle PM, Persechini A (1997) Detection in living cells of Ca2+-dependent

changes in the fluorescence emission of an indicator composed of two green fluorescent protein variants linked by a calmodulin-binding sequence. A new class of fluorescent indicators. J Biol Chem 272:13270–13274

9. Grynkiewicz G, Poenie M, Tsien RY (1985) A new generation of Ca2+ indicators with greatly improved fluorescence properties. J Biol Chem 260:3440–3450
10. Zhang J, Campbell RE, Ting AY, Tsien RY (2002) Creating new fluorescent probes for cell biology. Nat Rev Mol Cell Biol 3: 906–918
11. Dulla C, Tani H, Okumoto S, Frommer WB, Reimer RJ, Huguenard JR (2008) Imaging of glutamate in brain slices using FRET sensors. J Neurosci Methods 168:306–319
12. Hires SA, Zhu Y, Tsien RY (2008) Optical measurement of synaptic glutamate spillover and reuptake by linker optimized glutamate-sensitive fluorescent reporters. Proc Natl Acad Sci U S A 105:4411–4416
13. Laxman B, Hall DE, Bhojani MS, Hamstra DA, Chenevert TL, Ross BD, Rehemtulla A (2002) Noninvasive real-time imaging of apoptosis. Proc Natl Acad Sci U S A 99:16551–16555
14. Ai HW, Hazelwood KL, Davidson MW, Campbell RE (2008) Fluorescent protein FRET pairs for ratiometric imaging of dual biosensors. Nat Methods 5:401–403
15. Klarenbeek JB, Goedhart J, Hink MA, Gadella TW, Jalink K (2011) A mTurquoise-based cAMP sensor for both FLIM and ratiometric read-out has improved dynamic range. PLoS One 6:e19170
16. van der Krogt GN, Ogink J, Ponsioen B, Jalink K (2008) A comparison of donor-acceptor pairs for genetically encoded FRET sensors: application to the Epac cAMP sensor as an example. PLoS One 3:e1916
17. Vincent P, Gervasi N, Zhang J (2008) Real-time monitoring of cyclic nucleotide signaling in neurons using genetically encoded FRET probes. Brain Cell Biol 36:3–17
18. Zhang J, Hupfeld CJ, Taylor SS, Olefsky JM, Tsien RY (2005) Insulin disrupts beta-adrenergic signalling to protein kinase A in adipocytes. Nature 437:569–573
19. Allen MD, Zhang J (2006) Subcellular dynamics of protein kinase A activity visualized by FRET-based reporters. Biochem Biophys Res Commun 348:716–721
20. Depry C, Allen MD, Zhang J (2011) Visualization of PKA activity in plasma membrane microdomains. Mol Biosyst 7:52–58
21. Gervasi N, Hepp R, Tricoire L, Zhang J, Lambolez B, Paupardin-Tritsch D, Vincent P (2007) Dynamics of protein kinase A signaling at the membrane, in the cytosol, and in the nucleus of neurons in mouse brain slices. J Neurosci 27:2744–2750
22. Hu E, Demmou L, Cauli B, Gallopin T, Geoffroy H, Harris-Warrick RM, Paupardin-Tritsch D, Lambolez B, Vincent P, Hepp R (2011) VIP, CRF, and PACAP Act at distinct receptors to elicit different cAMP/PKA dynamics in the Neocortex. Cereb Cortex 21: 708–718
23. Castro LR, Gervasi N, Guiot E, Cavellini L, Nikolaev VO, Paupardin-Tritsch D, Vincent P (2010) Type 4 phosphodiesterase plays different integrating roles in different cellular domains in pyramidal cortical neurons. J Neurosci 30:6143–6151
24. Ehrengruber MU, Lundstrom K, Schweitzer C, Heuss C, Schlesinger S, Gahwiler BH (1999) Recombinant Semliki Forest virus and Sindbis virus efficiently infect neurons in hippocampal slice cultures. Proc Natl Acad Sci U S A 96:7041–7046
25. Nguyen QT, Tsai PS, Kleinfeld D (2006) MPScope: a versatile software suite for multiphoton microscopy. J Neurosci Methods 156: 351–359
26. Tsien RY, Harootunian AT (1990) Practical design criteria for a dynamic ratio imaging system. Cell Calcium 11:93–109
27. DiPilato LM, Cheng X, Zhang J (2004) Fluorescent indicators of cAMP and Epac activation reveal differential dynamics of cAMP signaling within discrete subcellular compartments. Proc Natl Acad Sci U S A 101: 16513–16518
28. Nikolaev VO, Bunemann M, Hein L, Hannawacker A, Lohse MJ (2004) Novel single chain cAMP sensors for receptor-induced signal propagation. J Biol Chem 279: 37215–37218
29. Ponsioen B, Zhao J, Riedl J, Zwartkruis F, van der Krogt G, Zaccolo M, Moolenaar WH, Bos JL, Jalink K (2004) Detecting cAMP-induced Epac activation by fluorescence resonance energy transfer: Epac as a novel cAMP indicator. EMBO Rep 5:1176–1180
30. Norris RP, Ratzan WJ, Freudzon M, Mehlmann LM, Krall J, Movsesian MA, Wang H, Ke H, Nikolaev VO, Jaffe LA (2009) Cyclic GMP from the surrounding somatic cells regulates cyclic AMP and meiosis in the mouse oocyte. Development 136:1869–1878
31. Börner S, Schwede F, Schlipp A, Berisha F, Calebiro D, Lohse MJ, Nikolaev VO (2011) FRET measurements of intracellular cAMP concentrations and cAMP analog permeability in intact cells. Nat Protoc 6:427–438

32. Komatsu N, Aoki K, Yamada M, Yukinaga H, Fujita Y, Kamioka Y, Matsuda M (2011) Development of an optimized backbone of FRET biosensors for kinases and GTPases. Mol Biol Cell 22:4647–4656

33. Thevenaz P, Ruttimann UE, Unser M (1998) A pyramid approach to subpixel registration based on intensity. IEEE Trans Image Process 7:27–41

Chapter 15

Optical Calcium Imaging Using DNA-Encoded Fluorescence Sensors in Transgenic Fruit Flies, *Drosophila melanogaster*

Shubham Dipt, Thomas Riemensperger, and André Fiala

Abstract

The invention of protein-based fluorescent biosensors has paved the way to target specific cells with these probes and visualize intracellular processes not only in isolated cells or tissue cultures but also in transgenic animals. In particular, DNA-encoded fluorescence proteins sensitive to Ca^{2+} ions are often used to monitor changes in intracellular Ca^{2+} concentrations. This is of particular relevance in neuroscience since the dynamics of intracellular Ca^{2+} concentrations represents a faithful correlate for neuronal activity, and optical Ca^{2+} imaging is commonly used to monitor spatiotemporal activity across populations of neurons. In this respect *Drosophila* provides a favorable model organism due to the sophisticated genetic tools that facilitate the targeted expression of fluorescent Ca^{2+} sensor proteins. Here we describe how optical Ca^{2+} imaging of neuronal activity in the *Drosophila* brain can be carried out in vivo using two-photon microscopy. We exemplify this technique by describing how to monitor odor-evoked Ca^{2+} dynamics in the primary olfactory center of the *Drosophila* brain.

Key words Optical Ca^{2+} imaging, Neuronal activity, *Drosophila melanogaster*, Two-photon microscopy, G-CaMP, Olfactory coding

1 Introduction

The invention of DNA-encoded fluorescence sensor proteins designed to report correlates of physiological processes within cells has brought forward physiological research remarkably [1, 2]. This progress is most evident if one goes beyond the investigation of single cultured cells or tissue preparations and aims at analyzing cellular processes in tissues of intact, living animals, e.g., in correlation with behavior. For a long time, analyzing intracellular, biochemical processes, e.g., second messenger-mediated signaling, has relied on biochemical assays that require mechanical tissue preparation while monitoring changes in membrane potential typically involves electrophysiological techniques. In both cases one can either investigate only very few cells with high temporal preci-

Jin Zhang et al. (eds.), *Fluorescent Protein-Based Biosensors: Methods and Protocols*, Methods in Molecular Biology, vol. 1071, DOI 10.1007/978-1-62703-622-1_15, © Springer Science+Business Media, LLC 2014

sion and high specificity with respect to cell types, or investigate larger pieces of tissue (e.g., parts of a brain) consisting of many diverse types of cells.

This situation has changed with the emergence of optical imaging techniques. The spatiotemporal dynamics of second messenger synthesis, kinase activity, vesicle release or membrane potential can now be monitored in real-time across larger arrays of cells using fluorescence dyes and appropriate microscopic setups [3]. The advantage of DNA-encoded fluorescence sensors over "conventional" synthetic fluorescence dyes is obvious. Perfusing tissues with a fluorescent dye relies on physical parameters, e.g., the diffusion properties of the dye or the tissue density at the site at which the dye is injected. The exact cell type to be monitored is, therefore, difficult to specify. Conversely, if the dye is injected directly into the cells under investigation, the number of cells that can be simultaneously analyzed remains limited. On the contrary, DNA-encoded fluorescence proteins can be expressed by cells under the control of regulatory DNA elements (e.g., enhancers or promoters) that specify the cells with respect to a common genetic identity and, potentially, a common function. This advantage is most obvious in nervous tissue, as information processing is typically accomplished by networks of diverse types of interconnected nervous cells. Neuronal activity is primarily characterized by changes in membrane potential and subsequent transmitter release at chemical synapses, and the modulation of neuronal activity often involves second-messenger cascades. DNA-encoded fluorescence probes that are sensitive to these parameters indeed exist, which enables one to monitor dynamic changes in the synthesis of cyclic nucleotides [4], membrane potential [5] or intracellular Ca^{2+} [6]. Unfortunately, the DNA-encoded voltage sensor proteins designed for optically recording changes in membrane potential that are currently available do not yet provide sufficient signal-to-noise ratios for in vivo applications. On the contrary, sensors reporting synaptic vesicle release [7] have been very successfully used in the nervous system of largely intact animals. However, the most commonly measured parameter as a correlate of neuronal activity is the concentration of intracellular Ca^{2+} [6]. Whereas Ca^{2+} serves as a signaling molecule in various neuronal subcompartments [8], its overall cytosolic concentration represents a good correlate of membrane depolarization, albeit with limited temporal resolution. Since the first descriptions of a genetically encoded calcium indicator (GECI) [9, 10], a large variety of GECIs have been created that differ in their principal mode of action (Förster resonance energy transfer between two chromophores or direct shifts in emission intensity of a single chromophore), their Ca^{2+} affinity and several physical parameters, e.g., sensitivity to environmental cellular conditions (for reviews *see* refs. 2, 3, 11). The DNA of these GECIs can be readily transferred into cells by electroporation, transfection reagents, or somatic transfections, e.g., mediated by viruses. The advantages of DNA-encoded sensor proteins are most evident in the context of transgenic animals. Indeed, germ line transformation can now be accomplished in a variety of organisms. Currently, a repertoire of sophisticated genetic techniques that enable researchers to restrict the expression of transgenes, in this case a Ca^{2+} sensor protein, to small and selected subsets of cells or at defined developmental stages are available for several model organisms, e.g., in the round worm *Caenorhabditis elegans*, the zebrafish *Danio rerio*, the house mouse *Mus musculus* or the fruit fly *Drosophila melanogaster*. Here we focus on *Drosophila melanogaster* as a model organism, and in particular on optical in vivo Ca^{2+} imaging in the brain [12]. For a protocol describing germ line transformation in *Drosophila* please refer to [13].

Besides the availability of versatile genetic tools by which transgenes can be selectively expressed [14], a prominent feature of *Drosophila* is its relatively large behavioral repertoire relative to *C. elegans*, despite having a relatively simple central nervous system (when compared to vertebrates). Technically, a major strength of *Drosophila* relies on bipartite expression systems by which the spatial and temporal expression of the GECI (or any other transgene) is achieved by crossing a "driver strain" expressing a transactivating molecule to a second transgenic "reporter strain" that carries the DNA sequence of the GECI whose expression is controlled by the transactivator [14]. The most commonly used binary expression system is the Gal4-UAS system [15], but several alternative systems have been described with a number of variations that help restrict the expression of the transgene precisely to neurons of interest [12, 14].

Here, we describe how optical Ca^{2+} imaging of neuronal activity in the intact brain of an entire transgenic animal can be practically performed using the olfactory system of the *Drosophila* brain as an example. Fruit flies perceive odors with ~1,200 olfactory sensory neurons located on the third antennal segments and the maxillary palps [16]. Each olfactory sensory neuron expresses one or very few olfactory receptors, each of which detects a variable range of odorants [17]. These sensory neurons project with their axons to the antennal lobe, the primary olfactory center of the fly's brain, and extend their terminal arborizations into spherical structures called glomeruli [16]. Since each sensory neuron expresses a limited number of olfactory receptors with a specific ligand-binding profile [17], odor information is subsequently represented within the antennal lobes as spatiotemporal patterns of glomerular activities [18]. To determine the spatiotemporal activation of neuronal structures in the *Drosophila* brain, we use two-photon microscopy because the application of wide-field microscopy is limited by light scattering across the z-axis of extended pieces of tissue. The invention of two-photon microscopy [19] has helped to improve spatial resolution by restricting the excitation of chromophores to defined focal planes. In summary, we intend to demonstrate how in vivo optical Ca^{2+} imaging in the brain of an intact transgenic animal can be achieved by providing a protocol for monitoring odor-evoked changes in Ca^{2+} concentration in olfactory sensory neurons at the level of the antennal lobes of *Drosophila* using two-photon microscopy.

2 Materials

2.1 Materials for the In Vivo Preparation

2.1.1 Transgenic Drosophila Strains

1. The "driver strain" Or83b-Gal4 [20].
2. The "reporter strain" UAS-G-CaMP 3.0 [21].

A variety of different transgenic fly lines carrying the DNA of diverse GECIs have been described (reviewed by 12, 23). For preparations in which slight movements of the specimen cannot be completely avoided, we prefer FRET-based sensors because shifts in fluorescence intensity caused by movements can be corrected through the intrinsic ratiometric properties of these sensors [12, 18, 23, 24]. For other purposes, single chromophore sensors, e.g., G-CaMP 3.0 [21], are advantageous because only one emission wavelength needs to be detected. In order to express G-CaMP 3.0 in specific neuronal subsets, a "driver strain" that specifies the expression pattern must be chosen. Several binary expression systems have been described to date (reviewed by 12, 14, 23). The most commonly used system is the bipartite Gal4-UAS expression system [15], which we have used in our example here. To express G-CaMP 3.0 in olfactory sensory neurons, we have chosen a well described Gal4 driver line that expresses Gal4 under the control of a promoter fragment for the olfactory receptor co-receptor "Or83b" [20]. Both strains are available at the Bloomington stock center (http://flystocks.bio.indiana.edu/). The F1 progeny from the cross of these two strains will express G-CaMP 3.0 specifically in a large population of olfactory sensory neurons.

2.1.2 Materials for the In Vivo Brain Preparation

1. Microscope slides (~76 mm × 26 × mm 1 mm).
2. Fine stainless steel metal mesh (diameter of wire ~0.3 mm).
3. Regular transparent adhesive tape.
4. 0.1 mm breakable razor blades and blade holder.
5. A pair of very fine forceps (tip < 0.1 mm).
6. A syringe with a very fine diameter (0.5 mm).
7. Ringer's solution: 5 mM HEPES-NaOH, pH = 7.3, 130 mM NaCl, 5 mM KCl [17], 2 mM $CaCl_2$, 2 mM $MgCl_2$, and 36 mM sucrose [22].

8. Binocular stereomicroscope.
9. Low melting agarose with a congealing temperature of 26–30 °C.

First one has to build a preparation chamber in which the animal can be positioned and immobilized for the surgery and the optical imaging. To accomplish this, cut a piece of the metal mesh about the width of the objective slide. Put the metal mesh pieces over the microscope slide and fix it by wrapping it around the microscope slide using adhesive tape (Fig. 1a, step 1). Stick several layers of adhesive tape (about the thickness of a fly, i.e., ~1.5 mm) on top of the mesh. Then use a splint of a breakable razor blade and the blade holder to cut a small chamber of about the size of a fly (~3 × 1 × 1.5 mm^3) into the layers of adhesive tape (Fig. 1a, step 2). In front of the chamber, cut a passage through the layers of the adhesive tape for the odor application (Fig. 1a, step 2). The fly can now be positioned into this chamber for the preparation and optical imaging (Fig. 1a, steps 3–5).

2.2 Equipment for Optical Calcium Imaging

2.2.1 Optical Imaging Microscope

Optical Ca^{2+} imaging in the brain of *Drosophila* can be performed using a regular wide-field fluorescence microscope equipped with a sensitive CCD camera and appropriate filters. In fact, in many cases a wide-field microscope provides advantages over laser scanning microscopes if the *z*-axis resolution is not important [23, 24]. If one intends to differentiate between focal planes in the *z*-axis, however, an upright two-photon microscope is beneficial. User-friendly two-photon microscopes are available from several companies. In any case, the microscope and the laser should be installed on a well-damped vibration-isolated table on air supports. For two-photon excitation of the fluorophore, a mode-locked titanium:sapphire laser is needed as a light source (e.g., Chameleon lasers from Coherent or Mai Tai/Tsunami lasers from Spectra-Physics). These lasers generate femtosecond pulses of infrared light in the range between ~700 and 1,000 nm. The microscope should be equipped with a sensitive photomultiplier for detection and an *X/Y*-scanning unit with high scanning speed (at least 5 frames/s). For using the Ca^{2+} indicator G-CaMP 3.0, an excitation wavelength of 920 nm is appropriate. To separate excitation and emission light, a dichroic mirror (690 nm long pass) is required. The G-CaMP emission can be detected using a regular GFP band-pass filter. Importantly, the microscope must be equipped with a water immersion objective appropriate for infrared imaging, in our case a 20× objective (NA = 1.0). A xenon lamp as a conventional light source and a regular GFP filter set is helpful to locate the structures of interest. Software for data acquisition and the control of the components of the microscope is, of course, commercially available according to the microscope system used. Data analysis, however, is recommended to be done with specialized image processing software, e.g., MetaMorph (Molecular Devices) or ImageJ.

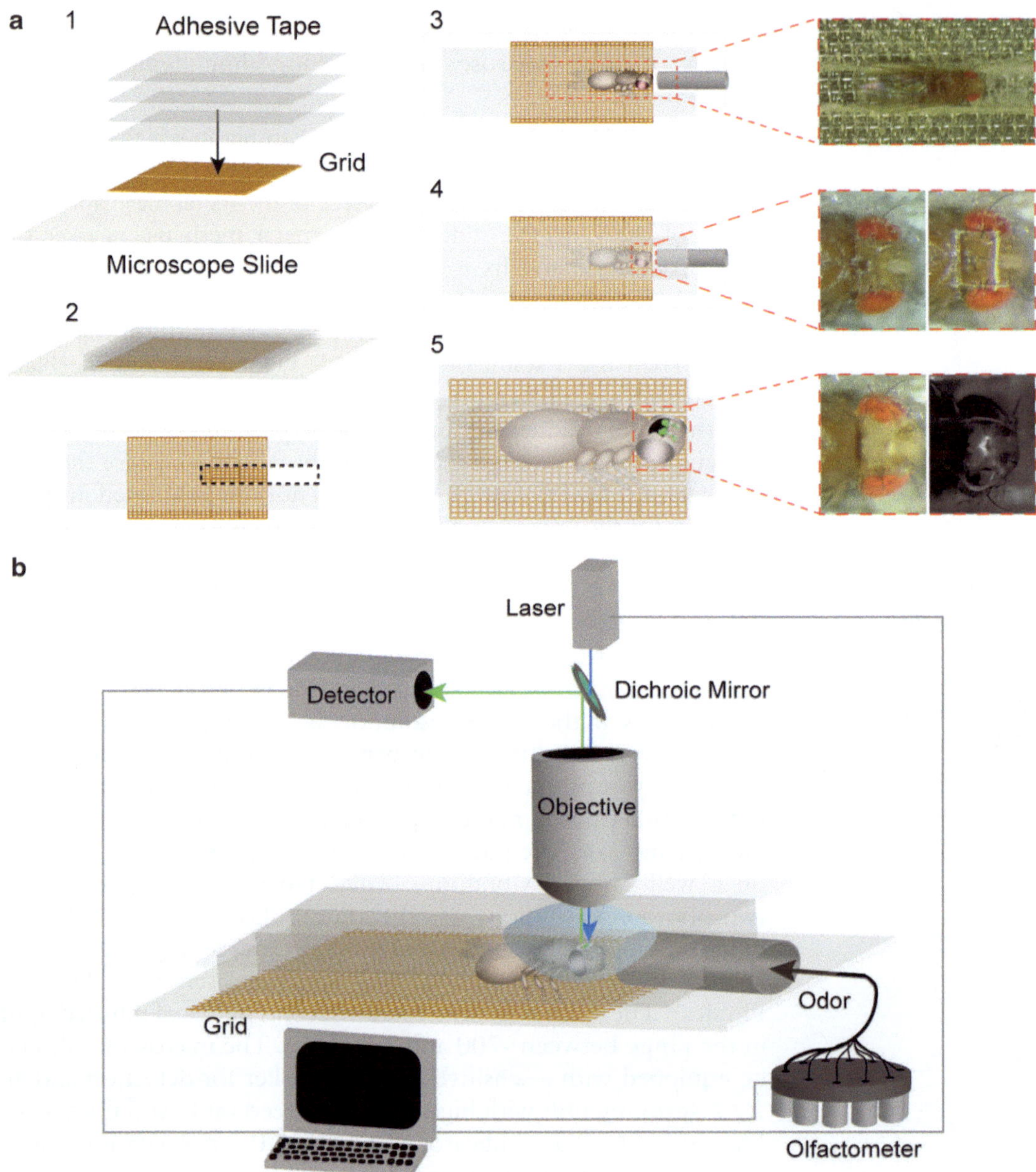

Fig. 1 (**a**) Schematic illustration of the fly brain in vivo preparation: (*1*) To restrain the fly in a small chamber, a metal grid is placed on a microscope slide and fixed with several layers of adhesive tape. Altogether the sticky tape layers should measure approximately the height of a fly. (*2*) A small passage is cut into the layers of the adhesive tape, leaving enough space for the fly and the tubing for the odor application. (*3*) The fly is placed into this corridor with the tubing directed towards its antennae. The chamber is then sealed from above with sticky tape. (*4*) In the next step a small window is cut into the adhesive tape, providing access to the head capsule such that the cuticle can be cautiously removed (*5*). (**b**) Schematic illustration of the optical imaging setup. The exposed brain is covered with Ringer's solution and placed under the microscope equipped with a water immersion objective. The Ca^{2+} sensor (e.g., G-CaMP 3.0) is excited by a 2-Photon laser, for G-CaMP at 920 nm with a maximum emission wavelength of ~509 nm. Odorants can be applied via a custom-built olfactometer that ensures a constant air flow rate for each odorant

2.2.2 Device for Odor-Application (Only for Odor-Stimulation Experiments)

For odor stimulation, a custom-built stimulus device ("olfactometer") is used to apply the air-borne odors at constant airflow rates and the desired concentrations. In our case, the air is pumped by a regular aquarium pump into the olfactometer and separated into different channels controlled by computer-triggered electronic valves. The software controlling the microscope should be capable of controlling these additional devices. The individual airstreams are guided into Teflon tubings with a constant air flow rate of~1 ml/s and from there to small (20 ml) odorant vials containing the odorants of choice. Under non-stimulation conditions, air is guided through a Teflon tube containing only the diluent, e.g., air or mineral oil. Airflows from different odor vials converge to a single outlet in front of the animal's antennae. To directly eliminate the odors from the experimental setup, an exhaust tube behind the animal is necessary to avoid odor contamination of the air. It is, of course, most important that contamination of all parts of the olfactometer with odorants is prevented and that any mechanosensory artifacts, such as might be caused by an air pressure decrease, are avoided. A variety of very similar olfactometer systems has been described (e.g., [25]).

3 Methods

3.1 In Vivo Preparation of a Drosophila Brain

1. Flies selectively expressing the G-CaMP 3.0 sensor [21] in neurons of interest can be generated by crossing the appropriate Gal4 line to the transgenic UAS-G-CaMP 3.0 line. In our experiments, we have crossed UAS-G-CaMP 3.0 with Or83b-Gal4 [20], which drives expression in a majority of olfactory sensory neurons (*see* **Note 1**).
2. Immobilize the flies by cooling in an empty vial for no longer % min in ice (*see* **Note 2**).
3. Use a small brush to gently put a single fly into the preparation chamber (*see* Fig. 1a step 3) described above and quickly seal it on top with a piece of adhesive tape before the animal recovers from anesthesia such that the fly remains restrained and the head capsule is in contact with the sticky surface of the adhesive tape.
4. Cut a small window through the adhesive tape and the cuticle of the head capsule with a splinter of a breakable razor blade. Appropriate blade holders are commercially obtainable (*see* Fig. 1a steps 4 and 5) (*see* **Note 3**).
5. To ameliorate optical access to the brain, remove cuticular structures as well as possible, then remove trachea and any accumulated fat from the head capsule using fine forceps. Proceed very carefully to avoid harming the brain tissue.

6. Remove the hemolymph surrounding the brain carefully using small pieces of tissue paper and then inject 2 % low melting agarose into the head capsule using a fine syringe of 0.5 mm diameter. Of course, the agarose should be close to the congealing point and not too warm. This step avoids desiccation and it physically immobilizes the tissue. The preparation is subsequently covered with Ringer's solution and the preparation chamber containing the fly is placed under the microscope.

3.2 Optical Ca^{2+} Imaging

1. Focus on the structure of interest (in our case the antennal lobes) using blue (488 nm) light from a regular light source, e.g., a xenon lamp, either by eye through the oculars or using a camera. Keep the illumination time as short as possible to avoid bleaching of the Ca^{2+} sensor.
2. Switch to the laser-scanning mode and refocus the specimen using the appropriate laser wavelength and band-pass filter (for G-CaMP 3.0 a two-photon excitation wavelength of 920 nm and a band-pass filter for GFP emission can be used). At this step fast scanning at low resolution is advisable to minimize bleaching.
3. Once the focus is set, choose a region of interest within which to scan the specimen. To maximize the scanning speed while maintaining the highest possible image resolution, the region of interest should be chosen as small as possible but large enough to cover all structures of interest.
4. Adjust the laser power and gain such that the laser power is low enough to avoid photo-toxicity, bleaching and heat production, but high enough that the baseline fluorescence is sufficiently visible.
5. Depending on the particular application and the expression level of the Ca^{2+} sensor, the scan speed may need to be adjusted. The adjustment of the frame rate may depend on the signal intensity of the baseline fluorescence, but frame rates from 3 to 10 Hz are reasonable for most physiological experiments. Lower frame rates increase photon collection per pixel at the cost of temporal resolution.
6. For the measurement itself, it is necessary to keep the number of acquired images as low as possible to avoid bleaching, but high enough to determine a reliable baseline as well as the onset and offset of the stimulus-evoked Ca^{2+} influx. For example, in the experiments shown in Fig. 2, a physiological stimulation with an odor of 2 s duration was used. The dynamics of intracellular Ca^{2+} concentration were observed using a recording protocol lasting for 85 frames at 5 Hz frame rate during which the odor stimulus was presented between frames 25–35. Between each measurement of physiological stimulations, a

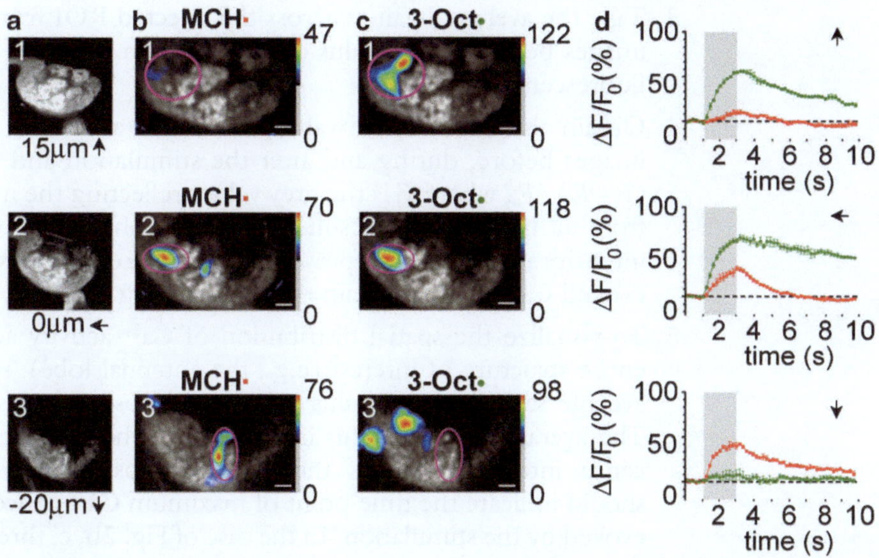

Fig. 2 Optical imaging of odor-evoked Ca^{2+} dynamics in olfactory sensory neurons of *Drosophila* at the level of the antennal lobe. (**a**) Baseline fluorescence of the G-CaMP 3.0 sensor expressed in a large population of sensory neurons. The *numbers* indicate the relative distances between different focal planes. Please note that olfactory glomeruli are very easily detectable. (**b**, **c**) *False-color* coded image of the spatial distribution of odor-evoked Ca^{2+} signals in the antennal lobe evoked by a 2 s stimulation with either 4-methylcyclohexanol (MCH, 1:750) (**b**) or 3-octanol (3-Oct, 1:500) (**c**) as odorants diluted in mineral oil. Recordings were taken at the three different focal planes of the antennal lobes shown in (**a**). *Warm colors* represent regions of high calcium activity while *cold colors* represent regions of low or no calcium increase. The *false-color* coded regions of Ca^{2+} increase are superimposed onto the grey scaled baseline fluorescence of the G-CaMP 3.0 sensor. The *numbers* indicate maximum relative changes in fluorescence (F/F_0) across the entire image. The *magenta circles* indicate regions in which the temporal dynamics of Ca^{2+} changes is determined (**d**) Scale bars: 10 μm. (**d**) Temporal dynamics of intracellular Ca^{2+} increase in regions of high activity evoked by MCH (**a**) or 3-Oct (**b**). The duration of the odor stimulus is indicated as *grey bars*. Relative changes in fluorescence are indicated in *red* for MCH and in *green* for 3-Oct. *Traces* indicate means ± SEM of three odor stimulations within the same animal. Note that the two odors evoke increases in intracellular Ca^{2+} in different regions of the antennal lobe and with different time courses

break period should be included to avoid continuous excitation of the fluorophore and potential photo-toxic effects. The acquisition is repeated several times at the desired focal planes.

3.3 Data Analysis

1. If there are slight movements of the structures under investigation during image acquisition, it is possible to align all the acquired images to reduce the movement effects using specialized algorithms (*see* **Note 4**).
2. In order to quantify potential dynamics in emission intensity reflecting intracellular Ca^{2+} dynamics using image processing software, arrange the acquired image files as a stack of images. Mark a region of interest (ROI) around the structure to be analyzed and obtain the raw grey values from the pixels within the ROI (*see* Fig. 2).

3. Take the averaged values across the selected ROI from several images before the stimulus onset to obtain a reliable baseline fluorescence value (F_o).
4. Obtain the fluorescence values from the same ROI from all images before, during and after the stimulation and calculate $(F_i - F_o)/F_o$, where F_i is the grey values reflecting the intensities from all images. As a result, the relative changes in emission intensity are obtained, representing the time course of stimulus-evoked Ca^{2+} activity within the ROI (*see* Fig. 2d).
5. To visualize the spatial distribution of Ca^{2+} activity across the entire structure of interest (e.g., the antennal lobe), separately average several pre-stimulus images and post-stimulus images. The averaged pre-stimulus images reflect the baseline fluorescence intensity whereas the averaged post-stimulus images should indicate the time-point of maximum Ca^{2+} concentration evoked by the stimulation. In the case of Fig. 2b, c, three images before the odor onset were selected as pre-stimulus images (F_{pre}) and the images at around 2 s after the onset were chosen as post-stimulus images (F_{post}). The averages of both were subjected to a mean filter in which several pixels around each pixel as a center (in our case 5×5) are averaged to remove noise. $(F_{post}/F_{pre}) - 1$ was calculated which equals $(F_{post} - F_{pre})/F_{pre}$. To shift all pixels to positive values (negative values cannot be depicted in an image), an arbitrary number is added which then reflects the baseline. To present the data in percentage values, multiply it by a factor of 100. The resulting images are then shown in false-colors in which the different colors cover the range of signal intensities. One can superimpose these images onto the grey scaled baseline fluorescence images if one skips the lowest values, thereby showing only highly activated regions in correlation with the entire anatomy of the structure (Fig. 2b, c).

4 Notes

1. If expression level of the GECI is too low, it is helpful to create flies that are homozygous for both the UAS construct and the Gal4 construct. Raising the animals at higher temperatures (e.g., 29 °C) also may help to increase expression levels of the GECI.
2. Anesthetizing the animals using CO_2 or chemicals is not recommended because it may affect the central nervous system.
3. Be careful that no Ringer's solution leaks through the window over the head capsule so that the antennae remain dry.
4. For aligning the images obtained by two-photon microscopy, we use a custom-written JAVA script executable by ImageJ which is based on the plugin "TurboReg" [26]. This rigid body algorithm uses landmarks present in a target image and aligns source images

to target images such that the mean-squared difference between the source and the target images are minimized.

Acknowledgments

This work was supported by the Deutsche Forschungsgemeinschaft (SPP1392, FI 821/2-1, and SFB 889/B4) and the German Federal Ministry for Education and Research via the Bernstein Center for Computational Neuroscience Göttingen B01, grant number 01GQ1005A.

References

1. Miyawaki A (2003) Fluorescence imaging of physiological activity in complex systems using GFP-based probes. Curr Opin Neurobiol 13:591–596
2. Looger LL, Griesbeck O (2012) Genetically encoded neural activity indicators. Curr Opin Neurobiol 22:18–23
3. Miyawaki A (2005) Innovations in the imaging of brain functions using fluorescent proteins. Neuron 48:189–199
4. Willoughby D, Cooper DM (2008) Live-cell imaging of cAMP dynamics. Nat Methods 5:29–36
5. Mutoh H, Perron A, Akemann W, Iwamoto Y, Knöpfel T (2011) Optogenetic monitoring of membrane potentials. Exp Physiol 96:13–18
6. Tian L, Akerboom J, Schreiter ER, Looger LL (2012) Neural activity imaging with genetically encoded calcium indicators. Prog Brain Res 196:79–94
7. Dreosti E, Lagnado L (2011) Optical reporters of synaptic activity in neural circuits. Exp Physiol 96:4–12
8. Berridge MJ (1998) Neuronal calcium signaling. Neuron 21:13–26
9. Miyawaki A, Llopis J, Heim R, McCaffery JM, Adams JA, Ikura M, Tsien RY (1997) Fluorescent indicators for Ca2+ based on green fluorescent proteins and calmodulin. Nature 388:882–887
10. Romoser VA, Hinkle PM, Persechini A (1997) Detection in living cells of Ca2+-dependent changes in the fluorescence emission of an indicator composed of two green fluorescent protein variants linked by a calmodulin-binding sequence. A new class of fluorescent indicators. J Biol Chem 272:13270–13274
11. Mank M, Griesbeck O (2008) Genetically encoded calcium indicators. Chem Rev 108:1550–1564
12. Riemensperger T, Pech U, Dipt S, Fiala A (2012) Optical calcium imaging in the nervous system of *Drosophila melanogaster*. Biochim Biophys Acta 1820:1169–1178
13. Bachmann A, Knust E (2008) The use of P-element transposons to generate transgenic flies. Methods Mol Biol 420:61–77
14. Venken KJ, Simpson JH, Bellen HJ (2011) Genetic manipulation of genes and cells in the nervous system of the fruit fly. Neuron 72:202–230
15. Brand AH, Perrimon N (1993) Targeted gene expression as a means of altering cell fates and generating dominant phenotypes. Development 118:401–415
16. Vosshall LB, Stocker RF (2007) Molecular architecture of smell and taste in *Drosophila*. Annu Rev Neurosci 30:505–533
17. Hallem EA, Carlson JR (2006) Coding of odors by a receptor repertoire. Cell 125:143–160
18. Fiala A, Spall T, Diegelmann S, Eisermann B, Sachse S, Devaud JM, Buchner E, Galizia CG (2002) Genetically expressed cameleon in *Drosophila melanogaster* is used to visualize olfactory information in projection neurons. Curr Biol 12:1877–1884
19. Denk W, Strickler JH, Webb WW (1990) Two-photon laser scanning fluorescence microscopy. Science 248:73–76
20. Larsson MC, Domingos AI, Jones WD, Chiappe ME, Amrein H, Vosshall LB (2004) Or83b encodes a broadly expressed odorant receptor essential for *Drosophila* olfaction. Neuron 43:703–714
21. Tian L, Hires SA, Mao T, Huber D, Chiappe ME, Chalasani SH, Petreanu L, Akerboom J, McKinney SA, Schreiter ER, Bargmann CI, Jayaraman V, Svoboda K, Looger LL (2009) Imaging neural activity in worms, flies and mice with improved G-CaMP calcium indicators. Nat Methods 6:875–881
22. Estes PS, Roos J, van der Bliek A, Kelly RB, Krishnan KS, Ramaswami M (1996) Traffic of

dynamin within individual *Drosophila* synaptic boutons relative to compartment-specific markers. J Neurosci 16:5443–5456

23. Strutz A, Völler T, Riemensperger T, Fiala A, Sachse S (2012) Calcium imaging of neural activity in the olfactory system of *Drosophila*. In: Martin JR (ed) Genetically encoded functional indicators. Springer Neuromethods 72:43–70
24. Fiala A, Spall T (2003) In vivo calcium imaging of brain activity in *Drosophila* by transgenic cameleon expression. Sci STKE (174):PL6
25. Olsson SB, Kuebler LS, Veit D, Steck K, Schmidt A, Knaden M, Hansson BS (2011) A novel multicomponent stimulus device for use in olfactory experiments. J Neurosci Methods 195:1–9
26. Thévenaz P, Ruttimann UE, Unser M (1998) A pyramid approach to subpixel registration based on intensity. IEEE Trans Image Process 7:27–41

Chapter 16

A Multiparameter Live Cell Imaging Approach to Monitor Cyclic AMP and Protein Kinase A Dynamics in Parallel

Nwe-Nwe Aye-Han and Jin Zhang

Abstract

Parallel detection of signaling activities allows us to correlate activity dynamics between signaling molecules. In this review, we detail a multiparameter live cell imaging method to monitor 3′,5′-cyclic adenosine monophosphate (cAMP) levels and protein kinase A (PKA) activities in parallel.

Key words Parallel imaging, FRET, cAMP, Protein kinase A, Signal transduction

1 Introduction

Signal transduction is a way for cells to communicate internally or externally to respond to various stimuli. At the heart of such communication are the signaling nodes that integrate diverse signals into specific outputs [1], resulting in desired functional responses. Rather than being linear, signaling pathways exist as structured networks, with multiple layers of regulation that ensure signal specificity and amplification. Signaling molecules involved in these pathways are, thus, coupled to one another in spatiotemporally synchronized manners. Knowledge of such dynamic relationships can be critical in assessing the information flow within the cells [2] or in identifying new therapeutic targets [3]. To comprehend inter-regulation of multiple signaling molecules at a systems level, it is important to be able to watch them in parallel, over the course of their cellular actions, and to correlate their activity dynamics.

1.1 cAMP/PKA Signaling

Protein kinase A (PKA), a ubiquitous protein kinase, is central to many signaling pathways. It is involved in regulating major biological processes, such as metabolism, differentiation and growth. Upon ligand binding and activation of seven transmembrane G-protein coupled receptors (GPCRs), associated G-proteins

Jin Zhang et al. (eds.), *Fluorescent Protein-Based Biosensors: Methods and Protocols*, Methods in Molecular Biology, vol. 1071, DOI 10.1007/978-1-62703-622-1_16,

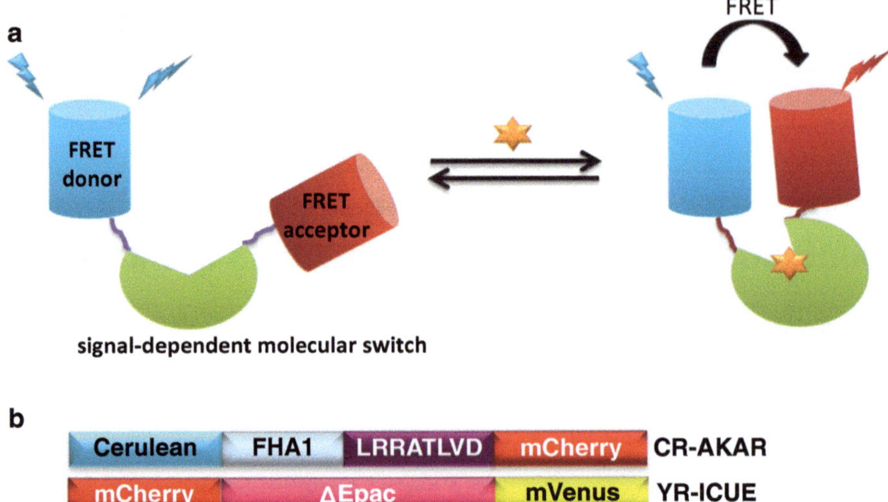

Fig. 1 (**a**) General scheme of FRET-based activity reporters. A signal-dependent conformational change induces a change in FRET. We have constructed red-shifted reporters, in which CR-FRET and YR-FRET change can be efficiently detected. (**b**) Domain structures of CR-AKAR and YR-ICUE. These reporters can be conveniently generated by replacing a fluorescent protein from the CY-based reporters with mCherry

activate adenylyl cyclases that produce cyclic 3′,5′-monophosphate (cAMP). cAMP then binds to PKA in its tetrameric holoenzyme form, triggering the dissociation of catalytic subunits from the latter's regulatory subunits [4]. The catalytic subunits then phosphorylate diverse substrates, altering various aspects of cellular physiology. Although PKA is the most well-known mediator of cAMP signaling, it is not the sole effector downstream of cAMP. Exchange proteins activated by cAMP (Epacs) represent another family of major cAMP effectors that channels most of the PKA-independent cAMP signaling inside cells [5].

1.2 FRET-Based Reporters

Genetically encodable fluorescent protein (FP)-based reporters are useful tools for monitoring signaling activities inside living cells [6]. In these reporters, fluorophores from two spectrally distinct FPs serve as a pair of donor and acceptor for fluorescence resonance energy transfer (FRET). These FPs are fused to a signal-dependent molecular switch constructed via semirational protein engineering (Fig. 1a). Upon signal activation, conformational changes in the molecular switch lead to a shift in relative distance and orientation of the fluorophores from the two FPs, generating a change in FRET. This allows FRET change to be used as a readout to detect signaling dynamics in real time and in the native cellular environment.

1.3 Orthogonal FRET Pairs with a Shared FRET Acceptor

A typical and widely used FRET pair contains a cyan fluorescent protein (CFP) as a donor and a yellow fluorescent protein (YFP) as an acceptor. Emergence of fluorescent proteins across the visible spectrum made parallel FRET detection possible when spectrally distinct donors and acceptors were put together to generate orthogonal FRET pairs [7]. Various laboratories have developed methods to monitor two or more signaling activities, taking advantage of these newly developed FPs (reviewed in refs. 7, 8).

Recently, we have reported a parallel imaging method, where mCherry [9], a red fluorescent protein (RFP) derived from *Discosoma* sp., serves as a shared FRET acceptor for two FRET donors, CFP and YFP [10]. This method relies on the distinct absorption peaks of CFP and YFP. More importantly, our experiments showed that FRET was efficient between CFP and RFP [11], and YFP and RFP with miminal cross-excitation of RFP upon YFP excitation, which can be easily corrected (*see* Subheading 3.7 for correction methodology). The advantages of the shared acceptor imaging are several-fold. First, FPs from any established CY-based FRET reporters can be easily replaced with mCherry, requiring minimal reporter characterization. In addition, imaging can be conveniently achieved by addition of an RFP filter set in an existing CY-FRET protocol.

Using the common acceptor approach, we have constructed a cyan and red FP-based PKA activity reporter called CR-AKAR (*C*FP/*R*FP-based *A*-*k*inase *A*ctivity *R*eporter) based on a previously developed, widely used CY-based AKAR [12]. In CR-AKAR, a phospho-threonine binding domain, forkhead associated domain 1 (FHA1) and a surrogate PKA substrate motif serve together as a signal-dependent switch that Cerulean (a CFP) and mCherry are flanking (Fig. 1b). Upon PKA activation and phosphorylation of the surrogate substrate, a phosphorylation-dependent conformational switch results in an increase in cyan to red FRET. We have also engineered a yellow and red FP-based cAMP sensor called YR-ICUE (*Y*FP/*R*FP-based *I*ndicator of *c*AMP *u*sing *E*pac), by replacing the CFP in the original CY-ICUE biosensor [13] with mCherry. In this reporter, the cAMP sensing domain of exchange protein activated by cAMP-1 (Epac1) is sandwiched between Venus (a YFP) and mCherry (Fig. 1b). Conformational changes in Epac1 domain, upon cAMP binding, result in a FRET decrease from Venus to mCherry.

Expressing both of these reporters in single living cells, we observed differential dynamics of cAMP and PKA upon stimulation with different G-protein coupled receptor agonists (Fig. 2). This has opened up the possibility to study and characterize the pathway parameters such as feedback loops and cross-regulation in a more systematic approach. Below, we outline the detailed method for parallel monitoring of cAMP and PKA activity dynamics using YR-ICUE and CR-AKAR.

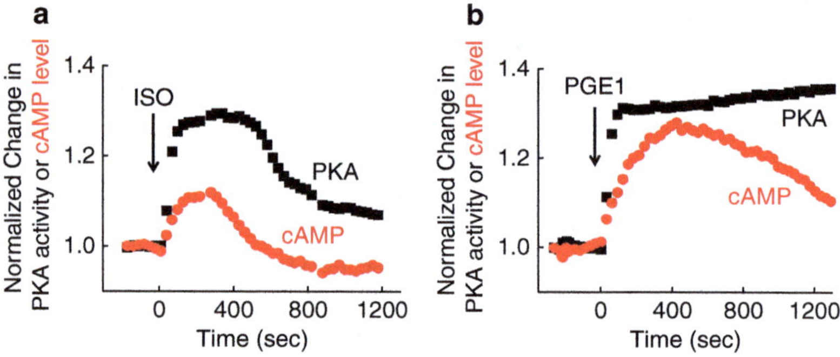

Fig. 2 Parallel detection of differential cAMP and PKA dynamics upon a GPCR-agonist stimulation. Representative timecourses of PKA activity (*black*) and cAMP level dynamic (*red*) in HEK293T cells upon stimulation with (**a**) isoproterenol (ISO) and (**b**) prostaglandin E1 (PGE1)

2 Materials

2.1 Cell Culture and Transfection

1. Cell lines: Human Embryonic Kidney with SV40 T Antigen (HEK293T).
2. Dulbecco's phosphate-buffered saline without Mg^{2+} and Ca^{2+} (DPBS).
3. T-25 cm^2 tissue culture flasks.
4. Imaging dish: 35 mm glass bottom petri dishes for live cell imaging (MatTEK).
5. HEK293T culture medium: Dulbecco's Modified Eagle's Medium (DMEM, low glucose) supplemented with 10 % fetal bovine serum (FBS) and 1 % penicillin–streptomycin to culture HEK293T cells. Other suitable tissue culture medium for additional cell lines of interest.
6. Solution of trypsin (0.05 %) and ethylenediamine tetraacetic acid (EDTA, 0.53 mM) or relevant trypsinization reagents.
7. Constructs: CR-AKAR and YR-ICUE biosensors.
8. Calcium phosphate-mediated transfection reagents: 2×HBS (50 mM HEPES, 10 mM KCl, 12 mM dextrose, 280 mM NaCl, 1.5 mM Na_2PO_4), pH adjusted to 7.05 using KOH; and 2 M $CaCl_2$; both filter-sterilized with 0.22 μm filters.

2.2 Preparation for Imaging

1. Hanks' Balanced Salt Solution for Imaging (HBSS*): 1× Hanks' Balanced Salt Solution (Gibco) with 2.0 g/L D-glucose; pH-adjusted to 7.4 using NaOH and filter sterilized using a 0.22 μm filter. Store at 4 °C and bring to room temperature prior to imaging.

2. 1,000× stock of stimuli: forskolin (FSK; Calbiochem), prostaglandin E1 (PGE1; Sigma), ritodrine (RITO; Sigma), isoproterenol (ISO; Sigma), and H89 (Sigma) (*see* **Note 1**) are prepared in DMSO and stored at −20 °C.

2.3 Epifluorescence Microscopy

1. Microscope: Axiovert 200M microscope; 40×/1.3NA oil-immersion objective lens (Zeiss).
2. Camera: MicroMAX BFT512 cooled charge-coupled device camera (Roper Scientific).
3. Xenon lamp: XBO 75W (Zeiss).
4. Neutral density (ND) filters 0.6 and 0.3 (Chroma Technology).
5. Filter sets for individual channels (All from Chroma Technology):

   CR-FRET—420DF20 excitation filter, 450DRLP dichroic mirror, 653DF95 emission filter.

   CFP—420DF20 excitation filter, 450DRLP dichroic mirror, 475DF40 emission filter.

   RFP—568DF55 excitation filter, 600DRLP dichroic mirror, 653DF95 emission filter.

   YFP—495DF10 excitation filter, 515DRLP dichroic mirror, 535DF25 emission filter.

   YR-FRET—495DF10 excitation filter, 515DRLP dichroic mirror, 653DF95 emission filter.
6. Lambda 10-2 filter changer (Sutter Instruments).
7. Immersol® 518F fluorescence free immersion oil (Zeiss).

2.4 Image Acquisition and Data Analysis

1. METAFLUOR 6.2 software (Molecular Devices).
2. Microsoft Office Excel.

3 Methods

3.1 Cell Culture

HEK293T cells are maintained in T-25 cm^2 flasks at 37 °C with 5 % CO_2. Upon reaching about 80 % confluency (about every 2–3 days), the cells were subject to passage in the following steps:

1. Wash cells twice with DPBS buffer (*see* **Note 2**). This step removes general debris as well as facilitates cell detachment.
2. Trypsinize the cells with 0.05 % trypsin (+EDTA) at room temperature (RT) for 1 min.
3. Gently tap the flask and neutralize the trypsin with desired volume of cell culture medium (*see* **Note 3**).
4. Split a desired volume of cells to a new flask or imaging dishes containing fresh medium.

3.2 Preparation for Transfection

Bring solutions to room temperature before use.

1. Aliquot 100 μL of 2× HBS (solution 1).
2. Mix 500 ng each of CR-AKAR and YR-ICUE in another eppendorf tube and add 8 μL of 2 M $CaCl_2$ into the mixture. Bring the volume to 100 μL with deionized, distilled water (solution 2).
3. Add solution 2 drop-wise onto solution 1 while the latter is being vortexed and incubate for 2–40 min at RT (*see* **Note 4**).

3.3 HEK293T Cells Transfection

1. (*see* **Note 5**) Replace the medium in the imaging dishes with fresh medium. Gently add the DNA complex drop-wise into the center of imaging dish (*see* **Note 6**).
2. Culture for additional 18–48 h before imaging.

3.4 Preparation for Imaging (Microscope)

1. Turn on the lamp, microscope, filter changer, camera and computer sequentially. Place a drop of immersion oil on the 40× objective.
2. Load METAFLUOR software and set a protocol for image acquisitions in CR-FRET, CFP, RFP, YFP and YR-FRET channels every 30 s. Check to make sure proper filters are in place for each channel.

3.5 Preparation for Imaging (Cells)

All steps are performed at room temperature unless otherwise noted.

1. For each imaging experiment, prepare a 2 μL aliquot of the desired stimulus in a new 1.5 mL eppendorf tube and leave it on ice until ready to use.
2. Aspirate the culture medium from an imaging dish and wash the cells twice gently with HBSS* to remove dead cells and general debris (*see* **Note 7**).
3. Add 2 mL of HBSS* into the dish and mount the dish securely on the microscope for imaging.

3.6 Image Acquisition and Data Analysis

1. Focus the cells properly using the bright field setting.
2. Switch to fluorescence mode and look for healthy (*see* **Note 8**), co-transfected cells with optimal CFP and YFP levels (*see* **Note 9**).
3. Draw a region of interest (ROI) on each of the fluorescent cells in the field. Also choose a region of interest on a region with an untransfected cell or no cell for background correction.
4. Log the intensity data and save the images from all five channels for each acquisition (*see* **Note 10**).
5. Acquire a series of images to establish the baseline before stimulus addition.

6. Stimulate the cAMP/PKA pathway by addition of a desired activator (FSK) or a GPCR agonist (ISO, PGE or RITO) (*see* **Notes 11** and **12**). Draw ~500 μL of imaging buffer from the dish and add it to the eppendorf tube prepared during **step 1** of Subheading 3.5. Mix thoroughly and pipette drop-wise into the periphery of the imaging area. Gently pipette up and down several times without touching or disrupting the imaging area. Resume image acquisition immediately afterward.
7. Export the logged data to an excel file and calculate the FRET emission ratio change using the following formulae for each construct.

For CR-AKAR,

$$\text{Emission Ratio} = \frac{(\text{CR-FRET channel intensity of ROI} - \text{CR-FRET channel intensity of background})}{(\text{CFP channel intensity of ROI} - \text{CFP channel intensity of background})}$$

For YR-ICUE,

$$\text{Emission Ratio} = \frac{(\text{YFP channel intensity of ROI} - \text{YFP channel intensity of background})}{(\text{YR-FRET channel intensity of ROI} - \text{YR-FRET channel intensity of background})}.$$

8. Set the time point immediately after stimulus addition to time 0 and adjust subsequent time points accordingly using simple subtraction. Plot the emission ratio versus time on a graph (*see* **Note 13**). Emission ratio should increase upon signal activation.

3.7 Correction for Cross-Excitation of RFP Following YFP Excitation

If necessary, the contribution from cross-excitation of RFP can be corrected using the following steps:

1. Use an RFP construct (for example mCherry in pcDNA3.1) to obtain the correction factor, defined by YR/RR, where YR and RR represent readings in the RFP intensity following excitation of YFP and RFP, respectively;
2. Acquire the RFP emission to obtain the total RFP intensity in cells where correction is required (RR′) (*see* **Note 14**); and
3. Multiply the correction factor (YR/RR) by the total RFP direct emission (RR′) to determine the contribution of cross-excited RFP following YFP excitation (YR_x),
4. Subtract YR_x from the experimental reading in the YR-FRET channel (YR_{exp}). Thus, after correction, the actual YR-FRET signal (YR_{actual}) can be calculated as

$$YR_{actual} = YR_{exp} - YR_x,$$

where $YR_x = RR' \times (YR/RR)$.

4 Notes

1. Additional stimuli can be prepared at desired concentration (preferably 1,000× for easier addition during imaging; *see* **Note 11**).
2. Gentle washing is required since HEK293T cells are very easy to detach. Use a 10 mL pipette to gently load the wash buffer in a corner of the flask.
3. Typically, three volumes of serum-containing culture medium should completely neutralize one volume of trypsin.
4. 200 μL of total DNA–calcium phosphate solution should be prepared for each imaging dish.
5. About 20–40 % confluency of cells is desired at this point to achieve optimal transfection efficiency.
6. Check for even distribution of DNA complex (as tiny black particles) on top of cells under the microscope.
7. HEK293T cells are very easy to detach. Thus, extra caution is necessary at this step so as not to disturb the attached HEK293T cells. Gently rock the dish side to side.
8. Healthy cells have a distinct morphology compared to cells under stress. Healthy cells will adhere well and have a flat appearance.
9. Optimal intensity for each FP varies under different microscope settings. Thus, it is necessary to adjust the neutral density (ND) filters and exposure times of individual channels accordingly in order to achieve the desired intensity for each FRET pair and to avoid photobleaching of each FP. Maintain the same exposure time for donor and FRET channels of each FRET pair.
10. The saved images can later be used to reanalyze the data, if necessary, or to make pseudocolor images or movies for visual purposes.
11. The final concentration should be 50 μM and 1 μM for FSK and ISO respectively, and should be 10 μM for both PGE1 and RITO. If a higher volume of stock solution is required, prepare the stock solution using imaging buffer (HBSS*). Avoid adding high volume of DMSO into the imaging dishes to minimize DMSO-induced cytotoxicity.
12. Pre-incubate cells with the PKA inhibitor H89 to check for pathway specificity. Be sure to use H89 containing imaging buffer for the subsequent steps.

13. Sometimes it is desirable to normalize the emission ratio at a particular time point to the ratio at time 0 so as to easily analyze the percent change in signal. For example, normalized ratio change at time point A can be calculated as,

$$\text{Normalized Emission Ratio (at time point } A) = \frac{\text{Emission Ratio (at time point } A)}{\text{Emission Ratio (at time 0)}}.$$

14. It is necessary to use the same setting for the microscope, camera, ND, and exposure time in **steps 1** and **2**.

References

1. Jordan JD, Landau EM, Iyengar R (2000) Signaling networks: the origins of cellular multitasking. Cell 103:193–200
2. Howe AK (2011) Cross-talk between calcium and protein kinase A in the regulation of cell migration. Curr Opin Cell Biol 23:554–561
3. Papa S, Bubici C, Zazzeroni F, Franzoso G (2009) Mechanisms of liver disease: cross-talk between the NF-kappaB and JNK pathways. Biol Chem 390:965–976
4. Kim C, Cheng CY, Saldanha SA, Taylor SS (2007) PKA-I holoenzyme structure reveals a mechanism for cAMP-dependent activation. Cell 130:1032–1043
5. Seino S, Shibasaki T (2005) PKA-dependent and PKA-independent pathways for cAMP-regulated exocytosis. Physiol Rev 85:1303–1342
6. Zhou X, Herbst-Robinson KJ, Zhang J (2012) Visualizing dynamic activities of signaling enzymes using genetically encodable FRET-based biosensors from designs to applications. Methods Enzymol 504:317–340
7. Carlson HJ, Campbell RE (2009) Genetically encoded FRET-based biosensors for multiparameter fluorescence imaging. Curr Opin Biotechnol 20:19–27
8. Schultz C, Schleifenbaum A, Goedhart J, Gadella TW Jr (2005) Multiparameter imaging for the analysis of intracellular signaling. Chembiochem 6:1323–1330
9. Shaner NC, Campbell RE, Steinbach PA, Giepmans BN, Palmer AE, Tsien RY (2004) Improved monomeric red, orange and yellow fluorescent proteins derived from *Discosoma* sp. red fluorescent protein. Nat Biotechnol 22:1567–1572
10. Aye-Han NN, Allen MD, Ni Q, Zhang J (2012) Parallel tracking of cAMP and PKA signaling dynamics in living cells with FRET-based fluorescent biosensors. Mol Biosyst 8:1435–1440
11. Allen MD, Zhang J (2008) A tunable FRET circuit for engineering fluorescent biosensors. Angew Chem Int Ed Engl 47:500–502
12. Allen MD, Zhang J (2006) Subcellular dynamics of protein kinase A activity visualized by FRET-based reporters. Biochem Biophys Res Commun 348:716–721
13. DiPilato LM, Zhang J (2009) The role of membrane microdomains in shaping beta2-adrenergic receptor-mediated cAMP dynamics. Mol Biosyst 5:832–837

Chapter 17

FRET and BRET-Based Biosensors in Live Cell Compound Screens

Katie Herbst Robinson, Jessica R. Yang, and Jin Zhang

Abstract

Live cell compound screening with genetically encoded fluorescence or bioluminescence-based biosensors offers a potentially powerful approach to identify novel regulators of a signaling event of interest. In particular, compound screening in living cells has the added benefit that the entire signaling network remains intact, and thus the screen is not just against a single molecule of interest but against any molecule within the signaling network that may modulate the distinct signaling event reported by the biosensor in use. Furthermore, only molecules that are cell permeable or act at cell surface receptors will be identified as "hits," thus reducing further optimization of the compound in terms of cell penetration. Here we discuss a detailed protocol for using genetically encoded biosensors in living cells in a 96-well format for the execution of high throughput compound screens and the identification of small molecules which modulate a signaling event of interest.

Key words Live cell compound screening, Genetically encoded biosensor, Cell signaling, Fluorescence resonance energy transfer, Bioluminescence

1 Introduction

High throughput compound screening, the process in which thousands to millions of compounds are assayed against a particular molecule of interest, is a common method to identify novel small molecule modulators of a given cellular target. After further testing and optimization, ultimately such compounds can be used as chemical tools to probe the function of a target protein within a cell in a dynamic fashion, a feature that is not provided through traditional genetic manipulation [1]. Moreover, when implemented in a particular fashion, a high throughput compound screen can serve as the initial phase of the drug discovery process. In this case, a compound library is screened against a particular disease target with the intention of identifying novel therapeutic agents or new uses for existing drugs [2]. Prior to initiating a high throughput screen against a target of interest, however, it is critical

Jin Zhang et al. (eds.), *Fluorescent Protein-Based Biosensors: Methods and Protocols*, Methods in Molecular Biology, vol. 1071, DOI 10.1007/978-1-62703-622-1_17,

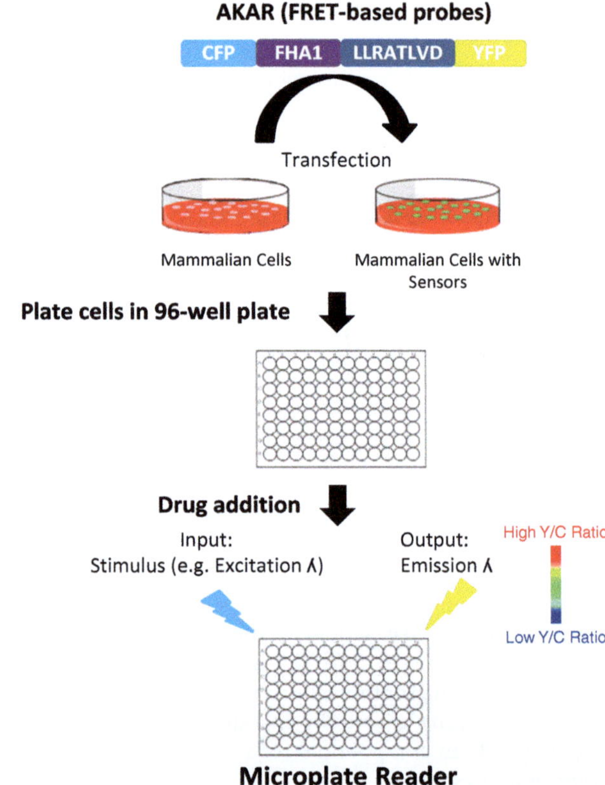

Fig. 1 Schematic representation of high throughput screen using genetically encodable FRET and BRET-based biosensors (e.g., AKAR)

to develop an assay which can rapidly and accurately report the effect of a small molecule on a target protein [1]. Specifically, in order for an assay to be used in a high throughput compound screen, it must be easily adapted to a 96-well format, or with the increasing availability and size of compound libraries, to 384- or 1,536-well format. Moreover, to be efficient in a high throughput format, the assay must be sensitive and reproducible (i.e., high signal-to-noise and low variability) and be able to be employed within a reasonable timeframe [3]. When compared to in vitro biochemical assays, achieving suitable sensitivity and reproducibility for a high throughput screen, often measured by the Z factor [4], is more challenging in cell-based assays. However, as genetically encoded biosensors can sensitively monitor signaling dynamics in living cells, when adapted accordingly, they have been proven to be a suitable assay format for live cell compound screens (Fig. 1) [5].

1.1 The cAMP/PKA Signaling Pathway

Throughout this protocol, biosensors that monitor activities in the cAMP-mediated signaling pathway will be used as an illustrative example of how to use a genetically encoded biosensor in a high throughput screen. The classical route of the production of cAMP, a canonical second messenger, is initiated when an extracellular stimuli binds to a G_s-coupled G protein coupled receptor (GPCR) on the cell surface, inducing the activation of adenylyl cyclase (AC), the enzyme which catalyzes the conversion of intracellular ATP into cAMP [7]. Once produced, cAMP can activate its effector molecules, which include the cAMP-activated protein kinase (PKA) and Exchange Protein activated by cAMP (Epac), before being degraded by phosphodiesterase (PDE). Collectively, the molecular players involved in the cAMP-mediated signaling pathway operate in a highly regulated fashion to coordinate numerous cellular functions, including insulin secretion and long term memory formation [8–10]. Furthermore, pharmacological manipulation of the molecular players involved in the cAMP-mediated signaling pathway is implicated in the treatment of various diseased states. As such, there is great effort to probe the regulation of the cAMP-mediated signaling in living cells.

1.2 Genetically Encoded Biosensors for Live Cell Compound Screens

This article focuses on the application of live cell compound screens as a tool for identifying various modulators of particular signaling pathway of interest. Compared to in vitro compound screens, live cell compound screens have two additional benefits. First, since the entire signaling network remains intact, the screen will identify any compound that modulates any molecule within the signaling pathway, not just the target molecule under study. For instance, when screening for molecules that regulate intracellular levels of cAMP, a compound that acts at the levels of the GPCR, G protein, AC, or PDE could be identified. Second, the screen will only identify compounds which are cell permeable or work at the level of the cell surface, reducing the need for further optimization in terms of cell permeability [5].

Genetically encoded biosensors have been used in live cell compound screens. A class of such biosensors is the fluorescence resonance energy transfer (FRET)-based reporters which are able to detect complex spatiotemporal dynamics of second messengers, enzyme activationctivities and protein-protein interactions in living cells (Fig. 2a) [6]. As a proof of concept, FRET-based biosensors designed to report cAMP dynamics and PKA activity were expressed in a population of cells in a multi-well format, where a microplate reader was used to detect changes in fluorescence signal over time (Fig. 2b). In doing so, these biosensors accurately identified molecules which modulated cAMP levels and the activity of PKA within the cell [5]. Similarly, FRET-based biosensors which report the activity or activation state of other promising drug targets, such as proteases and GPCRs, serve as promising tools for live cell high throughput compound screens [1, 11, 12]. An alternative design is based on bioluminescence resonance energy transfer (BRET) where a bioluminescent protein and an appropriate fluorescent protein (FP) typically serve as the donor

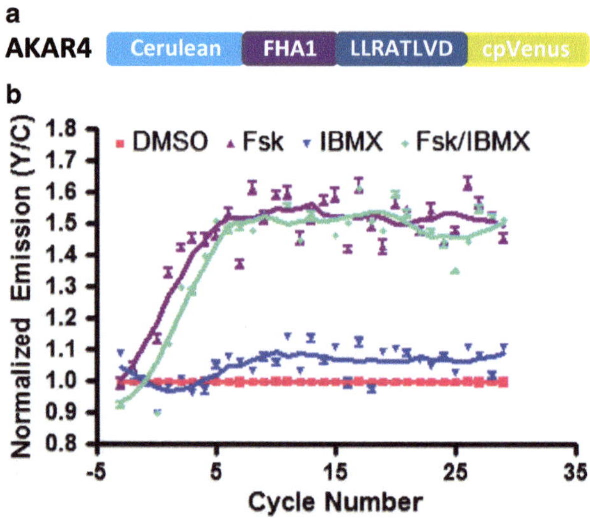

Fig. 2 High throughput activity assay based on an improved AKAR (AKAR4) (*a*). Design of AKAR4 (*b*). Representative time-course curves depicting changes in emission ratios (Y/C) of HEK293T cells expressing AKAR4, treated with DMSO (1 %), Forskolin (50 μM), IBMX (100 μM), and a combination of Forskolin and IBMX (50 μM, 100 μM respectively). 1 cycle = 16 s. Error bars represent standard deviation (*n* = 4)

and acceptor, respectively [12, 13]. Because BRET-based assays do not require exogenous illumination, they are not susceptible to autofluorescence and can achieve higher sensitivity than fluorescence-based assays [14, 18]. This frequently makes BRET-based assays more desirable than FRET-based methods in high throughput compound screens [15, 16]. For example, genetically encoded biosensors based on BRET have been used in a high throughput manner to detect signaling events such as second messenger dynamics and GPCR activation [12, 13]. It is also important to note that both FRET- and BRET-based assays are susceptible to compound interference. Specifically, in FRET-based screens, the inherent fluorescence of small molecules present in the compound libraries may interfere with the assay readout. Similarly, several small molecules have been shown to be direct modulators of luciferase activity, and thus would affect the readout from a BRET-based screen [19, 20]. As both FRET- and BRET-based screens have their respective strengths and limitations, care should be taken when choosing an assay for a high throughput screen [1].

1.3 Biosensors Applicable to This Protocol

Here, we present a detailed protocol for using genetically encodeable biosensors, based on either FRET or BRET, for high throughput compound screens in living cells. As an example, we use A-kinase activity reporter 4 (AKAR4), a FRET-based biosensor designed to report PKA activity [17]. We begin by discussing strategies for transfection of the biosensor cDNA before providing instructions on passing the cells into a 96-well format. Then we outline the steps for performing the compound screen and end with tips for data analysis and identification of "hits." The details provided are specific for HEK293T cells, but also provide a detailed outline which can be applied to numerous cell lines and compound libraries.

2 Materials

2.1 Cell Culture

1. Dulbecco's Modified Eagle's Medium (DMEM) supplemented with 10 % fetal bovine serum (FBS) and 1 % penicillin–streptomycin.
2. 10 cm tissue culture dish.
3. 96 Well Assay Plate, White Plate, Clear Bottom with Lid, Polystyrene, Sterile (Costar, No. 3610).
4. 96 Well Assay Plate, Black Plate, Clear Bottom with Lid, Polystyrene, Sterile (Costar, No. 3603).
5. Poly-D-lysine: 0.1 mg/mL.

2.2 Transfection Materials

1. Biosensor DNA.
2. Dulbecco's Phosphate Buffered Saline—without Mg^{2+} and Ca^{2+} (DPBS).
3. DMEM supplemented with 10 % FBS and 1 % penicillin–streptomycin.
4. Calcium phosphate transfection: prepare 2× HBS buffer (0.28 M NaCl, 0.05 M HEPES, 0.0015 M Na_2HPO_4, pH = 7.05), and a separate solution of 1 M $CaCl_2$.

2.3 Harvesting and Counting

1. Solution of trypsin (0.25 %) and ethylenediaminetetraacetic acid (EDTA, 0.53 mM).
2. Hemocytomer.
3. Trypan blue.

2.4 Imaging Materials and Devices

1. Phosphate Buffer Saline (PBS; 10×): 0.385 M $Na_2HPO_4 \cdot 7H_2O$, 0.17 M $NaH_2PO_4 \cdot H_2O$, 0.342 M NaCl, pH 7.4.
2. Hank's Balanced Salt Solution (HBSS): 0.15 M NaCl, 4 mM KCl, 1.2 mM $MgCl_2 \cdot 6H_2O$, 55.5 mM Glucose, 20 mM HEPES, pH 7.2, Store at 4 °C.
3. FLUOstar OPTIMA Fluorescence microplate reader (BMG Labtechnologies, Inc.).
4. Filters: one 420DF20 excitation filter, two emission filters (470DF40 for cyan and 535DF25 for yellow).
5. Compound library.

2.5 Analyzing Images and Data

1. Spreadsheet application (e.g., Microsoft Office Excel).

3 Methods

3.1 Cell Culture and Transfection

1. Grow two 10 cm dishes of HEK293T cells to 60–70 % confluency (*see* **Note 1**).
2. Transfect the cells via calcium phosphate-mediated transfection.
   (a) For each dish, mix 5 μg DNA with 55 μL of 2 M $CaCl_2$ and add ddH_2O to a final volume of 500 μL.
   (b) Add the DNA solution dropwise to 500 μL of 2× HBS while the latter is being vortexed. Vortex the final solution well and incubate for 30–40 min at room temperature. The solution should turn opaque.
   (c) Mix again and add the DNA solution dropwise onto the cells, being careful not to disrupt the cells.
3. Check the transfection efficiency after 24 h. Proceed only if there is greater than 70 % transfection efficiency (*see* **Note 2**).

3.2 Harvesting, Counting, and Plating

1. Wash each dish once with 5 mL of PBS.
2. Add 500 μL of 0.05 % trypsin to each dish. Add 9 mL of fresh DMEM to the first dish to harvest cells. Harvest the cells in the second dish using the same 9.5 mL so that the cells from both dishes are collected in the same 10 mL total volume.
3. Spin cells at 1,500 × *g* for 6–8 min.
4. Remove the media and resuspend the pellet in 1 mL of fresh media (*see* **Note 3**).

   Count cells: dilute 7.5 μL of cells two fold with 7.5 μL of trypan blue, add the solution to the hemocytometer and count all of the cells in one large square. Determine the number of cells per μL according to the following equation (*see* **Note 4**):

$$\text{Cells per }\mu\text{L} = \frac{(\#\text{ cells}) \times (\text{dilution factor, in this case 2})}{0.1\text{ mm}^3}.$$

5. Determine the number of cells that should be plated so that 24 h later the cells form a monolayer in the well. For HEK293T cells, this is about 30,000 cells per well (*see* **Note 1**). Calculate the volume of media needed to dilute the cell suspension so that the desired number of cells can be plated in a volume of 100 μL per well (*see* **Note 5**). For example, HEK293T cells should be diluted to 300 cells per μL so that 30,000 cells will be added to each well in 100 μL. Importantly, some of the control wells should be left either without cells or with non-transfected cells.
6. Place the 96-well plate in the CO_2 incubator and incubate overnight.

3.3 Live Cell Compound Screen

1. Wash the cells once with 100 μL of PBS (*see* **Note 5**).
2. Add 190 μL of HBSS to each well (*see* **Notes** 7 and **8**).
3. Insert the plate into the microplate reader
4. Set up an imaging protocol such that both the CFP (excitation 420 ± 10 nm, emission 480 ± 10 nm) and FRET (excitation 420 ± 10 nm, emission 535 ± 10 nm) channels are acquired for each well (*see* **Note 8**).
5. Establish a three point baseline reading by reading all necessary channels for the entire plate three times.
6. Add compounds from the compound library of choice using a multichannel pipette. Add 10 μL of each compound to each well of the plate, except for the control wells (*see* **Note 6**). Mix by pipetting and be sure to change tips after each addition (*see* **Note** 7). If a robot system is available, it can be programmed to add the compounds.
7. Insert the plate into the plate reader and acquire data for each channel for the entire plate for a time period appropriate for the signaling events under study. For cAMP-mediated signaling, this is about 20 min.
8. Optional step: Perform an antagonist screen by adding a subsaturating dose of compound to each well in order to activate the signaling event under study, including half of the vehicle control wells. For example, when screening molecules which antagonize PKA activity using AKAR4, a 100 nM dose of the β2-adrenergic receptor agonist is added to each well.

3.4 Data Analysis

1. Using Excel, determine the average reading from the wells which had no cells or non-transfected cells for both the donor and acceptor channels. Subtract this value from the respective value for each well at each time point. These values serve as the background corrected values.
2. For each time point, take the ratio of the acceptor to donor.

$$\text{Yellow to Cyan Emission Ratio} = \frac{\text{FRET channel Emission Intensity} - \text{FRET channel Emission Intensity of Background}}{\text{CFP channel Emission Intensity} - \text{CFP channel Emission Intensity of Background}}.$$

3. Normalize the calculated ratio at each time point to the value of the respective well just before drug addition.
4. Also, normalize the calculated ratios to the vehicle control wells.
5. For a time course, plot the normalized ratio over time.
6. Calculate the standard deviation of the vehicle control wells. A "hit" is a compound that falls 4–5 standard deviations above or below the average of the vehicle control wells.

4 Notes

1. The details provided are using HEK293T cells as an example. Optimization of cell culture, transfection techniques, and plating density may be needed for other cell types.
2. If transfection efficiency is not greater than 70 %, the signal-to-noise ratio may not be high enough to achieve high sensitivity. As an alternative to transient transfection, a cell line stably expressing the biosensor of interest could be generated and used.
3. Be sure not to aspirate off the pellet. It is best to turn the centrifuge tube horizontally to aspirate the media off of the side of the tube, keeping the tip away from the pellet.
4. The volume of one square on the hemocytometer is 0.1 mm^3 or 0.1 μL, so the quick calculation is: cells/μL = #cells × Dilution Factor × 10.
5. For cell lines which do not adhere well to plastic, such as HEK293T, it is beneficial to plate each well of the 96-well plate with poly-D-lysine beforehand. To do this, add 50 μL of 0.1 mg/mL poly-D-lysine to each well of a 96-well plate and incubate at RT for 1 h. Aspirate off the solution and wash once with PBS. Allow the plate to dry for 30 min in the laminar flow hood before plating cells.
6. Controls for the experiment will be a set of naïve cells, meaning no drugs or vehicle control is added to the cells. The second set of controls will be cells treated with only the vehicle control.
7. How much media is added to each well depends on the composition of the compound library. This example is using a library of stock concentration of 200 μM such that when 10 μL of compound is added to a well containing 190 μL of HBSS the final screening concentration is 10 μM.
8. If using a BRET-based biosensor, there will be some differences in the protocol for using a FRET-based biosensor. In particular, when using a BRET-based biosensor:
   (a) Cells do not need to be washed with PBS.
   (b) A white walled plate should be used.
   (c) The desired concentration of luciferase substrate, e.g., coelenterazine-h, must be added to HBSS.
   (d) Acquisition of emission wavelengths must be adjusted accordingly and no excitation light is required.

References

1. Inglese J, Johnson RL, Simeonov A, Xia M, Zheng W, Austin CP, Auld DS (2007) High-throughput screening assays for the identification of chemical probes. Nat Chem Biol 3(8): 466–479
2. Herzberg RP (2009) Design and implementation of high-throughput screening assay. Methods Mol Biol 565:1–32
3. Mayr L, Bojanic D (2009) Novel trends in high-throughput screening. Curr Opin Pharmacol 9(5):580–588
4. Zhang JH, Chung TDY, Olderburg KR (1999) A simple statistical parameter for use in evaluation and validation of high throughput screening assays. J Biomol Screen 4(2)
5. Allen MD, DiPilato LM, Rahdar M, Ren YR, Chong C, Liu JO, Zhang J (2006) Reading dynamic kinase activity in living cells for high-throughput screening. ACS Chem Biol 1(6): 371–376
6. Ni Q, Titov DV, Zhang J (2006) Analyzing protein kinase dynamics in living cells with FRET reporters. Methods 40(3):279–286
7. Adjobo-Hermans MJW, Goedhar J, van Weeren L, Nijmeijer S, Manders EMM, Offermanns S, Gadella TWJ (2011) Real-time visualization of heterotrimeric G protein Gq activation in living cells. BMC Biol 9(1):32
8. Tasken K, Aandahl EM (2004) Localized effects of cAMP mediated by distinct routes of protein kinase. Physiol Rev 84:137–167
9. Williams C (2004) cAMP detection methods in HTS: selecting the best from the rest. Nat Rev Drug Discov 3:125–135
10. Zhang J, Hupfeld CJ, Taylor SS, Olefsky JM, Tsien RY (2005) Insulin disrupts beta-adrenergic signalling to protein kinase A in adipocytes. Nature 437:569–573
11. Tian H, Ip L, Luo H, Chang DC, Luo KQ (2007) Br J Pharm 150(3):321–324
12. Thomsen W, Frazer J, Unett D (2005) Functional assays for screening GPCR targets. Curr Opin Biotechnol 16(6):655–665
13. Jiang LI, Collins J, Davis R, Lin KM, DeCamp D, Roach T, Hsueh R, Rebres RA, Ross EM, Taussig R, Fraser I, Sternweis PC (2007) Use of a cAMP BRET sensor to characterize a novel regulation of cAMP by the sphingosine 1-phosphate/G13 pathway. J Biol Chem 282(14):10576
14. Boute N, Jockers R, Issad T (2002) The use of resonance energy transfer in high-throughput screening: BRET versus FRET. Trends Pharmacol Sci 23(8):351–354
15. Fan F, Binkowski BF, Butler BL, Stecha PF, Lewis MK, Wood KV (2008) Novel genetically encoded biosensors using firefly luciferase. ACS Chem Biol 3(6):346–351
16. Fan F, Wood KV (2007) Bioluminescent assays for high-throughput screening. Assay Drug Dev Technol 5(1):127–136
17. Depry C, Allen MD, Zhang J (2011) Visualization of PKA activity in plasma membrane microdomains. Mol Biosyst 7:52–58
18. Herbst KJ, Allen MD, Zhang J (2011) Luminescent kinase activity biosensors based on a versatile bimolecular switch. J Am Chem Soc 133(15):5676–5679
19. Herbst KJ, Allen MD, Zhang J (2009) The cAMP-dependent protein kinase inhibitor H-89 attenuates the bioluminescence signal produced by *Renilla* luciferase. PLoS One 4(5)
20. Thorne N, Inglese J, Auld DS (2010) Illuminating insights into firefly luciferase and other bioluminescent reporters used in chemical biology. Chem Biol 17(6):646–657

Chapter 18

Integrating Fluorescent Biosensor Data Using Computational Models

Eric C. Greenwald, Renata K. Polanowska-Grabowska, and Jeffrey J. Saucerman

Abstract

This book chapter provides a tutorial on how to construct computational models of signaling networks for the integration and interpretation of FRET-based biosensor data. A model of cAMP production and PKA activation is presented to provide an example of the model building process. The computational model is defined using hypothesized signaling network structure and measured kinetic parameters and then simulated in Virtual Cell software. Experimental acquisition and processing of FRET biosensor data is discussed in the context of model validation. This data is then used to fit parameters of the computational model such that the model can more accurately predict experimental data. Finally, this model is used to show how computational experiments can interrogate signaling networks and provide testable hypotheses. This simple, yet detailed, tutorial on how to use computational models provides biologists that use biosensors a powerful tool to further probe and evaluate the underpinnings of a biological response.

Key words Computational modeling, FRET biosensors, Virtual cell, Cell signaling, cAMP

1 Introduction

Fluorescence microscopy of genetically encoded fluorescence resonance energy transfer (FRET)-based biosensors is a powerful technique for understanding signaling kinetics in living cells. These biosensors allow researchers to visualize and quantify the spatiotemporal distribution of signaling molecules within the cell. As discussed in previous chapters, a variety of FRET biosensors have been developed to directly detect changes in intracellular concentrations or activation of signaling molecules in real time [1]. It is often the goal to understand how these biochemical dynamics are modulated by the overall signaling network, particularly in response to pharmacologic or genetic perturbations. The complexity of signaling networks often hinders attempts to relate biosensor data directly to the molecular mechanisms underlying dynamic cell responses. Computational models allow integration of diverse biochemical and biosensor data

Jin Zhang et al. (eds.), *Fluorescent Protein-Based Biosensors: Methods and Protocols*, Methods in Molecular Biology, vol. 1071, DOI 10.1007/978-1-62703-622-1_18,

into a common quantitative framework. These models can be used to quantitatively evaluate the plausibility of current hypotheses with existing data or to computationally generate new hypotheses that can be tested in the wet lab. This book chapter provides experimental biologists an accessible tutorial on computational modeling, such that they may begin to analyze the system-level implications of their fluorescence-based biosensor data.

2 Materials

1. The Virtual Cell Software (version 5.1 or greater).
2. Neonatal Cardiomyocyte Isolation Kit, Cellutron cat# nc-6031.
3. 35 mm Glass-Bottom Culture Dishes, MatTek Cat# P35GC-1.5-14-C.
4. SureCoat, Cellutron, Cat# sc-9035.
5. Cell Culture Media: Dulbecco's Modified Eagle Medium (DMEM) supplemented with 10 % heat-inactivated fetal bovine serum (FBS), 4 mM L-glutamine, 20 U/mL penicillin, and 20 U/mL streptomycin.
6. Tyrode's buffer: 155 mM NaCl, 5 mM KCl, 2 mM $CaCl_2$, 1 mM $MgCl_2$, 2 mM NaH_2PO_4, 10 mM HEPES, and 10 mM glucose, pH 7.2.
7. Lipofectamine 2000 Kit, Invitrogen.
8. ICUE, Indicator of cAMP accumulation Using Epac, FRET biosensor DNA plasmid, GenScript.
9. AKAR, A-Kinase Activity Reporter, FRET biosensor DNA plasmid, GenScript.
10. Forskolin (FSK), an AC agonist.
11. IBMX, a general PDE inhibitor.
12. Isoproterenol (ISO), a β-adrenergic receptor agonist.
13. Olympus wide-field inverted microscope IX81.
14. Olympus UPlanSApo 10×/1.3 numerical aperture (NA) objective.
15. Lambda DG4 excitation filter, Sutter Instrument Company.
16. Lambda 10-3 emission filter wheel changer with a Smart-Shutter.
17. ECFP/YFP-ET filter set, Chroma Technology (Rockingham, VT).
18. 12-bit Hamamatsu digital C9300-21 digital camera (Hamamatsu Photonics, Bridgewater, NJ).
19. Image acquisition software, such as Metamorph or IP-LAB.
20. Image J software.

3 Methods

3.1 Model Generation

3.1.1 Model Definition

This book chapter focuses on the generation of computational models that incorporate experimental data from FRET-based biosensors. The main goal of computational models is to help answer questions that cannot be directly answered using biosensor responses and intuition alone. Computational models provide a unified understanding of how a signaling pathway acts as a whole by integrating experimental information about the different aspects of the signaling cascade. There are many issues that models address, such as quantitatively examining the viability of a hypothesis or evaluating competing hypotheses for explaining an observed phenomenon. These models are often then probed to make predictions and generate testable hypotheses about particular biological responses, such as the action of drugs or gene silencing.

When building a computational model, the most important consideration is the biological question that will be addressed. The biological question determines the size and scope of the model needed and dictates the expected output of the model. As a prototypical signaling pathway, this chapter will examine the production of the second messenger cyclic adenosine monophosphate (cAMP) and the subsequent activation of cAMP-dependent Protein Kinase A (PKA). PKA participates in many diverse and integral signaling pathways and thus understanding the dynamics of PKA activation is important in analyzing these pathways [2, 3]. An example of a biological question that could be asked about PKA activation might be: "What is the rate-limiting step for PKA activity in response to receptor stimulation: cAMP production or PKA activation?" This question defines the scope of the model by saying that the desired output is PKA activity and the model must include both cAMP production and PKA activation.

Now that scope and objective of the model system have been specified by the biological question, the specifics of the signaling pathway must be defined. This is done by sifting through the information presented in literature, as well as databases, to define the most accurate signaling network [4–6]. The model considered here (Fig. 1b) incorporates cAMP production by direct activation of adenylyl cyclase (AC) upon stimulation with forskolin (FSK) or stimulation of AC by β-adrenergic receptor agonist isoproterenol (ISO). After synthesis, cAMP is degraded by cAMP-specific phosphodiesterase (PDE). The cAMP biosensor ICUE is directly incorporated into this model, binding cAMP to form an ICUE-cAMP complex [7]. PKA holoenzyme consists of two catalytic and two regulatory subunits. Upon full activation by cAMP binding to the regulatory subunits, PKA's catalytic subunits dissociate and phosphorylate its downstream targets. In this simplified model, PKA is represented as having just one regulatory and one catalytic

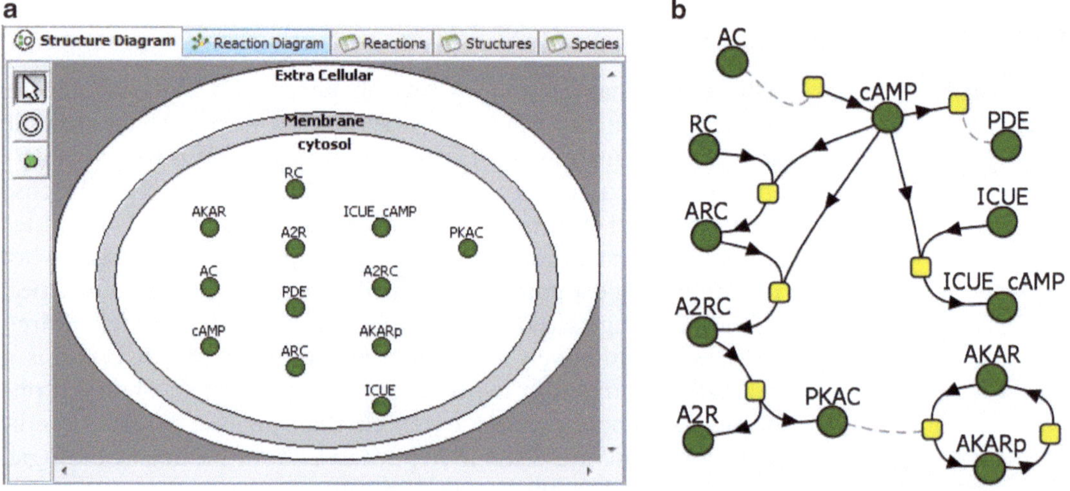

Fig. 1 (**a**) Virtual Cell structure diagram with the cell compartment and model species defined. (**b**) Reaction diagram of model. *Green circles* represent model species, *yellow boxes* and *arrowed lines* are reactions and *dotted lines* represent enzyme catalysis

subunit, which switch from an inactive to active state by the binding of two cAMP molecules followed by the dissociation of the regulatory subunits from the catalytic subunits (RC, ARC, A2RC, A2R and PKAC in order from inactive to active catalytic subunits). PKA activity can be monitored by the FRET biosensor A-Kinase Activity Reporter (AKAR) [8, 9]. Activated PKA catalytic subunits phosphorylate AKAR, inducing a conformational change in the reporter that generates FRET signal. Both the AKAR phosphorylation and subsequent FRET signal are reversed by protein phosphatases.

It is a good rule of thumb to start with the simplest model possible and then expand as needed to explain the observed data. When developing any model, assumptions must be made explicit when publishing or disseminating the model. For example, the model presented here makes the simplifying assumption that activation of AC by stimulation of the β-adrenergic receptor by isoproterenol (ISO) can be approximated as direct AC activation in place of explicitly modeling the receptors themselves and activity of G proteins. Another assumption is that PKA activation, which requires the binding of two cAMP molecules, is not affected by which site the cAMP binds to (i.e., binding order is independent). These assumptions will affect the model and will be discussed later.

The last piece of information needed to build the model is the kinetic rate constants. This can be one of the more difficult aspects of model building because it requires extensive literature mining. It is often useful to look at published computational models, as authors will typically publish the original literature sources for the kinetic parameters [2, 3, 10]. However, many parameters are unavailable in the literature. In this case, you can start by using order-of-magnitude estimates of that parameter [11]. These gaps in information will be filled later in this chapter when we use the experimental biosensor data to constrain the model and fit these parameters.

3.1.2 Model Implementation in Virtual Cell

The computational models discussed here are formulated mathematically as Ordinary Differential Equations (ODEs). ODE models can be implemented and computationally solved using numerous programming languages (e.g., MATLAB, C++), but for this book chapter we will use the user-friendly Virtual Cell software (http://vcell.org/) [12]. A key advantage of Virtual Cell is that the math is performed "behind the scenes", so the user can focus on the biochemical reactions and biology of interest. Virtual Cell is free to use and was developed by Les Loew and colleagues at the National Resource of Cell Analysis and Modeling to be a simple yet powerful tool to allow students and biologists with relatively little math background to perform computational modeling [12]. Virtual Cell also has advanced capabilities including stochastic and spatial modeling; however, this book chapter will focus on its basic ODE modeling features. More detailed instruction on the use of Virtual Cell, as well as additional tutorials and resources, can be found on the Virtual Cell webpage (http://vcell.org). The following steps provide a tutorial walk-through for creating your own simple cAMP/PKA pathway model in Virtual Cell.

Step 1. Creating a model. After loading Virtual Cell, create a user account if you don't have one already. To create a new model, select *File>New>Biomodel.*

Step 2. Define compartments. Virtual Cell can incorporate signaling pathways that occur in separate cellular compartments. Compartments can be physical compartments within the cell, such as the nucleus or mitochondria, but they can also represent regions of the cell that are separated by diffusional barriers within the cell, such as the dyadic cleft which is separated from the cytosol in cardiac myocytes. To create a compartment, use the "Compartment Tool" (empty circle icon) in the "Structure Diagram" tab of the Physiology section of the model, and click in the graph area to create a compartment. If sub-compartments are needed within the cell, click the "Compartment Tool" inside the cell to create individual sub-compartments. Our example uses one compartment to represent the cell, where the inside of the cell has been labeled "cytosol" (Fig. 1a). Compartments can be renamed using the Object Properties tab in the lower window.

Step 3. Define model species. To define the proteins and second messengers, referred to as species, use the "Species Tool" (green circle icon) and click in the compartment in which they will be reacting. If implementing a multi-compartment model, species that can interchange between compartments must be defined in all relevant compartments. Figure 1a presents all of the needed species, shown as small circles, inside the cytosol compartment. Each "species" has been labeled to define what it represents. It is important to note that complexes between two reactants, such as cAMP bound ICUE (ICUE_cAMP), or different phosphorylation states, such as AKAR and phosphorylated AKAR (AKARp), need to be defined as separate species. Again, rename species using the Object Properties tab.

Step 4. Define reactions. In the "Reaction Diagram" tab of the Physiology section, all of the species should be present in their assigned compartment before connecting reactants to products. Reactants are connected to products using the "RX connection tool". When this tool is selected, click on a reactant species and drag it to the product species. A line with arrows pointing from reactant to product and a yellow box in the middle should be created (Fig. 1b). If multiple reactants are combining to form a product, each additional reactant can be added to the reaction by dragging from the reactant to the yellow box of the desired reaction. For example, to define the cAMP binding to the inactive PKA holoenzyme to form the intermediate "ARC", a reaction is dragged from cAMP to ARC and then another reaction line is dragged from RC to the newly formed yellow box. Similarly, if multiple products are formed in the reaction, click on the yellow box of the desired reaction and drag to the additional product. Finally, enzymes can be connected to the reaction that they catalyze using the "set a catalyst" tool and dragging from the enzyme to the yellow box of the reaction. Once all of the reactions are created, your "Reaction Diagram" should look similar to Fig. 1b.

Once the Reaction Diagram is created, a kinetic type and rate constants must be defined for each of the reactions in the diagram. To define these, select the reaction (yellow box) and the equations and parameters for that reaction will be shown below (*see* **Note 1**). The initial parameters used in our example model are listed in Table 1.

For binding or dissociation reactions, use the "Mass Action" kinetic type. Mass action kinetics allows the reaction to be reversible. Virtual Cell will then define the reaction rate equation based on how the species have been connected to it and create two variables, K_f and K_r, which represent the forward and reverse rate constants, respectively. If the forward and reverse rate constants are not directly available in the literature, the dissociation constant, K_D, can be used to infer the rate constants using the following equation (*see* **Note 2**).

$$K_D = K_r / K_f$$

Table 1
Initial parameter values defined for the example model. See public model "ecg5pc: Greenwald MIMB 2012—Base Model" to relate reaction and parameter names to the model fluxes

Reaction	Parameter	Value	Unit
cAMP_synth	ATP	5,000	μM
	Km	860	μM
	Km_Iso	315	μM
	kcat_iso†	0.75	s^{-1}
	Kd_iso	0.1	μM
	kfsk†	7.3	s^{-1}
	Iso_stim	1	μM
	t_iso	200	s
	Kd_fsk	860	μM
	FSK_stim	50	μM
	t_FSK	600	s
cAMP_deg	Km	1.305	μM
	kcat	5	s^{-1}
	Ki	30	μM
	IBMX_stim	100	μM
	t_IBMX	600	s
ICUE_bind	Kf†	5	$\mu M^{-1} s^{-1}$
	Kr	10	s^{-1}
PKA_bind1	Kf‡	1,000	$\mu M^{-1} s^{-1}$
	Kr‡	9,140	s^{-1}
PKA_bind2	Kf‡	1,000	$\mu M^{-1} s^{-1}$
	Kr	1,640	s^{-1}
PKA_act	Kf	4,375	s^{-1}
	Kr	1,000	$\mu M^{-1} s^{-1}$
AKAR_phos	Km‡	21	μM
	kcat	54	s^{-1}
AKAR_dephos	kcat‡	8.5	$\mu M^{-1} s^{-1}$
	PPase‡	2.14	μM

Parameters that were used to fit the model to ICUE (Fit 1) or AKAR (Fit 2) data are identified by † and ‡, respectively

An estimate of either K_r or K_f will allow for the calculation of the other constant such that its relative magnitude agrees with experimental data. The above equation can be rearranged and typed into the expression section for the associated rate constant and Virtual Cell will create a new variable, K_d, where you can input the dissociation constant value.

For enzyme catalyzed reactions, such as phosphorylation of AKAR (AKARp) by the catalytic subunit of PKA (PKAC), use the Henri–Michaelis–Menten (Irreversible) kinetic type.

Virtual Cell will then create a reaction rate that has two variables, K_m and V_{max}. K_m is the Michaelis constant and is often calculated when enzyme kinetics are measured. V_{max} is the "maximum velocity" of the enzyme catalysis, which can be described by $V_{max} = k_{cat} \times E_{tot}$, where k_{cat} is the catalytic rate constant and E_{tot} is the concentration of the active enzyme. Put this equation into the expression column for V_{max} and in place of E_{tot} put the species name given to the relevant enzyme that catalyzes this reaction, e.g., PKAC for AKAR phosphorylation. Virtual Cell will automatically create a new variable for k_{cat} where you can define the catalytic rate constant for this reaction. For enzymatic reactions that do not have an explicit reactant defined (e.g., production of cAMP), Virtual Cell is not able to automatically define the Michaelis–Menten rate equation. Therefore, for this reaction, you will need to choose the general kinetic type and manually input the Michaelis–Menten reaction rate expression.

Finally, for reactions that do not fall into these two categories, the General kinetic type, where the user defines the equation for the reaction rate manually, can be used. One common expression used in models is the Michaelis Type equation, which has the following general form:

$$\frac{A}{\mathrm{EC}_{50} + A},$$

where A is the species of interest and EC_{50} is the concentration of species A at which half of the maximal activity is achieved. This form of the equation is used in our model to approximate ISO stimulation of cAMP production.

Public models can be accessed by selecting *file>open>biomodel* and selecting the model of interest from the Public Biomodels folder. Use our public Virtual Cell model (model name "ecg5pc: Greenwald MIMB 2012—Base Model") to verify that the reactions and rate parameters are specified correctly in your model. In particular, examine the cAMP synthesis reaction because this reaction required manual definition and incorporates Michaelis type equations to define stimulus activation strength.

Step 5. Model simulation. Once the model has been defined in Virtual Cell, the behavior of the model is determined by numerical solution of the model equations. To accomplish this, first create a new application by right clicking on "Applications", selecting "Add New" and select "Deterministic". This will create a new application to numerically solve the ODE model you defined in Virtual Cell. If creating a multi-compartment model, you will need to specify the volume of each compartment. If using a single compartment, as is used in the example, it is usually not

Table 2
Initial concentrations used in the example model

cAMP	0 μM
AKAR	1.25 μM
AC	0.05 μM
RC‡	1 μM
A2R	0 μM
PDE	0.014 μM
ARC	0 μM
ICUE_cAMP	0 μM
A2RC	0 μM
AKARp	0 μM
ICUE	0.15 μM
PKAC	0 μM

‡ identifies parameters that were allowed to vary to fit AKAR data (Fit 2)

necessary to change the size definitions away from the default. Next, define parameters for each of the species in the model. Under the "Specifications" tab will be a list of all the species in the model. This is where you define the initial concentrations of each species under the "Initial Condition" column. The initial concentrations used in our model are defined in Table 2. It is also possible in this section to specify a species as having a fixed concentration over time by checking the box under the "Clamped" column. For example, ATP concentration is often assumed to be constant because of its high concentration and strict regulation within the cell.

You are now ready to run the simulation. Under the "Simulations" tab, click the "New Simulation" button. Select the simulation that was just created and push the "Edit Simulation" button. From this window you can specify what changes you wish to make for this simulation. This can be used to simulate the addition of a stimulus, such as changing the concentration of ISO from 0 to 1 μM at the start of the simulation which is done by entering the value 1 μM in the "New Value/ Expression" column for ISO. It is often necessary to be able to apply a stimulus later in a simulation in order to compare simulation results to experimental data where multiple perturbations are applied in series. To add time delays, change the parameter in the reaction diagram to have the following form,

$$p = p_{\text{new}} \times (t > t_1)$$

where p is the parameter that you wish to change, p_{new} is the new value for that parameter, and t_1 is the time at which the parameter will change. Then you can define p_{new} and t_1 at the beginning of the simulation as described above.

On the "Solver" tab of the edit simulation window, you can specify the time period over which the simulation will be performed, as well as other properties such as the maximum time step and how many samples will be recorded. Since the experimental data which are used for our example were collected for 1,200 s at 10 s intervals, we set the ending time in the model to 1,200 and set the output interval to 10 s. Finally, with the fully defined simulation selected, the simulation can be run by pushing the "Run and Save Simulation" button. This will submit the simulation to be solved remotely on the Virtual Cell servers. When the model is solved, the "Running Status" column will read "completed". To view the results, push the "Simulation Results" button. This will bring up a window where you can view the simulation results for each of the different species (for error checking, *see* **Note 3**). Also, the simulation values can be seen in table format by clicking the "Show Data" button in the bottom right corner, which can be useful for extracting data for further analysis or plotting. Figure 2 shows the results of the example model with addition of 1 μM ISO at 200 s and the addition of the direct AC agonist forskolin (FSK, 50 μM) plus PDE inhibitor (IBMX, 100 μM) at 600 s.

3.2 Collection of FRET Data for Model Integration

Computational models of signaling dynamics often use FRET biosensors because of their ability to collect real-time signaling dynamics. Here we explain how to collect and analyze data obtained from FRET biosensors such that they can be used to evaluate and inform computational models. Various methods can be applied to measure FRET from the changes in donor and acceptor emission. FRET signal can be detected when an excited donor fluorophore transfers energy to an acceptor fluorophore in close proximity (<10 nm) [1]. A standard way to measure FRET is by quantifying the acceptor emission upon donor excitation (acceptor-sensitized emission). This method is usually corrected for spectral bleed-through caused both by leakage of the donor and acceptor emission into the FRET channel and by acceptor photobleaching. This method is also referred to as the 3-image FRET technique since it requires the measurement of three different intensities: acceptor emission, emission from the donor into the acceptor channel (due to FRET), and direct excitation of the acceptor. Our example uses two FRET biosensors, ICUE [7] and AKAR [8], which both utilize the Cyan and Yellow Fluorescence Proteins, CFP and YFP, a common and well-established FRET pair [13]. Since both ICUE and AKAR are unimolecular reporters, with fixed stoichiometry between donor and acceptor, the changes in donor emission ratio directly correlate

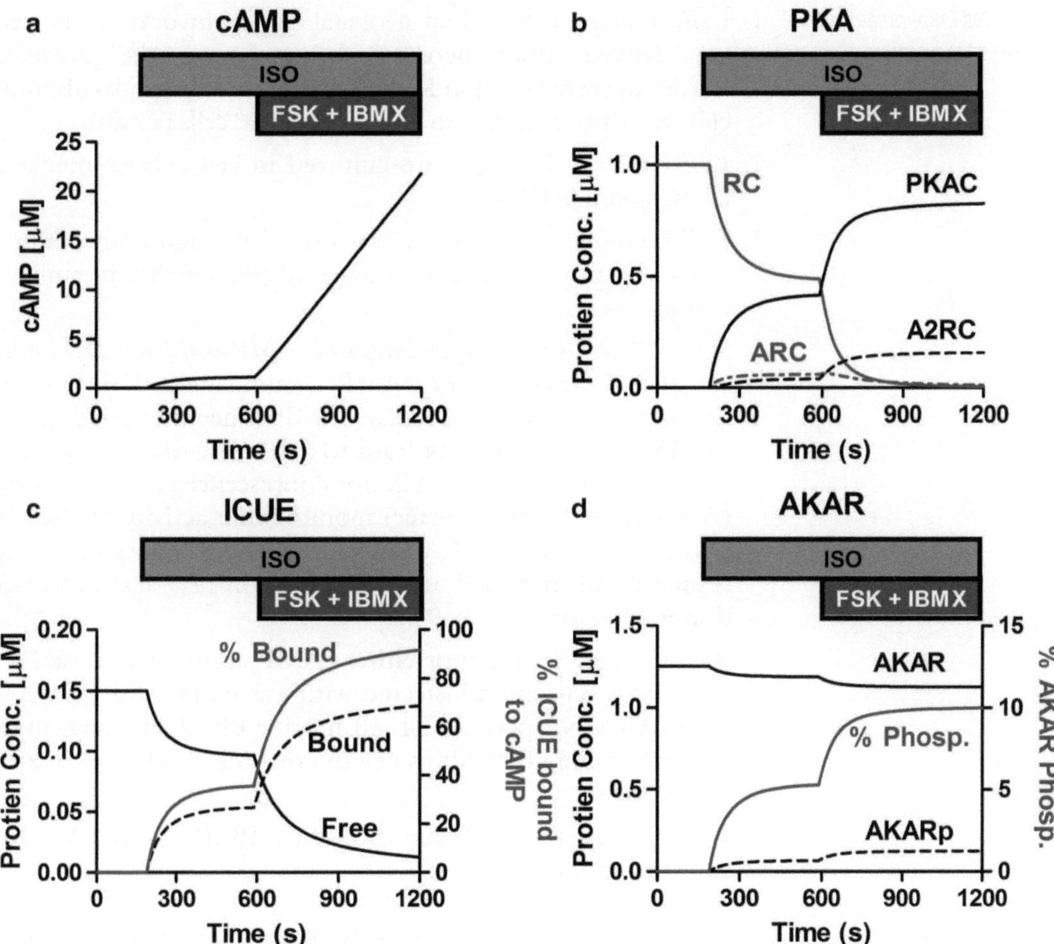

Fig. 2 Simulation results with 1 μM ISO applied at 200 s then 50 μM FSK and 100 μM IBMX applied at 600 s. (**a**) cAMP concentration over time where the concentration grows unbounded in the model because cAMP generation becomes faster than the inhibited cAMP degradation. (**b**) The four activation states of PKA. (**c**) [*Left axis*] Concentrations of ICUE that are bound to cAMP (*bound*), and ICUE that does not have cAMP bound to it (*free*). [*Right axis*] The percent of total ICUE that has cAMP bound. (**d**) [*Left axis*] Concentration of AKAR and AKARp. [*Right axis*] Percent of total AKAR that is phosphorylated

to changes in FRET signal. The main purpose of this part of the chapter is to describe how to obtain data from FRET experiment which would be used for validation of the model predictions.

3.2.1 FRET Experimental Data Collection and Normalization

The generation of FRET experimental data for use in models consists of two interdependent parts: (A) collection of experimental data using FRET-based biosensors and (B) normalization of collected experimental data for model validation and integration. One can also use the model to predict the biological response of the signaling pathway to new perturbations and experimentally test these predictions to gain new insight into the modeling network behavior.

3.2.2 Cell Preparation for FRET Measurement

1. *Cells.* Primary cultured rat neonatal cardiac myocytes. The cells are derived from neonatal hearts using the Neonatal Cardiomyocyte Isolation Kit and plated on 35 mm glass-bottom culture dishes coated with SureCoat at 10^6 cells per dish.
2. *Cell Culture.* The cells are cultured in cell culture media at 37 °C and 5 % CO_2.
3. *Cell Imaging.* Tyrode's buffer is used for imaging since it provides low autofluorescence and good pH stability in ambient conditions.
4. *FRET Based Reporters for Imaging cAMP and PKA Dynamics.*
   ICUE (Indicator of cAMP accumulation Using Epac) shows changes in intracellular cAMP concentration. Binding of cAMP to this biosensor leads to a decrease of acceptor emission (I_A) and an increase of donor fluorescence (I_D) [7]. *AKAR* (A-Kinase Activity Reporter) monitors the activity of PKA by phosphorylation of PKA substrate. AKAR phosphorylation results in an increased acceptor emission (I_A) and decreased donor emission (I_D) [8, 9].
5. *Transfection.* To monitor either cAMP accumulation or PKA activation, cells are transfected with 1.2 μg per dish of ICUE or AKAR DNA plasmid 24–48 h prior FRET imaging, using Lipofectamine 2000 Kit as per the manufacturer's instruction (*see* **Note 4**).
6. *Cell Treatment.* Forskolin (50 μM), IBMX (100 μM) and isoproterenol (1 μM).

3.2.3 Image Acquisition

Controls for Normalized FRET Ratio

It is important to collect controls for minimal and maximal FRET signal to be able to normalize the changes in the FRET ratio and to determine fractional activation. Fractional activation is the fraction of the total biosensor that is activated (where 1 means the biosensor is fully activated and 0 means that biosensor is inactive). This is necessary because changes in the FRET ratio are a relative measure of changes in biological responses and normalization allows the absolute quantification of the biosensor response that is necessary for model validation and fitting. The minimal biosensor signal (R_{min}) is defined as FRET signal measured under conditions where basal enzyme activity or basic protein formation is inhibited by the appropriate inhibitors. In practice, determination of the minimal FRET signal (R_{min}) depends on the amount of basal activity or concentration of the protein of interest in the imaged cells. For example, protein kinase C (PKC) has a significant amount of basal activity, thus to measure the minimal FRET signal PKC inhibitors must be used [14]. In cardiac cells, we determined that cAMP levels are low in the resting state and that PKA has minimal basal activity, thus it is assumed that unstimulated cells exhibit the minimal biosensor signal without inhibition of baseline activity. To establish this baseline FRET for each AKAR or ICUE experiment, we usually collect data for the first 2–5 min without addition of any drug. In contrast, the maximal biosensor signal (R_{max}) is determined by stimulating the signaling pathway in such a way that the FRET biosensor is maximally activated. This is usually done by applying a strong stimulator and inhibition of negative regulators of the signal response. To establish a maximal biosensor signal for cAMP or PKA, we add a mixture of 50 μM FSK and 100 μM IBMX 2–10 min before the end of each experiment.

To visualize the reporters, we use the wide-field Olympus motorized inverted microscope IX81 equipped with an Olympus UPlanSApo 10X/1.3 numerical aperture (NA) objective. The Lambda DG4 excitation filter changer with Xenon arc lamp can be used as an illumination system designed for rapid wavelength change. Since the microscope for 3-color FRET must be capable of collecting donor fluorescence, acceptor fluorescence, and FRET emission, the Lambda 10-3 emission filter wheel changer with a Smart-Shutter is applied for automatic and fast switch between emission filters (for CFP and YFP). To filter excitation and emission wavelengths, we use an ECFP/YFP-ET filter set. The set contains beam splitter (CFP/YFP) and the following filter sets: for the YFP cube: excitation filter 500/20 nm and emission filter 535/30 nm; for the CFP cube: excitation filter 430/24 nm and emission filter 470/24 nm and for the FRET cube (CFP/YFP): excitation filter 430/24 nm and emission filter 535/30 nm. To record the fluorescent images, we recommend using high-sensitivity cameras, as they will minimize the duration of excitation required to obtain the images, thereby reducing photobleaching of the sample and photodamage to the cell. We use a 12-bit Hamamatsu digital C9300-21 digital camera. Automated acquisition of images can be achieved using a variety of software packages, such as Metamorph or IP-LAB (*see* **Note 5**).

3.2.4 Image Analysis

To extract data obtained from an imaging experiment, open the recorded experiment and load the raw images using Image J software. Figure 3a shows an example of CFP images for both ICUE and AKAR in cardiac myocytes. Select the cells with similar intensity levels of CFP and YFP and draw ROIs both on the cell and on the background (Bkgr). The Image J software will extract the mean intensity values of I_A, I_D, and I_F from the ROIs drawn. Save the experiment data as an .xls or .txt file.

To correct for background fluorescence, subtract the mean emission intensity of the background ROI from the corresponding mean emission intensity of the cell ROI for appropriate channels. Emission ratios are calculated from the collected data using the following formula for each time point:

$$\text{Emission Ratio}\left(\text{AKAR}\right)=\frac{I_F - I_{F\text{-Bkgr}}}{I_D - I_{D\text{-Bkgr}}}.$$

$$\text{Emission Ratio}\left(\text{ICUE}\right)=\frac{I_D - I_{D\text{-Bkgr}}}{I_F - I_{F\text{-Bkgr}}}.$$

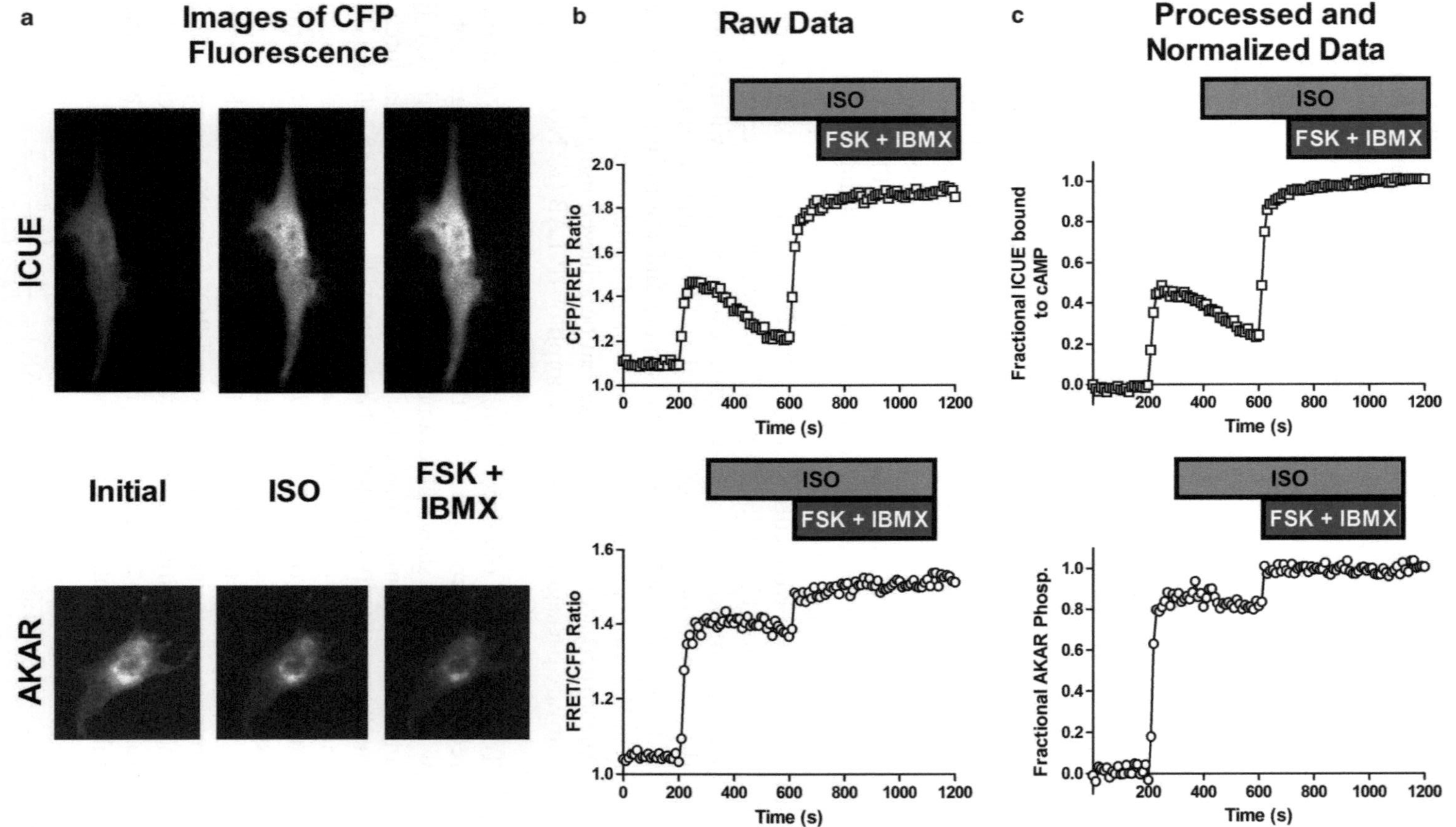

Fig. 3 Examples of experiments showing measurements of ICUE and AKAR responses to changes in intracellular cAMP levels in neonatal cardiac myocytes. Cardiac myocytes were first stimulated with the β-adrenergic receptor agonist isoproterenol (ISO, 1 μM) and subsequently to reach maximal biosensor responses with mixture of IBMX (100 μM) and forskolin (FSK, 50 μM). (**a**) Raw FRET donor (CFP) images of cardiac cells before stimulation ("Initial") and after stimulation with ISO followed by addition of FSK and IBMX. The increase in CFP fluorescence in the ICUE probe is directly correlated with elevation of intracellular cAMP. In contrast, increased activation of PKA is correlated with decrease in CFP fluorescence in the AKAR probe. (**b**) Representative emission ratio time courses of ICUE3 (*upper panel*) and AKAR3 (*bottom panel*) stimulated with 1 μM ISO followed by 50 μm FSK and 100 μM IBMX. CFP to FRET (CFP/FRET) emission ratio (ICUE, *upper panel*) and FRET to CFP (FRET/CFP) emission ratio (AKAR, *bottom panel*) in a single cell. Ratios were calculated with background subtraction for both channels. (**c**) Normalized FRET ratio

These data are plotted as changes in the corrected CFP/FRET and FRET/CFP emission ratios for ICUE and AKAR, respectively (Fig. 3b). Spectral bleed-through correction in FRET biosensors can be done using the method described by Periasamy et al. [15]. For unimolecular FRET-based biosensors, such as ICUE or AKAR, the bleed-through correction can be omitted because of the fixed stoichiometry between the FRET pairs. In this case, the FRET data is expressed as direct fluorescence emission ratios from donor and FRET channels only with background (Bkgr) correction as indicated in step 2 (Fig. 3b). Photobleaching can cause a progressive decrease in fluorescence and signal strength during image collection, which may interfere with FRET signal quantification. In addition, YFP bleaching results in an artificial CFP increase. Thus, illumination intensities and duration of cell illumination (exposure or acquisition time) should be kept as low as possible for minimizing photobleaching and phototoxicity. To correct for photobleaching, multiply the FRET channel intensity (I_F) by the ratio of the acceptor intensity (I_A) measured on direct acceptor excitation at the beginning of the experiment (I_{A_0}) to the intensity at the corresponding time point during the experiment (I_{A_t}) [16]. Finally, calculate the average emission ratio of the minimal ($R_{\min}$) and maximal biosensor signal ($R_{\max}$) and normalize the raw emission ratio (EM) with the following formula:

$$\text{Normalized Emission Ratio} = \frac{\text{EM} - R_{\min}}{R_{\max} - R_{\min}}.$$

The normalized FRET data can be plotted on the scale from 0 to 1 (Fig. 3c).

3.3 Integrating Experimental Data

3.3.1 Model Validation

Now that the computational model is built and some data has been acquired, the model needs to be compared with experimental data obtained using biosensors to validate that it captures the properties of the real biological system. The inclusion of experimental controls for minimal and maximal FRET biosensor signals allows us to directly relate the FRET signals to the fraction of the total biosensors that are active. If the control conditions do not span the entire activity range (0–100 %) the interpretation of the experimental data may be affected. To compare the computational results to the experimental data, the biosensor response must be normalized in terms of fractional activation. Fractional activation is calculated using the following algebraic equations:

$$\text{ICUE Ratio} = \frac{\text{ICUE\_cAMP}}{(\text{ICUE} + \text{ICUE\_cAMP})}$$

$$\text{AKAR Ratio} = \frac{\text{AKARp}}{(\text{AKAR} + \text{AKARp})}$$

To have Virtual Cell calculate these algebraic equations, new global parameters need to be defined. In the "Parameters and Functions" section, click the "Add New Global Parameter" button which creates a new global parameter. Rename this parameter (e.g., ICUE_ratio) and in the expression column enter in the algebraic functions as shown above. Now you will be able to see the fractional activation of each biosensor in simulation results. Figure 4 shows the comparison between the base model that was created using the best initial guesses for kinetic rate parameters. This shows that the model in its current state does not accurately recapitulate the experimental evidence. This discrepancy can be caused by inaccuracies or uncertainty in kinetic parameters and protein concentrations. Therefore some parameter values will be adjusted so that the model predictions better fit the experimental data.

3.3.2 Fitting to Experimental Data

Gaps and uncertainty in information about model parameters can create discrepancies between model results and experimental data. To try to better capture the experimental data, we will vary some of

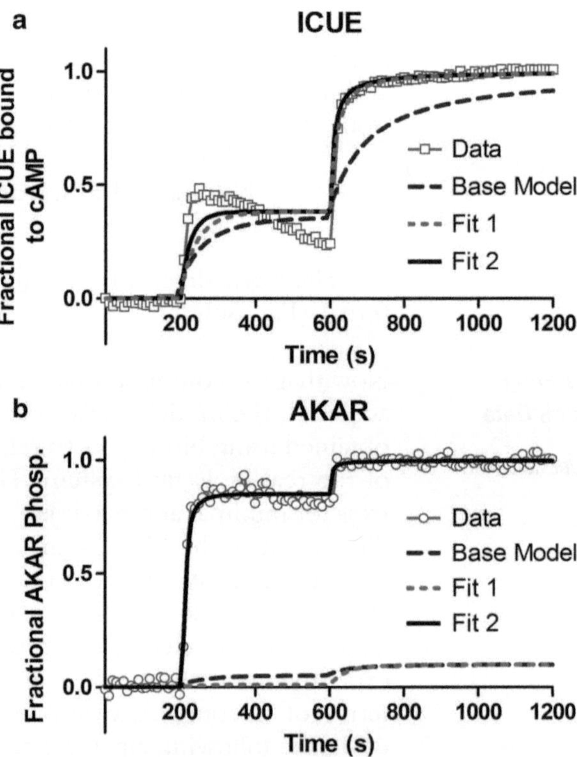

Fig. 4 (**a**) Normalized ICUE FRET data (*squares*) is compared to the ICUE ratios for computational models in different stages of the fitting process. (**b**) Normalized AKAR FRET data (*circles*) is compared to AKAR ratios for computational models in different stages of the fitting process. Base model (*black dash*) is the simulation results from the computational model that uses kinetic parameters from the literature. Fit 1 (*grey dotted*) is the simulation results after fitting kinetic parameters involved in cAMP production. Fit 2 (*black solid*) is the simulation results after fitting kinetic parameters from Fit 1 that are involved in PKA activation

the model parameters to fit to the experimental data. When fitting data, one must choose which parameters to fit based on the strength of evidence that was used to define that parameter. The goal is to fit as few parameters as possible but still accurately fit the biological response. When fitting parameters to data that monitor different aspects of a signaling network, such as ICUE monitoring cAMP concentration and AKAR reporting PKA activity, it can be more effective to fit different modules of the model separately. In this model, parameters involved in cAMP generation will be fitted first using the ICUE data (noted in Table 1 with †). Then, when a good fit of cAMP production is achieved, parameters involving PKA activation will be fit against both ICUE and AKAR data. To use Virtual Cell to fit parameters, use the following steps:

Step 1. Fitting is done in the "Parameter Estimation" tab of the Application that was defined earlier for your model. The first task is to define the parameters that will be varied to fit the experimental data. In the "Parameters" tab, click the add parameter button (green plus) and a list of all the model parameters will be shown. Select the parameters that will be varied and select "ok," which will put those parameters into the parameter tab. For each of the parameters that were selected, lower and upper bounds must be defined. One rule of thumb for the bounds is to allow one order of magnitude in both directions. This allows a fair amount of parameter variation but hopefully keeps parameters within a biologically relevant range.

Step 2. To bring the experimental data into Virtual Cell, it must be saved in a comma separated values (.csv) format (which can be done using "save as" in Excel). The data and associated time values need to be put into columns with labels in the first row of the column. In the "Experimental Data Import" tab, click the "Import from CSV file" button and select the file where your data is saved. The graph should show the data from the file you just imported. The normalized data for this example can be extracted from the public model "ecg5pc: Greenwald MIMB 2012—Fit 2". The data can be found in the "Experimental Data Import" tab of the Parameter Estimation section of the application. The data can be copied by selecting the "show data" button (blue grid), selecting all the data and copying it to a .csv file. When fitting to the ICUE data alone, do not import the AKAR data because Virtual Cell will try to fit the model to all data imported into that model.

Step 3. In the "Experimental Data Mapping" tab, you need to relate the experimental data to Virtual Cell variables. When using biosensor data it will be necessary to use global variables to define fractional activity as explained earlier. For each data set in the file you imported, there will be a row that will be labeled with the title given in the data file. Select each row and push the "Map Experimental Data..." button. For that experimental data set, choose the appropriate global variable or species concentration to which it relates. For the normalized ICUE data, select the global parameter, as defined earlier, ICUE_ratio.

Step 4. Finally, the fitting optimization can be done by selecting the "Run Task" tab. There are many methods to find the optimal parameters. We use the Levenberg–Marquadt method because it is an efficient and deterministic solution method. Once a solution method is selected, press the "Solve by Copasi" button and Virtual Cell will run the fitting algorithm. Once a solution is found, the fitted values for the parameters will be shown on the right. To inspect how well the fit agrees with the experimental data, you can push the "Plot" button. Also, the "save solution as new simulation" button is useful to record the results of the fit as well as provide the ability to examine this fit later.

The fit of the cAMP production, called Fit 1, is shown in Fig. 4 and the model now has fairly good agreement with experimental ICUE FRET data. This model is not able to explain the decrease in ICUE response following ISO stimulation because the model did not include negative feedback loops in the β-adrenergic receptor pathway, including receptor desensitization and increased PDE activation by PKA-mediated phosphorylation. Such discrepancies between models and data can help identify what parts of a signaling network are necessary to capture a response. At this point, it may be useful to save a different copy of your model (*File>Save As*...). After fitting the model to the ICUE data, we saved our model as the public model "ecg5pc: Greenwald MIMB 2012—Fit 1". In this new copy of the model, the parameters that were fit in the previous iteration (for this model, cAMP generation) can be put into the model directly so that other components can be fit to the model data. With the "Fit 1" version of the model, we can now fit the parameters relating to PKA activity to the normalized AKAR data. The fit was done by again following steps 1–4 described above, except in step 2 we use both the normalized ICUE and AKAR data. Parameters chosen were both involved in PKA activation and catalytic activity as well as phosphatase activity, because the amount of AKAR phosphorylation is determined by the balance of PKA and phosphatase activity (parameters chosen indicated in Tables 1 and 2 by ‡). As it is shown in Fig. 4, Fit 2 now has good agreement with both ICUE and AKAR data. This model is available as the public model "ecg5pc: Greenwald MIMB 2012—Fit 2". To validate that the fitted model accurately represents the biological response, the model needs to be compared to an independent set of experimental data, distinct from the data used to fit the model.

Congratulations! Now you have successfully built a computational model, normalized experimental data to compare to model results and fit the model to the experimental data. Now we will provide some examples of how models can be used to test hypotheses.

3.4 Interpreting Biosensor Data Using a Model

The purpose of building models is to understand or evaluate hypotheses as well as create new testable hypotheses. One application of our example model is to identify the rate limiting step during PKA activation. This question was addressed by Saucerman et al. using the AKAR biosensor and a more complex model, finding that PKA activation kinetics are rate-limited by cAMP generation (see the Public Model "jsaucer: AKARmyocyte") [17]. Additionally, they were able to use this model with Virtual Cell's spatial modeling capabilities (not discussed here) to examine molecular mechanisms underlying subcellular PKA phosphorylation gradients. Violin et al. also used computational modeling to interpret experimental FRET results involving cAMP signaling [18]. Using ICUE, they examined how the duration of the cAMP signal is regulated in β-adrenergic signaling. They were able to validate their interpretation of experimental data by building a model that incorporated receptor desensitization through both β-arrestin and G-protein coupled receptor kinases as well as PKA-dependent PDE activation. Both of these examples exhibit the use of computational models to quantitatively validate hypotheses generated by experimental data.

Models can be used to generate hypotheses by simulating the effects of changing kinetic parameters or protein concentrations. To showcase some examples of model perturbations, here we test how biosensors may introduce three artifacts (biosensor saturation, buffering and kinetics) that affect biological interpretation of biosensor data. The model simulations of these perturbations can be found in the public model "ecg5pc: Greenwald MIMB 2012—Perturb". One issue that can arise is that a strong biological response to a stimulus may saturate the biosensor. This can cause misinterpretation of biosensor data because the biological response may be greater than the biosensor is able to measure. We have used the model to simulate cases where ICUE responses can become saturated and no longer accurately reflect cAMP concentration. Figure 5a shows the ICUE response to different concentrations of the adenylyl cyclase agonist forskolin. As cAMP accumulates in cell, the ICUE response becomes saturated and unable to accurately report the cAMP concentration.

Biosensors can also have a direct impact on the biological system. For example, ICUE binds cAMP and thus can reduce the concentration of free cAMP. When a signaling molecule binds to a protein and removes it from solution, it is called buffering because it can maintain the concentration just like pH buffers can prevent the pH from changing. We used the model to show how changes in the ICUE concentration can affect the cAMP concentration in the cell (Fig. 5b). As the concentration of ICUE increases, cAMP concentration equilibrates more slowly because the cAMP is first being bound to ICUE and then contributing to the cytosolic concentration. This shows that using cells with low biosensor expression minimizes the disturbance of the natural signaling kinetics.

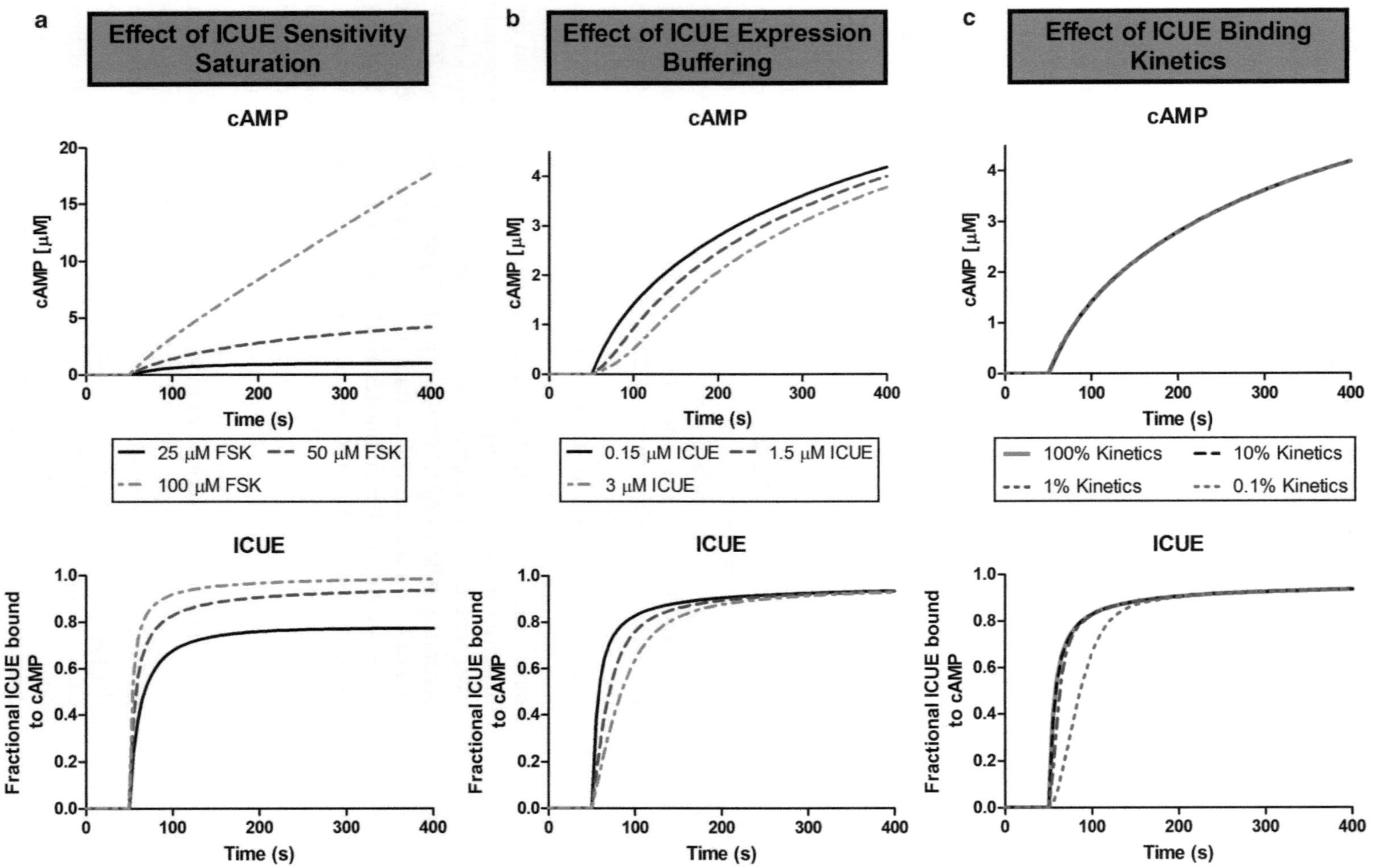

Fig. 5 (**a**) Stimulation with increasing concentrations of FSK correlates with increasing formation of cAMP but ICUE response can saturate. (**b**) Increasing concentration of ICUE biosensor in the cell can lead to buffering of cAMP concentration thus affecting the normal response of the signaling network. (**c**) ICUE binding rates are decreased and the biosensor is no longer able to accurately present the cAMP concentration

Finally, it is important to understand the effect of biosensor kinetics on the measured pathway dynamics. If a biosensor has slow activation kinetics it can falsely represent the actual kinetics that are being monitored. Lai Hock Tay et al. observed this with the Ca^{2+} FRET biosensor TN-L15 and determined a method to quantitatively correct for the slow activation kinetics [19]. We modeled examples where ICUE binding kinetics would be slow by reducing both the binding and dissociation rates by an equal amount to only change the speed of binding and not the relative amount of cAMP bound. Figure 5c shows that as the ICUE kinetics get slower, the biosensor response becomes more dissimilar to the actual cAMP concentration. This is an important consideration in biosensor selection as well as design.

3.5 Conclusions

Computational modeling is a powerful tool for understanding and dissecting cell signaling that is often underutilized in biological studies. These models integrate biosensor data with biochemical data to be able to analyze the signaling network as a whole. Models can be used to both test hypotheses as well as generate new hypotheses. Models can also be useful for identifying and minimizing the impact of potential artifacts. It is the hope of the authors that this chapter allows experimental biologists to extract additional insights from their biosensor data using computational models.

4 Notes

1. It is often useful to set kinetic parameters to have the value of 1 when setting up the model to be able to verify that each part of the model is working properly before taking the time to put specific values in for each parameter.
2. The dissociation constant can sometimes be presented as the association constant, which is the inverse of K_D.
3. The fluxes of each of the reactions can also be seen from this window, which can help to determine the source of errors if the simulation is not working as it should.
4. For primary cells, such as rat neonatal cardiac myocytes, transfection efficiency varies from 5 to 15 %.
5. It is critical that the same settings and exposure times are maintained for each filter set during image acquisition of all samples.

References

1. Aye-Han N-N, Ni Q, Zhang J (2009) Fluorescent biosensors for real-time tracking of post-translational modification dynamics. Curr Opin Chem Biol 13:392–397
2. Saucerman JJ, Brunton LL, Michailova AP, McCulloch AD (2003) Modeling beta-adrenergic control of cardiac myocyte contractility in silico. J Biol Chem 278:47997–48003
3. Sample V, Dipilato LM, Yang JH, Ni Q, Saucerman JJ, Zhang J (2012) Regulation of nuclear PKA revealed by spatiotemporal manipulation of cyclic AMP. Nat Chem Biol 8:375–382

4. Li C, Donizelli M, Rodriguez N, Dharuri H, Endler L, Chelliah V, Li L, He E, Henry A, Stefan MI, Snoep JL, Hucka M, Le Novère N, Laibe C (2010) BioModels database: an enhanced, curated and annotated resource for published quantitative kinetic models. BMC Syst Biol 4:92
5. Bader GD, Cary MP, Sander C (2006) Pathguide: a pathway resource list. Nucleic Acids Res 34:D504–D506
6. Lloyd CM, Lawson JR, Hunter PJ, Nielsen PF (2008) The CellML model repository. Bioinformatics 24:2122–2123
7. DiPilato LM, Cheng X, Zhang J (2004) Fluorescent indicators of cAMP and Epac activation reveal differential dynamics of cAMP signaling within discrete subcellular compartments. Proc Natl Acad Sci U S A 101: 16513–16518
8. Allen MD, Zhang J (2006) Subcellular dynamics of protein kinase A activity visualized by FRET-based reporters. Biochem Biophys Res Commun 348:716–721
9. Zhang J, Ma Y, Taylor SS, Tsien RY (2001) Genetically encoded reporters of protein kinase A activity reveal impact of substrate tethering. Proc Natl Acad Sci U S A 98:14997–15002
10. Bhalla US, Iyengar R (1999) Emergent properties of networks of biological signaling pathways. Science 283:381–387
11. Soltis AR, Saucerman JJ (2011) Robustness portraits of diverse biological networks conserved despite order-of-magnitude parameter uncertainty. Bioinformatics 27:2888–2894
12. Moraru II, Schaff JC, Slepchenko BM, Blinov ML, Morgan F, Lakshminarayana A, Gao F, Li Y, Loew LM (2008) Virtual cell modelling and simulation software environment. IET Syst Biol 2:352–362
13. Rizzo MA, Springer G, Segawa K, Zipfel WR, Piston DW (2006) Optimization of pairings and detection conditions for measurement of FRET between cyan and yellow fluorescent proteins. Microsc Microanal 12:238–254
14. Gallegos LL, Kunkel MT, Newton AC (2006) Targeting protein kinase C activity reporter to discrete intracellular regions reveals spatiotemporal differences in agonist-dependent signaling. J Biol Chem 281:30947–30956
15. Periasamy A, Wallrabe H, Chen Y, Barroso M (2008) Chapter 22: Quantitation of protein–protein interactions: confocal FRET microscopy. Methods Cell Biol 89:569–598
16. Zal T, Gascoigne NRJ (2004) Photobleaching-corrected FRET efficiency imaging of live cells. Biophys J 86:3923–3939
17. Saucerman JJ, Zhang J, Martin JC, Peng LX, Stenbit AE, Tsien RY, McCulloch AD (2006) Systems analysis of PKA-mediated phosphorylation gradients in live cardiac myocytes. Proc Natl Acad Sci U S A 103:12923–12928
18. Violin JD, DiPilato LM, Yildirim N, Elston TC, Zhang J, Lefkowitz RJ (2008) beta2-adrenergic receptor signaling and desensitization elucidated by quantitative modeling of real time cAMP dynamics. J Biol Chem 283: 2949–2961
19. Tay LH, Griesbeck O, Yue DT (2007) Live-cell transforms between Ca2+ transients and FRET responses for a troponin-C-based Ca2+ sensor. Biophys J 93:4031–4040

INDEX

Jin Zhang et al. (eds.), *Fluorescent Protein-Based Biosensors: Methods and Protocols*, Methods in Molecular Biology, vol. 1071, DOI 10.1007/978-1-62703-622-1, © Springer Science+Business Media, LLC 2014

Q

R

S

V

Z

MIX
Papier aus verantwortungsvollen Quellen
Paper from responsible sources
FSC® C105338

If you have any concerns about our products,
you can contact us on
ProductSafety@springernature.com

In case Publisher is established outside the EU,
the EU authorized representative is:
Springer Nature Customer Service Center GmbH
Europaplatz 3, 69115 Heidelberg, Germany

Printed by Libri Plureos GmbH
in Hamburg, Germany